CHEMISTRY AND TECHNOLOGY

OF WINES AND LIQUORS

CHEMISTRY AND TECHNOLOGY

OF

WINES AND LIQUORS

By

KARL M. HERSTEIN, F.A.I.C.

Consulting Chemist

and

THOMAS C. GREGORY

Consulting Chemist

NEW YORK

D. VAN NOSTRAND COMPANY, Inc.

250 FOURTH AVENUE

1935

Printed in U. S. A.

PRESS OF
BRAUNWORTH & CO., INC.
BOOK MANUFACTURERS
BROOKLYN, NEW YORK

PREFACE

It is hoped that the present volume will, in a sense, serve to mark the end of an era, and the beginning of a new one. Mankind has had certain arts from time immemorial. Weaving, smelting, pottery, and the production of alcoholic beverages are noteworthy among these. And they share, besides great age, the distinction of having reached a fairly high peak of perfection without that intensive application of scientific development which has been characteristic of the newer arts whose origin has been in the advance of scientific knowledge.

This is not to say that they have been untouched by science until the twentieth century. In particular the art of alcoholic beverages owes much to the workers of the nineteenth century. Pasteur, Hansen, Lavoisier, and many of the immortals of science have left their imprint and monuments in this field as well as in many others. More recently, but still apart from the modern age were the great investigations by the Royal Commission in Great Britain and President Taft's Board in this country into the question "What is Whiskey?" In our wine production, the work of the beloved Harvey W. Wiley culminating in the "American Wines at the Paris Exposition" had a far-reaching decisive effect. This summary cannot do more than pay its respects to the thousands of earnest workers here and abroad who by their labors have added vastly to our knowledge of the art of making alcoholic beverages and their composition.

The beverage art, however, has been distinguished in another way. It has had to suffer under the inherent conservative tendency of any old art, and also it has been specially hampered by various legal bedevilments. The era just past in the United States, prohibition, may be likened, by not a too strained analogy, to the Dark Ages in Europe from the fourth to the fourteenth century. During the prohibition period, the beverage art, under the necessity of continuing its existence to satisfy a demand which would not cease even at an official behest, and yet under the need of concealment to evade legal requirements, went through a curious semi-comatose state.

The time happened to coincide with a period in our national life when in all other arts, the sciences, especially the science of chemistry and the growing knowledge of chemical engineering, were finding broad new fields of extensive and intensive application. The vast results of these applications both in new products and in increased and improved productiveness are too well known to require illustration.

Hence the repeal of prohibition found the beverage art as a sort of stepchild. Chemical science was ready to step in. Chemical engineering had its techniques ready. But the art to which these were to be applied was demoralized. Bootlegging required very little of its product. A very bare resemblance to its prototype and a substantial "kick" were sufficient to satisfy the market. Quality of product was generally unattainable by bootleg manufacturer, and really unnecessary to his market. Economy of production was a relatively minor consideration when the liability to government seizure and the maintenance of an army of thugs and wholesale bribery constituted the larger items in the final selling price of the product.

In this historical background, the present volume is offered. The authors are unaware of any other summary of the art as it now exists which has been published recently and they feel that there may be a need for it. On this account the authors have felt it necessary to include between the same covers a wide diversity of material of varying degrees of technical density, and they have been thereby forced to an equal diversity of treatment. Those sections which are primarily descriptive are necessarily broad rather than detailed. On the other hand, in the analytical sections, for example, precision of detail has been a specific aim. With this rationale for its apparent lack of uniformity, the authors offer it to the scientists and engineers who may within the next few decades largely transform the beverage art, in the hope that for them it will prove a useful starting point.   To the larger number who may wish to have a general knowledge of existing techniques or to have handy a reference for special purposes, the volume is also introduced in the hope that they will find in it such information as they may need.

February, 1935.

CONTENTS

TABLES

TEXT FIGURES

xi

CHAPTER I

THEORETICAL CONSIDERATIONS.—SUGARS AND STARCH

General Statement.—The entire wine and liquor industry rests on the fact of nature that under suitable conditions sugar is transformed into potable alcohol, while at the same time the other materials in the sugar solution and the by-products resulting along with the alcohol lend various pleasant characteristics to the finished product. It follows, then, that the character of the finished product depends, first, on the raw material which furnishes the sugar, second, on the conditions of the transformation of the sugar into alcohol, and third, on the after treatment of the alcoholic solution. An exact knowledge of the effect of each of these factors is the key to the successful production of a uniformly palatable result.

Fermentation.—Basically, the transformation of sugar into alcohol is the one step which is common to all liquor production. This change, which is only one of a vast number of similar changes resulting from the action of living bodies on suitable organic (carbon, hydrogen and oxygen) compounds, is called alcoholic fermentation to distinguish it from the many similar processes which, starting with different chemicals, result in different products.

Sugars.—Alcoholic fermentation involves the transformation of a sugar, usually dextrose, into alcohol. Hence some discussion of sugars is the logical starting point. We have used the term sugar in a more generalized sense than it is used in lay language. To the chemist there are known many sugars, all of which are chemical compounds containing carbon, hydrogen and oxygen. The two latter elements are present always in the same ratio as in water, so that the sugars come into a broad classification of

3

chemical compounds called carbohydrates. Within this larger group the sugars are generally distinguished by their ability to form crystals. The chemist classifies sugars, first, according to the number of carbon atoms contained in their molecules, and second, according to the number of carbon atom chains which are present in their molecule. This system of classification leads to the following schematic terminology:

Monosaccharides

Trioses—$C_3H_6O_3$	Glycerose
Tetroses—$C_4H_8O_4$	Erythrose, etc.
Pentoses—$C_5H_{10}O_5$	Arabinose, Xylose, Rhamnose, etc.
Hexoses—$C_6H_{12}O_6$	Dextrose, Fructose, Mannose, Galactose, etc.
Heptoses—$C_7H_{14}O_7$	Manno-heptose, Gluco-heptose, etc.
Octoses—$C_8H_{16}O_8$	Gluco-octose, etc.

Disaccharides

Hexabioses—$C_{12}H_{22}O_{11}$	Sucrose, Maltose, Lactose, etc.

Trisaccharides

Hexatrioses—$C_{18}H_{32}O_{16}$	Raffinose, Melezitose, etc.

Poly-saccharides

$(C_6H_{12}O_6)_n$	Starch, Inulin, Cellulose, etc.

Within each group the sugars are distinguished from each other by differences in chemical structure which result in differences of such properties as solubility, sweetness, crystal form, optical rotating power, melting point, etc. In particular, it has been found that for each structure there are pairs of sugars which have equal but opposite optical rotating powers. Usually only one of each pair is of common occurrence. Further discussion of the chemistry of the sugars is beside our point here, which considers them merely as raw materials for the production of alcohol.

For this purpose only the hexoses are directly suitable. Such di- or poly-saccharides as can be converted readily into hexoses are, of course, also of primary importance. The common hexoses which are often encountered in the fermentation industry are, in order of decreasing importance:

Dextrose
d-Fructose, Laevulose
Galactose.

Of the disaccharides:

Maltose
Sucrose (Saccharose)
Lactose, are of importance.

Starch is the chief poly-saccharide encountered in the fermentation industry.

Hexoses.—*Dextrose* occurs naturally in the juice of fruits (grapes, etc.), from which is derived its common name, grape sugar; and in blood and many other sources. It is prepared commercially by the hydrolysis of starch by means of dilute acid and can be bought in crystal form of very nearly the same high degree of purity as cane sugar (sucrose). Its sweetness is slightly less than that of cane sugar. More often, however, dextrose is prepared directly in the wort (fermentation liquor) and converted into alcohol without isolation. In pure solution dextrose may be determined from the specific gravity or refractive index of the solution. In ordinary solutions dextrose may be separately determined by making use of its chemical reducing power, optical rotation, fermentability, etc. Ordinarily the fermentation industry is more interested in the total content of fermentable sugars than in any special one.

d-Fructose is usually also present in fruit juices, makes up very nearly half the sweetness of honey and is obtained in equal amounts as dextrose by the hydrolysis (so-called inversion) of cane sugar. It is somewhat sweeter to the taste than cane sugar. With dextrose, to which it is structurally a very close relative, it is the most readily fermentable of sugars.

Galactose.—This sugar occurs almost exclusively as a product of the hydrolysis of lactose, milk sugar. It is of rather minor importance except in the production of such beverages as koumiss, etc. by the fermentation of milk.

Hexabioses.—*Maltose.*—Almost the sole occurrence of this sugar is in the product of the hydrolysis of starch either by the action of enzymes or by acid hydrolysis. It is not isolated but is fermented in the solution in which it is prepared. The enzyme, maltase, is usually elaborated by the same yeasts as carry on the fermentation. This enzyme converts the maltose into two equivalents of dextrose which are then directly fermentable. When isolated, maltose forms hard white crystalline masses, very similar to grape sugar. It is determinable by the facts that its solutions have some reducing power (about two-thirds that of glucose), and that its solutions are strongly dextro-rotary.

Sucrose (*Saccharose*).—This is the sugar which is commonly meant when the word sugar is used. It occurs naturally in sugar-cane, beets, sugar maple, sorghum and in many other plants. Its production and purification are among the world's major industries. Cane-sugar crystallizes in large monoclinic prisms which are readily soluble in water, very sweet in taste, and, as marketed, represents probably the nearest approach to absolute purity of all materials sold in large bulk. It is determinable in pure solution by either the specific gravity, polarization, or refractive index of its solution. In impure solution, its optical rotatory power, lack of reducing property and the change in both these properties after hydrolysis (inversion) furnish means of determination.

The inversion of sucrose results either from the action of acid or of a special enzyme, invertase, which is present in yeast. In either case the process results in one equivalent each of dextrose and fructose. Sucrose, itself, does not ferment, that is, does not break up into alcohol and carbon dioxide under the influence of the enzyme, zymase. However, since most yeasts also contain invertase, the fermentation of sucrose proceeds as soon as this latter enzyme has had a little time to act.

Lactose.—Milk-sugar or lactose is found naturally, as its name indicates, in milk. It forms small white crystals which dissolve with some difficulty in water. Lactose, on hydrolysis, yields equivalent amounts of dextrose and galactose. After hydrolysis the dextrose and galactose can be fermented with the production of alcohol, and usually in practice, with some production also of lactic acid. To this change is due the special character of beverages like koumiss.

Starch.—While many fermented liquors obtain sugar for conversion into alcohol from sources indicated above; by far the largest single source of sugar, especially for distilled liquors, is the poly-saccharide, starch. Its importance arises from the fact that by suitable treatment almost 100% conversion of starch into fermentable sugars, dextrose, maltose, etc. can be obtained. Hence the general nature, occurrence and physical and chemical properties of starch are of major interest in the fermented liquor industry.

General Statement.—Starch is the compound in which all of the higher (green-leaved) plants store the sugar they need for food. Hence it occurs almost universally in their tissues; and in their special storage places, seeds and tubers, makes up the bulk of the solids. When pure, it is a fine white powder having a density of 1.6 and at ordinary temperature it is quite insoluble in water, alcohol, ether, and other common solvents. Under the microscope, starch appears as minute, white, translucent grains varying widely in size and shape according to its origin. In each case, however, the average size and shape of the starch grains is so characteristic that it is usually comparatively easy to determine their botanic origin. Morphologically, starch granules can be classified into the following groups:

(1) *The potato group*—large oval granules, showing concentric rings and a nucleus or hilum, eccentrically placed. This group includes the arrowroot and potato starches.

(2) *The legume group*—round or oval granules usually also showing concentric rings and with an irregular hilum. The starches of peas, beans and lentils belong in this group.

(3) *The wheat group*—round or oval granules with a central hilum. Wheat, barley, rye and acorn as well as the starches of many medicinal plants are found in this group.

(4) *The sago group*—round or oval granules truncated at one end. The group includes sago, tapioca and cinnamon starches.

(5) *The rice group*—small, angular, polygonal grains. Corn, rice, buckwheat and pepper starches are included in this group.

Within various groups the grain size may vary from 0.005-0.15 mm. or more.

Structure.—The structure of the granules is quite complex, but consists essentially of an envelope of rather condensed nature enclosing a colloidal substance of slightly more diffuse structure. The envelope constitutes approximately 2% of the substance of the granule.

Properties.—The outstanding physical property of starch is that of forming a paste when heated in the presence of water. It can be shown under the microscope, that what happens is that the granulose, the interior material of the granule, swells as it absorbs water, and finally bursts its shell. A similar result can be obtained without the use of heat if the starch is first ground in a ball mill to break the cell wall, or if it is treated with chemical reagents which destroy the cell wall. Dilute caustic alkalies and solutions of zinc chloride are among the reagents which produce this result.

The temperature range required to produce pastification and the viscosity of the resulting paste are highly characteristic of the variety of starch employed. For most starches, however, pastification does not commence below 70° C. (158° F.)

Classification.—Commercial starches are classified, according to the viscosity of the paste produced, as thick- or thin-boiling. Wheat starch is a typical thin-boiling starch, as a 5% mixture of wheat starch in water yields a thin, translucent syrup, scarcely gelatinous at boiling temperature. Corn starch, on the other hand, is a characteristic thick-boiling starch. Its 5% mixture with boiling water is practically non-fluid.

While it is now known that variations in the pasting qualities of a variety of starch can be induced by changes in the conditions of manufacture or by suitable treatments, these properties as well as the degree of gelatinization of the cooled paste are of great importance to many industries. In laundry practice and some branches of textile manufacture, for instance, it is essential that the starch paste be thin enough to penetrate the fabric when hot, without piling up on the surface, and at the same time that it have body enough to provide the necessary stiffness to the finished article. On the other hand, in paper-box making, to cite one example, a thick-boiling starch which will be adhesive without soaking into the stock is required.

In the fermentation industry the pasting qualities of a starch are only of minor importance since adaptations can always be made to provide for them. The essential value of starch to the fermentation industry is that by suitable treatment it yields progressively more soluble products and finally can yield almost 100% of dextrose.

Conversion.—The conversion of starch results first in the so-called soluble starches, then in dextrin, then in maltose and finally in dextrose. The earlier stages of the process are not sharply defined from a chemical point of view and they are controlled entirely with a view to the special qualities desired in the product. As the conversion continues, mixtures of an unfermentable gum, dextrin, with varying proportions of maltose and dextrose are obtained. By carrying the process further, a yield of almost pure dextrose can be secured. This, of course, is the object in the processing of starch for the fermentation industry. The conversion of starch results either by the action of the enzyme, diastase, by boiling with dilute acids or by gentle roasting. The first two methods are used in the fermentation industry, often in conjunction with each other.

Chemically the process is a hydrolysis. That is, by the addition of water to the starch molecules, the latter are split into more soluble materials. The products obtained depend upon the agent used, and, also, upon whether the action is allowed to go to completion. Many researches have been made in the study of this subject. C. O'Sullivan (J.C.S. 1872, *579;* 1876, *125*), showed that the products of diastatic action are maltose and dextrin, and that the proportion of maltose in the product decreases as the temperature of conversion is raised above 63°C. According to Brown, Heron and Morris (J.C.S. 1879, *596*), malt extract at room temperature converts starch paste into 80.9 parts of maltose and 19.1 parts of dextrin, and the same change occurs at all temperatures to 60°C. The intermediate dextrins were investigated by Brown and Morris (J.C.S. 1885, *527;*

1889, *449, 462*), and by Brown and Miller (J.C.S. 1889, *286*). Various views have been held as to the nature of the intermediate products, and even of the final products. Daish (J.C.S. 1914, *105, 2053, 2065*) and Nanji and Beazley (J.S.C. Ind., 1926, *215T*), state that prolonged treatment with mineral acid converts starch into dextrins and maltose and finally into *d*-glucose. Ordinary diastase, or amylase, a β-diastase, converts starch finally into dextrins and maltose, whereas takadiastase, which contains the enzyme maltase in addition to an α-diastase, yields *d*-glucose as final product. (Davis and Daish, B. C. Abs. 1914, *ii, 588*. Cf. Baker and Hulton, J.C.S. 1914, *105, 1529;* W. A. Davis, J. S. Dyers, 1914, *30, 249*). G. W. Rolfe (Rogers' Manual of Industrial Chemistry, 1921, *905-906*) states that since the discovery of the process of converting starch into dextrose by the action of heat and acids, dextrose in a crude form and known as starch sugar or grape sugar has entered into commerce, more or less. There is also a very pure dextrose commercially sold under the name "Cerelose." Its importance is small, however, as compared to that of "commercial glucose." He draws attention to the fact that there is some confusion of terms which associate this starch product with grape sugar and dextrose. It is quite true that dextrose (*d*-glucose) is an ingredient of commercial glucose, but the dextrose in the commercial glucose of today is the least important ingredient, both in quantity and for the qualities which it imparts to the product. He gives a diagram (See Figure 1) which shows the variation in proportion of the three primary constituents of commercial glucose; dextrin, maltose, and dextrose, present as acid hydrolysis of the carbohydrate matter proceeds. The progress of the hydrolysis is shown by the change in optical rotation from that of starch paste to that of dextrose. The diagonal dotted lines show the respective dextrin and maltose percentages obtained in starch products hydrolyzed by diastase (malt) conversion for the corresponding rotation values. These are corrected for the polarization influence of carbohydrates introduced in the malt, which do not come from starch hydrolysis. He states, further, that the stage of hydrolysis most favorable for the manufacture of commercial glucose for ordinary purposes lies between the rotation values, 120° and 140°, although glucose used for special brewing purposes may be somewhat outside these limits.

Maquenne and Roux (C.R., 1905, *140, 1303*), claim that starch is a mixture of two substances, amylose and amylopectin, the former in the interior portion of the granule, and the latter in the envelope. The amylose, obtained by reversion, or by heating starch with water under pressure and cooling, gives no

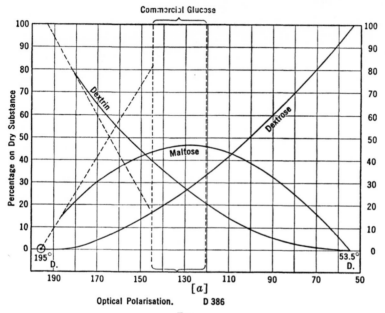

FIG. 1.

coloration with iodine in the solid state, is not readily attacked by diastase, and is scarcely soluble in water at 120° C. If, however, it is heated with water under pressure at 150° C. it dissolves fairly readily; the solution can be filtered, gives a blue coloration with iodine and is completely converted into maltose by malt extract at 56° C. It is probable that the amylose and amylopectin are not homogeneous (C.R. 1908, *146, 542*). Schryver and Thomas (Bio. J., 1923, *17, 497*), have found that certain starches contain small amounts of hemicelluloses, and Ling and Nanji (J.C.S. 1923, *123, 2666*), state the ratio of amylose to amylopectin is constant and is equal to 2:1, although their absolute percentage will vary according to the proportion of hemicellulose in the starch. For practical purposes, the progress of the hydrolysis or conversion of starch paste is shown by characteristic chemical and physical changes. The thick paste loses its colloidal nature and rapidly becomes more limpid, the concentration (*e.g.,* osmotic pressure) of the solution increases, although the dissolved carbohydrates become specifically lighter, and the solution becomes distinctly sweeter in taste. If tested with a weak aqueous solution of iodine, the deep sapphire blue given by the original starch paste changes as the hydrolysis proceeds, passing into violet, then to a rose red which in turn changes to a reddish brown which grows steadily lighter, until just before complete hydrolysis is reached, it disappears altogether. A few drops of the solution poured into strong alcohol give a copious white precipitate during the early stages of the conversion; as the hydrolysis continues the amount of the precipitate becomes less until near the end when no precipitate is produced.

If the conversion products are tested polariscopically, it will be found that there will be a progressive fall in specific rotation values from that of starch paste (202°) to that of dextrose (52.7°). The Fehling test shows no copper reduction with starch paste, at the beginning of the hydrolysis, but progressively increases till the maximum reducing power is reached when all of the converted products are finally transformed into dextrose.

CHAPTER II

THEORETICAL CONSIDERATIONS.—ENZYMES

At the beginning of the preceding chapter the statement was made that fermentation results from the action of living bodies on suitable organic compounds. The actual instruments of the conversion, however, are not living cells but constitute a group of special chemical compounds which are built up or secreted by the living cells and which are called enzymes. We know that these enzymes are not living bodies because the changes which they produce can be effected in the entire absence of any live cells. Yeast juice, for example, even when totally free of yeast cells can cause the fermentation of suitable sugar solutions. The activity of the juice diminishes in the course of time, and both in rate of fermentation and in total fermentation produced; the extract or juice is much less efficient than an equivalent amount of living yeast.

Description.—As a class, the enzymes are unstable, nitrogenous compounds of a colloidal nature and of great chemical complexity. Only two enzymes, urease and pepsin, have been isolated in a fairly pure state. The exact chemical composition of the enzymes is still unknown. They are not necessarily proteins although many of them have certain properties common to proteins. Their most important characteristic is the ability to produce chemical changes in other substances without themselves being changed. This is what chemists call catalytic power. Another name for the enzymes as a group is chemical or soluble ferments.

Specific Action.—When enzymes are considered separately rather than as a class it is found that their individuality manifests itself in two ways; a combination of one special enzyme with one special substrate (material to work on) is required for each particular result. However, it is possible to classify enzymes to some extent by the type of reaction which they produce, as follows:

GROUP	ACTION
1. Hydroxylases	Invertase hydrolyses cane sugar. Amylase hydrolyses starch.
2. Esterases	Hydrolyse esters, including the lipases which act specifically on fats.
3. Oxidases	Produce oxidation.
4. Reductases	Produce reduction as *e.g.* reduce aldehydes to alcohol.
5. Carboxylases	Split off carbon dioxide from organic acids.
6. Clotting enzymes	Thrombase, which clots blood. Rennin, which clots milks.

It is now customary to name enzymes after the compound on which they act with the addition of the ending "ase."

There are also known some cases in which enzymes tend to prevent rather than produce a reaction. In the majority of cases their action as catalysts is positive. That the action is truly catalytic is shown by the fact that the rate of reaction is directly proportional to the concentration of the enzymes, but the total amount of action is independent of the amount of enzyme, provided a sufficient time is allowed; and provided also that the enzyme is not decomposed by other means. As a rule a small amount of enzyme cannot decompose unlimited amounts of sub-strate, since most enzymes are relatively unstable. Hence, if only a small amount of enzyme is used the reaction slackens and finally ceases, owing to decomposition (autolysis) of the enzyme before all the substrate has reacted.

Conditions of Functioning.—Enzymes are sensitive to high temperatures, *e.g.*, when heated to below 100° C. their activity is completely destroyed. They are, however, resistant to many antiseptics which destroy protoplasm and kill fermenting organisms. Some germicides, such as formaldehyde, tend to destroy enzymes.

Preparation.—Enzymes are concentrated by precipitation from their solutions by the addition of alcohol or acetone, or by the addition of salts such as ammonium sulphate; by precipitation after adjustment of the solution to a definite pH; or by adsorption by such materials as alumina, silica gel, fullers' earth, etc. Often combinations of these procedures are used. The resulting products, however, are often contaminated with other enzymes and with inactive impurities, all of which makes the study of their reactions difficult.

Selectivity.—The action of enzymes is essentially selective, in this respect differing from the action of inorganic catalytic agents. The following tabular representation of the reactions of the trisaccharide, raffinose, will illustrate this:

SUGAR	CATALYST	PRODUCTS
Raffinose	Acid	Dextrose, fructose, galactose
Raffinose	Diastase	Melibiose, fructose
Raffinose	Emulsin	Galactose, sucrose

In general, esters, amides, carbohydrates, glucosides, etc. are all hydrolyzed by hydrochloric acid. Lipases will hydrolyze esters but not carbohydrates. Maltase will hydrolyze maltose but not sucrose. Even slight differences in the configuration of two sugars will be sufficient to affect their reactivity with a particular enzyme.

It will be seen that the activities of the hydrolytic enzymes are so specific that great care must be exercised to provide those enzymes which, working on the available materials, will produce the desired results. While this circumstance is of great theoretical importance, it enters into practical consideration rather rarely. As a rule each naturally occurring saccharide or other hydrolyzable substance is accompanied by its own specific hydrolytic enzyme so that it can be made available for natural utilization. The enzyme does not necessarily exist as such in the tissue, but may be present as a so-called zymogen, which liberates the enzyme under suitable conditions, such as a wound to the organism or the presence of an acid.

Co-enzymes.—Very often also there are required the presence of two factors, the enzyme and the co-enzyme to produce the fermentation. For example, it has been shown by Harden and Young (J.C.S. 1905 Abs. II; *109* and ibid. 1906, I; *470*) that yeast juice can be separated by dialysis into two fractions which when recombined are equal in activity to the original juice, although neither by itself will cause any fermentation. The dialyzable fraction, the co-enzyme, is resistant to boiling, but disappears from yeast juice during fermentation, or when the juice is allowed to undergo autolysis. It is decomposed by acid or alkaline hydrolyzing agents, by repeated boiling, and by the lipase of castor beans. The presence of both an enzyme and a co-enzyme has been found necessary in other fermentations, *e.g.*, in the action of lipase it has been found that a co-enzyme which is a salt of a complex taurocholic acid is required.

Alcohol-producing Enzymes.—As can be inferred from the above, the number of types of enzymes required to carry on a practical fermentation may vary from one to three or more, according to the substrate to be fermented.

Zymase, which carries on the final conversion of the monosaccharide to alcohol and carbon dioxide in accordance with the chemical equation:

$$\underset{\text{C}_6\text{H}_{12}\text{O}_6 \rightleftharpoons}{\textit{Dextrose (or fructose, etc.)}} \qquad \underset{\text{2C}_2\text{H}_5\text{OH}}{\textit{Alcohol}} \qquad \underset{+\ \text{2CO}_2}{\textit{Carbon Dioxide}}$$

is always necessary.

A hydroxylase such as maltase or invertase is usually needed to convert maltose or sucrose respectively into mono-saccharide sugar. This conversion follows the formulation

$$\underset{\substack{\textit{maltose, etc.)} \\ \text{C}_{12}\text{H}_{22}\text{O}_{11}}}{\textit{Disaccharide (Sucrose,}} \quad + \quad \underset{\text{H}_2\text{O}}{\textit{Water}} \quad \rightleftharpoons \underset{\substack{\textit{fructose, etc.)} \\ \text{C}_6\text{H}_{12}\text{O}_6 + \text{C}_6\text{H}_{12}\text{O}_6}}{\textit{Hexoses (Dextrose,}}$$

Finally another enzyme may be needed to saccharify, *i.e.*, hydrolyze, the starch into maltose. Successful fermentation depends very largely upon the supplying of the required enzymes at the proper stage in the operation, and in the necessary amount to produce the desired result. Further consideration of these topics will be found under the captions of Yeast and Malt.

CHAPTER III

THEORETICAL CONSIDERATIONS.—FERMENTATION

Alcoholic fermentation is a subject which has attracted the study of many chemists, even some of the greatest. Many days of research have been devoted to define completely the reactions which take place in this chemical transformation, and various theories have been advanced to explain the steps which occur. Despite all this work, our knowledge of its mechanism is still incomplete. We know that fermentation commences with sugars in the presence of certain other materials, ferments or enzymes; and we know most of the end products. But we do not know how these products result, why they result, or what are the intermediate steps in their formation. A completely satisfactory explanation still has not been found in answer to these questions.

General Requirements.—Sugar, water, the presence of a ferment, and a favorable temperature, usually 75° F.–85° F. and never over 90° F. are inescapable requirements. There are other limiting factors. The ferments or enzymes are known to be chemical compounds. They have not been analyzed completely, however, nor synthesized, and we are dependent for their production upon living plants, the yeasts. The reactions follow the law of Mass Action to the extent that as the concentration of the alcohol approaches 14-16% the reaction slows down and finally stops. The law of Mass Action states that in a reversible reaction the final state reached depends on the relation between the concentrations of the initial and end materials. However, attempts to reverse the process of fermentation have been successful only in the minutest degree.

Oxygen does not appear to enter into the fermentation reaction, but the presence of air and particularly aeration of the fermenting liquor, the substrate, do have a noticeable influence.

Essential Nature.—Whatever the starting point, in all cases the desired result is the presence of all of the sugars in a form suitable for conversion into alcohol. As previously stated, this process is called alcoholic fermentation. This change was believed by Lavoisier (1789) to follow the formulation:

$$\underset{\text{Hexose}}{C_6H_{12}O_6} \quad = \quad \underset{\text{Ethyl Alcohol}}{2C_2H_5OH} \quad + \quad \underset{\text{Carbon Dioxide}}{2CO_2}$$

It was shown by Pasteur (1857) that this formulation actually accounts only for about 95% of the sugar consumed. Glycerin, organic acids and traces of other by-products account for the balance. This figure of 95%, however, does represent the yield of alcohol which is obtainable under favorable conditions and represents also the upper limit of the result of good commercial practice in the production of alcohol, either potable or industrial.

Since their time, the combined labors of many students have, on the one hand, added somewhat to our knowledge of the mechanism of fermentation; and on the other hand, have defined some of the by-products as well as some of the minor prerequisites to success.

Products.—Pasteur found that the actual yield from the fermentation of 100 pounds of sugar was as follows:

Alcohol	48.55 lb.
Carbon Dioxide	46.74 lb.
Glycerol	3.23 lb.
Organic Acids	0.62 lb.
Miscellaneous	1.23 lb.
	100.37 lb.

The fact that the total weight of fermentation products exceeds slightly the weight of sugar fermented is explained by the absorption and fixation of small amounts of water to make certain of the by-products. According to Pasteur some sugar is also utilized by the yeasts in building new yeast cells.

In general, the chief products of vinous fermentation may be stated to be: alcohol and carbon dioxide (accounting for 94-95% of the sugar), glycerol 2.5-3.6%, acids 0.4-0.7%, and, in addition, an appreciable quantity of fusel oil (higher alcohols), some acetaldehyde and other aldehydes and some esters. Among the minor products of fermentation may be listed the following which have been identified:

> Formic Acid
> Acetic Acid
> Propionic Acid
> Butyric Acid
> Lactic Acid
> Ethyl Butyrate
> Ethyl Acetate
> Ethyl Caproate, etc.

Rate of Fermentation.—The ratio of carbon dioxide to alcohol produced and the ratio of yeast formed to alcohol produced both vary at different stages of the fermentation. They depend both on the age of the yeast and on the age of the fermentation. Slator, J.C.S. (1906), 89; 128, and ibid. (1908), 93; 217 has shown that the rate of conversion of dextrose into alcohol and carbon dioxide by yeast is exactly proportional to the amount of yeast present and, with the exception of very dilute solutions, is almost independent of the concentration of sugar. Slator and Sand, Trans. Ch. Soc. (1910) 97; 922-927 have further developed and explained this fact by showing that the diffusion of sugar into the yeast cell is so rapid even in dilute solutions that more sugar is present in the cell than can be fermented at any instant. Various yeasts will ferment levulose (fructose) at the same rate as that at which dextrose is fermented. Similarly, maltase-containing-yeasts will ferment maltose solutions at the same rate as if dextrose were being fermented.

Controlling Factors.—The factors which really govern the rate of fermentation, then, are two: the concentration of yeast and the temperature. The latter is of the greatest importance. Taking as a unit the amount of fermentation produced at 32° F., six times as much will result in the same time at 77° F., and twelve times as much at 95° F. However, the rates of formation of undesired by-products and of autolysis of the yeast also increase objectionably as the temperature is raised, hence it is usual to set an upper limit of 90° F. on the fermentation. Since heat is evolved during the process, this requires that the rate must be so controlled that either natural radiation or artificial cooling will keep the temperature within the desired limits.

Inorganic Requirements.—Various inorganic constituents must also be present in the fermenting liquor. Some of these, phosphates in particular, play a part in the mechanism of fermentation. Others are necessary to provide food for the yeast, nitrogen compounds, calcium, potassium and manganese, etc. In addition there is a still incompletely defined compound called "bios" which appears to be essential to success. Most of these are present in sufficient amounts in the raw materials of the fermentation industry although it is probable that close study of their occurrence might be rewarded by increased yields and/or improved products.

Mechanism.—The role of phosphates especially is of interest since it shows that the fermentation proceeds by steps instead of immediately in the sense of the equation

$$C_6H_{12}O_6 \rightleftharpoons 2C_2H_5OH + 2CO_2$$

Harden and Young, J. Ch. Soc. (1908) Abs. i, 590 were the first to show that the addition of phosphates (disodium phosphate) to a mixture of yeast juice and dextrose resulted in both an initial acceleration and in an increased total fermentation. They also showed that there was an optimum concentration of phosphate, deviations from which in either direction resulted in a diminished rate of fermentation. It is assumed that a hexose-di-phosphate is formed as follows:

$$C_6H_{12}O_6 + 2Na_2HPO_4 \rightarrow C_6H_{10}O_4(PO_4Na_2)_2 + 2H_2O$$

It is this compound which breaks down into alcohol and carbon dioxide and regenerates the sodium phosphate. The latter can then again combine repeatedly with the sugar, sensitizing it to the breaking down action of the enzyme until the fermentation is complete. It has been shown by careful experiments that during the period of increased fermentation the amounts of alcohol and carbon dioxide produced are stochimetrically related to the quantity of added phosphate in the ratio C_2H_5OH : Na_2HPO_4. Since the filtered enzyme plus phosphate will not of themselves induce fermentation, it follows that phosphate is not the co-enzyme. On the other hand, in the entire absence of phosphate no fermentation occurs even though both enzyme and co-enzyme are present. Arsenites and arsenates cause some acceleration of the fermentation, but they cannot be used in place of phosphates. They appear rather to act as accelerators in the decomposition of the dextrose phosphate.

It has been found by further investigation that there is apparently another step in the fermentation reaction in which the dextrose di-phosphate splits into two moles of triose mono-phosphate. It should be particularly noted that the immediately preceding study of the mechanism of the action of yeast juice is not directly applicable to the action of yeast. *E.g.*, Slator (loc. cit.) found that phosphates are without accelerating effect when living yeast cells are employed.

Many other suggestions have been made regarding the intermediate steps in the conversion of dextrose into alcohol and carbon-dioxide and the nature of the intermediate products. Büchner and Meisenheimer, B. (1905), 38; 620, suggested that lactic acid is the first product of the action of zymase on dextrose since it is known that this acid is formed in muscle tissue by the oxidation of glycogen, which is a polydextrose. They added to this theory the assumption of a second enzyme, lactacidase, which carries on the decomposition of the lactic acid into ethyl alcohol and carbon dioxide; cf. Bio. Z. (1922), 128; 144 and 132; 165. This suggestion was based on the observation that a concentrated solution of dextrose when treated with alkali yields about 3% of alcohol on exposure to sunlight. Under similar conditions a more dilute solution gives a 50% yield of lactic acid.

Another suggestion is that dihydroxy-acetone, $CO(CH_2OH)_2$ is the first result of the splitting of the hexose molecule. It has been shown by Büchner and Meisenheimer, B. 43; 1773 and by Lebedew, B., 44; 2932. Cf. also Franzen and Steppuhn, ibid., 2915 that dihydroxy-acetone is fermentable by yeast. It has also been shown that dihydroxy-acetone and glyceraldehyde can, under suitable conditions be condensed to form a hexose so that there is every probability that the reaction is reversible. Again, the suggestion has been made that glyceraldehyde, $C(HO) \cdot \dot{C}(HOH) \cdot C(H_2OH)$, by the loss of water yields an enolic compound $CH : C(OH) \cdot CHO$ called methyl-glyoxal; which, by the addition of water can yield either lactic acid or even ethyl alcohol and water. On the other hand, the above mechanisms, although plausible, appear improbable as important parts of the fermentation reaction since it has also been shown that glyceraldehyde is only slowly fermented while methyl-glyoxal and lactic acid are unacted upon by yeast juice or yeast.

Neuberg, Deut. Zuckerind. (1920) 45, 492, has shown that yeast contains an enzyme, carboxylase, which is capable of eliminating carbon dioxide from α-ketonic acids, and therefore suggests that pyruvic acid $CH_3 \cdot CO \cdot COOH$ represents the initial splitting product of fermenting hexose. The reaction, then, proceeds with the decomposition of the pyruvic acid into acetaldehyde and carbon dioxide, and the aldehyde is reduced by the yeast to ethyl alcohol. In support of this theory it has been shown that the addition of pyruvic acid to a fermenting liquor increases the yield of alcohol. The presence of glycerol, which may function merely as a yeast preservative, is required. It is also known that there is present in yeast an enzyme which is capable of reducing aldehyde. However, the conversion of the hexose into pyruvic acid is difficult to formulate logically.

Neuberg and Arinstein, Bio. Z. (1921), 117; *269* and 122; *307* suggest the following steps for the fermentation:

(1) Dextrose $\rightleftharpoons$ Water + Methyl-Glyoxal-Aldol

$$C_6H_{12}O_6 \rightleftharpoons 2H_2O + CH_3 \cdot CO \cdot CH(OH) \cdot CH_2 \cdot CO \cdot CHO$$

This keto form changes to the enol-form

$$CH_2 : C(OH) \cdot CH(OH) \cdot CH_2 \cdot CO \cdot CHO$$

(2) Methyl-Glyoxal-Aldol (keto form) + Water $\rightleftharpoons$
Glycerol + Pyruvic Acid

$$CH_2 : C(OH) \cdot CH(OH) \cdot CH_2CO \cdot CHO + 2HO \rightleftharpoons$$
$$CH_2(OH) \cdot CH(OH) \cdot CH_2(OH) + CH_3 \cdot CO \cdot COOH$$

Part of the pyruvic acid is converted by the action of carboxylase and reductase into alcohol and carbon dioxide.

(3) Methyl Glyoxal + Acetaldehyde + Water $\rightleftharpoons$
                                            Alcohol + Pyruvic Acid

$$CH_3CO \cdot CH : O + CH_3 \cdot CHO + H_2O \rightleftharpoons$$
$$CH_3CH_2(OH) + CH_3 \cdot CO \cdot COOH$$

In support of this exposition Neuberg cites the three different courses taken by the fermentation according to conditions:

(a) Normal in fairly acid media with over 90% yields

$$C_6H_{12}O_6 \rightarrow 2C_2H_5OH + 2CO_2$$

(b) In the presence of sulphites (to fix aldehyde)

$$C_6H_{12}O_6 \rightarrow C_3H_5(OH)_3 + CH_3CHO + CO_2$$

(c) In faintly alkaline conditions (in the presence of sodium bicarbonate)

$$2C_6H_{12}O_6 + H_2O \rightarrow 2C_3H_5(OH)_3$$
$$+ 2CO_2 + C_2H_5OH + CH_3CO_2H$$

In reply to the objection that acetaldehyde and pyruvic acid are not so readily fermentable as dextrose, Neuberg suggests that the reaction takes place with one of their tautomeric (enolic) forms.

By-products.—Glycerol.—The preceding formulations of the fermentation reaction indicate readily the production of glycerol. Büchner and Meisenheimer have shown that it is formed in sugar solutions even by the action of yeast juice without the requirement of living yeast cells. It is known also that by the addition of sodium carbonate or sulphite and by selection of the most suitable yeasts a yield of 25% or more of glycerol on the weight of sugar fermented can be obtained.

Fusel Oil and Succinic Acid, etc.—It has been shown that this group of by-products derives not from the sugar but from other materials present in the fermenting liquor. F. Ehrlich in many researches (1904-1910) has shown that the higher alcohols and aldehydes, which when mixed we call fusel oil, are formed by the deammination of amino acids resulting from the hydrolysis of proteins. Thus isoamyl alcohol, which is the chief constituent of fusel oil, is closely related to leucine, amino-isohexoic acid, and active amyl alcohol is similarly related to isoleucine, α-amino-β-

methyl-valeric acid. Both of these acids are formed in the hydrolysis of proteins and both, according to Ehrlich, are transformed into the corresponding amyl alcohols in the presence of sugar by the action of pure yeast cultures. The gross reaction is formulated as follows:

$$(CH_3)_2 \cdot CH \cdot CH_2 \cdot CH(NH_2) \cdot COOH + H_2O \rightleftharpoons$$
$$(CH_3)_2CH \cdot CH_2 \cdot CH_2(OH) + CO_2 + NH_3$$

In a similar manner the other amino acids react; tyrosine, β-p-hydroxy-phenyl-α-amino-butyric acid yielding p-hydroxy-phenyl-ethyl alcohol, tyrosol; while phenyl-alanine, α-amino-β-phenyl-propionic acid, yields phenyl-ethyl alcohol. Succinic acid, for example, which is of usual occurrence in fermented liquors is probably formed by a similar reaction from glutamic acid with the additional step of oxidation in the process.

These changes take place only by the action of yeast but not by the action of yeast juice. This fact points to the importance of these reactions in the life process of the yeast cell. A similar conclusion results from the fact that no free ammonia is found at the end of the reaction, all of it having been consumed in the building up of new yeast cells. This conclusion is further strengthened by the fact, also shown by Ehrlich, that if appreciable amounts of simple nitrogenous bodies, such as ammonium salts, are present in the fermenting liquor these will be used by the organisms in preference to decomposing the amino-acids. Ehrlich has found it possible to increase or diminish the amounts of fusel oil formed by diminishing or increasing the amounts of ammonium salt added to the wort; and also to increase the yield of fusel oil by adding larger amounts of amino-acids to the fermenting mixture. Practically all amino-acids formed in the hydrolysis of proteins can be similarly decomposed by yeast cells provided that sugar is present. As previously stated enormous amounts of research have been devoted to the mechanism of the fermentation reaction. Despite this, the complete definition of the reaction has not been arrived at. The point of interest here is that there are a large number of factors involved, many of them known and controllable and necessary to success.

CHAPTER IV

THEORETICAL CONSIDERATIONS.—RAW MATERIALS

General Classification.—The preceding chapters have been devoted to a brief discussion of the chemical background which is common to all of the fermentation industry. The various raw materials which, by the application of the principles developed, are converted into alcohol-containing beverages will be discussed in the present chapter. The diversity of these raw materials is very great, since anything which contains sugar or may be treated to yield sugar, will serve and probably has been applied, to produce intoxicating liquor. It is possible to classify these materials, therefore, in two ways. However any classification may err at times as industrial demands change. The classification may be:

I. According to their function as sugar suppliers or auxiliaries.
II. According to the products into which they enter.

On the first basis we find three groups:

A. Materials which supply preformed sugar for fermentation.
> Fruit and fruit juices
> Molasses
> Honey
> Sugar, etc.

B. Starch-containing materials yielding fermentable sugar on treatment.
> Cereal grains
> Potatoes, etc.

C. Auxiliary materials which contribute neither sugar nor starch.
> Spices and other Flavoring matters
> Coloring Agents
> Blending Agents, etc.

This classification is based on applicability or composition. On the basis of use the same materials may be reclassified:

A. Materials for the production of distilled spirits.

        Cereal grains
        Potatoes
        Molasses
        Flavoring and Coloring Agents, etc.

B. Materials for the production of wines.

        Fruit juices and fruits
        Sugar, etc.

C. Materials for the production of liqueurs and cordials.

        Alcohol
        Sugar
        Spices and Essential Oils
        Coloring Agents, etc.

Distilled Spirits.—The materials used for the production of distilled spirits are further classified according to the types of product in which they find employment:

American Whiskies

Rye
    Barley
    Rye

Bourbon
    Barley
    Corn
    Wheat

Scotch Whiskies

Pot Still Type
    Barley

Patent Still Type
    Barley
    Rye
    Corn
    Oats

Irish Whiskies

Pot Still Type
    Barley
    Rye
    Oats
    Wheat

Patent Still Type
    Barley
    Rye
    Corn
    Oats

Gin (*Distilled*)
    Barley
    Rye
    Corn

Kornbranntwein
    Barley
    Rye
    Corn
    Wheat

Vodka
    Barley
    Rye
    Corn (In cheaper grades)

Rum
    Molasses
    Cane Juice

Schnapps
    Potatoes

Brandy
    Grape juice and marc.
    Other fruits (Apples, prunes, cherries, etc.)

Since the entire spirit industry is designed to produce palatable products, the tabulation given above is not necessarily binding. Some of the materials listed are very seldom used. Practice may differ from plant to plant producing the same product or may be influenced from year to year by changed crop and economic conditions. Owing to the great variety of distilled liquors in flavor and general character, and the influence on these of both the materials used and the processes by which they are treated, it is impossible to be completely general. However, it may be stated that the important cereal grains to the liquor industry are in order: barley, rye, corn, oats, and wheat. A short discussion of these grains including both their botanical and chemical characteristics follows:

Cereal Grains.—*Barley.*—Barley takes the first place in importance in the spirit industry on account of its high production of the enzyme, diastase, when permitted to sprout (malt). There are several types of barley which are largely used. These include:

(1) a. Hordeum distichum, a two-row type which includes the well-known varieties Chevalier, Hallett, Hanna.

    b. Hordeum zeocritum, two-row fan-shaped barleys of which Goldthorpe is the leading variety.

(2) Hordeum vulgare, the ordinary six-row barley, such as Manchurian, Oderbrucker, Scotch, etc.

(3) Hordeum hexastichum, which likewise have their flowers on a spikelet fertile, but on account of the fact that the ears are wide, the appearance of the head is a hexagon when examined from the top. These are really six-row barleys. An example of these is the white barley of Utah and adjoining States.

The average composition of barleys of these three types, together with data as to weight, are given in the following tabulation:

TABLE I.—AVERAGE PERCENTAGE COMPOSITION OF THREE TYPES OF
MALTING BARLEYS (LE CLERC) [1]

| Type of barley | Moisture, % | Water-free basis | | | | | | Wt. per bushel, pounds | Wt. per 1,000 grains, grams |
		Ash, %	Protein, %	Fat, %	Fiber, %	Pentosans, %	Starch %		
Two-row...........	8.9	2.9	11.6	2.0	5.2	8.4	59.1	52	38
Ordinary six-row.....	8.7	3.0	11.9	2.0	5.8	9.6	58.9	47	27
Hexastichum.........	8.5	2.9	10.0	2.0	5.8	9.0	59.9	48	38

[1] U. S. Dept. Agr., Bureau of Chemistry, Bull. 124, Le Clerc and Wahl, Chemical studies of American barleys and malts.

The two-row barleys are chiefly grown in Europe, although they are raised in this country in Montana, Idaho, New York, etc., to a limited extent. On account of their relatively high carbohydrate content and low protein, they are particularly adapted to use in the brewery.

In this country, the ordinary six-row barley is the kind most abundantly produced, being raised extensively in the States of the Mississippi Valley. Its relatively high protein content causes it to produce malt of high diastatic power, and thus fits it especially to use in the distillery.

A good distiller's barley should be free of dirt and have good odor and color, small grains of uniform size, a high percentage of nitrogen, and high germinating capacity. With these characteristics and with proper treatment in the malt house, it is bound to yield a malt of good character.

Cleanliness is necessary, not only because dirt is not a source of alcohol, but because it is sure to carry large numbers of bacteria and molds, which interfere with the production of a good malt.

Barley is composed of about 12 per cent husks, 10 per cent bran, 2.5 per cent embryo, and the rest endosperm or the stored food for the plantlet. The husks and bran are merely protective. The germ is the seat of life; it consists of embryonic radicles or rootlets and the plumula or acrospire. The sprouting of this germ is essential to the production of malt since this serves to convert the starch of the endosperm into fermentable sugar. The endosperm is composed mostly of starch and protein, but both of these substances are insoluble and non-diffusible and can not be used directly in supplying food for the young plant. The agency which renders these soluble and diffusible is an enzyme or a series of enzymes secreted by the embryo during growth, one of which, the diastase, dissolves the starch and converts it into sugar and dextrin, while another, the peptase, acts on the protein. These enzymes are developed in the growing malt as the germ's need for food increases. The products of enzyme action are soluble and diffusible, and can be directly used as food by the growing embryo.

Barley contains about 65 per cent of fermentable matter; and at a weight of 48 pounds per bushel one ton should produce about 98 gallons of alcohol.

Barley, to be considered good, should show a germination of at least 97 per cent. If below this limit of vitality, it should be reduced in price, or rejected. Grains which are incapable of germination are not only useless, but harmful, because they act as carriers of micro-organisms which may infect the rest of the grain.

Rye.—Rye is, comparatively speaking, a minor crop in the United States. The only species under cultivation is the common rye: Secale cereale.

In structure and habits of growth, rye resembles wheat, and, like wheat, it is grown as both a spring and winter crop. It is a tall-growing, annual grass with fibrous roots, flat, narrow, bluish-green leaves, standing erect or decurved and having slender cylindrical spikes consisting of two or three-flowered spikelets. The flowering glumes are long-awned or bearded and lance-shaped, and are so firmly attached that little chaff results from the threshing. The individual grains are partly exposed and are longer, more slender and more pointed than wheat. They are dark, with a slightly wrinkled surface and are very hard and tough, requiring more power to mill than any other grain.

The following is an analysis of common rye:

TABLE II

Moisture	11.0%
Ash	2.0%
Protein (N × 6.25)	11.6%
Ether extract	1.7%
Crude fiber	2.0%
Pentosans	8.5%
Total sugar	4.0%
Other carbohydrates	59.2%
Wt. per 1,000 grains	25.0 grams
Wt. per bushel	56.0 lbs.

Rye is very largely used in distilleries which produce potable spirit such as whiskey, gin and vodka, and in the manufacture of compressed yeast. It is also used in relatively small amounts in the yeast mashes of alcohol distilleries. It is not suited to use as the chief ingredient of the mash in an alcohol distillery, on account of the tenacious quality of the mash which it forms. Furthermore, it gives a low yield in proportion to the amount of starch which it contains. Though it usually contains over 60 per cent of fermentable matters, it rarely produces over 85 gallons of alcohol to the ton.

Corn.—This valuable food stuff is the grain of a gigantic grass known to botanists as Zea Mays. It originated in America but subsequently was transplanted to many other countries, principally France, Hungary, Italy, Spain, Portugal, South Africa and Argentine.

There are over three hundred recognizable varieties, some of which are only a few inches in height, while others are giants of six feet or more; some come to maturity in two months, while others require three or four times as long before their cobs ripen. There is also great variety in the shape, size, and color of the actual grain. Some are white as, for example, the Cuzco maize, others are yellow, red, purple, or even striped, and the varieties differ among themselves in chemical composition.

The most important maize growing country of the world is the United States. Many varieties are cultivated, but the chief may be roughly grouped into four classes. First come the "Flint" varieties which are most commonly met with east of Lake Erie and north of Maryland, and the "Dent" varieties which are most popular west and south of these localities. The "Horsetooth," which passes insensibly into the above forms, is grown chiefly in the South. Lastly, the "Sweet" varieties are extensively cultivated for the green grain. These are boiled and used as a vegetable and seldom allowed to mature into the ripened grain.

For the making of whiskey the finest white "Flint Corn" is preferred. The Kentucky distilleries are extremely careful in their selection of the raw material. Indian Corn that is grown along the Ohio and the Kentucky Rivers is especially sought after by the distillers as being peculiarly suitable. Corn which has been injured by frost, heat, moisture or mold can readily be used in the manufacture of industrial alcohol as such damage does not affect the fermentable content of the grain and it can be bought at a low price, but it is unsuitable for the liquor industry.

A typical American maize should have the following composition:

TABLE III

Weight of 100 kernels	38 grams
Moisture	10.75%
Proteids	10 %
Ether extract	4.25%
Crude fiber	1.75%
Ash	1.50%
Carbohydrates other than crude fiber	71.75%

Dry corn of good quality should yield at least 65% of sugars and starch and should yield from 98 to 105 gallons of 180° alcohol per ton (shelled).

Oats.—Oats is the grain or seed of the cereal grass Avena sativa. It forms one of the most valuable sources of food for both man and beast, the nutritive value of the grain being very high. It is extensively grown in Great Britain, Continental Europe, Russia and North America.

Practically four-fifths of the oat crop of the United States is produced in the thirteen states extending from New York and Pennsylvania westward to North Dakota, South Dakota, Nebraska and Kansas. Each of these states devotes more than a million acres to oats. The average yield in the six northernmost states, New York, Michigan, Wisconsin, Minnesota, North Dakota and South Dakota is 31.68 bushels per acre while their total production is slightly less than one-third of the oat crop of this country. The average yield of the other seven states, Pennsylvania, Ohio, Indiana, Illinois, Iowa, Nebraska and Kansas is only 29.23 bushels per acre yet they produce more than half of the entire crop. The difference in yield of nearly two and one-half bushels to the acre between these two groups of states is due largely to the fact that the climatic conditions of the northern group are better suited to the production of the crop. There is no material difference in soil composition or other factors affecting the yield.

Oats are grown in the corn belt, which includes all the states of the second group, largely because a small-grain crop is needed in the rotation and because the grain is desired for feeding to work stock. Spring wheat is seldom satisfactory in this district and winter crops often do not fit well into a rotation which ordinarily includes corn, a small grain and grass. Under these conditions oats are generally grown as the best crop between corn and grass. This is particularly true in Illinois and Iowa, the two states producing the greatest quantity of both corn and oats.

There are many factors which reduce the yield of oats in the corn belt. In general, those varieties of oats which mature earliest are best adapted to the belt, for early maturing often enables a crop to escape hot weather, injury from storms, and attacks of plant diseases. The early varieties also usually produce less straw and for that reason are less likely to lodge than the ranker growing late varieties. A number of years ago the Early Champion and the Fourth of July varieties came into prominence but they are not now extensively grown, for, although early in maturing, their yield is often unsatisfactory. Burt is a very early variety much used for spring seeding south of the Ohio River but little known elsewhere. The type of early oats now most largely grown in this country is represented by the Sixty-Day and the Kherson varieties, two comparatively recent introductions from Europe.

The following is a typical analysis of a variety of oats:

TABLE IV

Moisture	11.6%
Ash	3.4%
Protein (N × 6.25)	11.5%
Ether extract	4.7%
Crude fiber	11.0%
Pentosans	12.0%
Total sugar	1.5%
Other carbohydrates	44.3%
Wt. per 1,000 grains	29.2 grams
Wt. per bushel	32.0 lbs.

This grain is not too well suited for distillery use because of the glutinous nature of the mixture which is formed when it is treated with hot water. It contains about 50 per cent of fermentable substance and might be made to yield about 70 gallons of alcohol per ton.

Wheat.—Wheats have been cultivated by man since before the dawn of history, and nothing is now known of the original wild forms from which they are descended. In old legends and ancient manuscripts wheat is spoken of as familiarly as at the present day. Nor do we know with any certainly in which country it was first found; but it seems probable that Central Asia was the original home of the wild forms from which the cultivated species have sprung.

Wheat belongs to the grass family, Pouceas (Germineae), and to the tribe called Hordeae, in which the 1 to 8 flowered spikelets are sessile and alternate on opposite sides of the rachis, forming a true spike. Wheat is located in the subtribe Triticeae and in the genus Triticum where the solitary two-to-many flowered spikelets are placed sidewise against the curved channeled joints of the rachis.

Great diversity is shown by the varieties of wheat grown in the United States. More than 200 distinct varieties are grown. Clark ("Classification of American Wheat Varieties" U. S. Department of Agriculture, Dept. Bull. 1074, 1922) divides American grown wheats into the following groups:

(1) Common
(2) Club
(3) Poulard
(4) Durum
(5) Emmer
(6) Spelt
(7) Polish
(8) Einkorn
(9) Unidentified

Within these groups are numerous types of which the following more important are arranged in approximate order of production and use:

Common
Turkey, Marquis, Fultz, Red Wave, Poole, Fulcaster.

Club
Hybrid (Grown only on Pacific coast)

Durum
In Dakotas and adjoining states

Emmer
  Minnesota, Dakotas and adjoining states

Polish
  New Mexico and Wyoming
  (Not grown much)

Spelt
  Not grown to any extent commercially.

The following is an analysis of two typical wheats:

TABLE V

	Soft wheat	Hard Wheat
Moisture	12.0%	12.0%
Ash	1.9%	1.8%
Protein (N × 6.25)	9.0%	12.4%
Ether extract	1.7%	1.7%
Crude fiber	2.5%	2.5%
Pentosans	7.0%	7.0%
Total sugar	2.7%	2.7%
Other carbohydrates	63.2%	59.9%
Wt. per 1,000 grains	38.7 grams	38.7 grams
Wt. per bushel	60.0 lbs.	60.0 lbs.

What has been said regarding the yield of alcohol to be obtained from rye applies also to wheat and therefore, the latter has not been much used in distilleries, even at times when it was relatively cheap.

Wine Materials.—The wine industry differs from the spirit industry, among other things, in that the distinguishing characteristics of the finished product, wine, are much more closely related to the individual character of the raw material. Although almost all wines are made from grapes, the character of a single batch of wine will depend not only on the variety of grape used, but even on the special plot of ground on which it was grown and the climatic conditions in the year of its growth. For this reason a classification of materials for the wine industry in the same manner as that given for the spirit industry is not possible. On the other hand, a means of classification is possible based on the geographical distribution of the grape varieties which are successfully grown for wine production.

TABLE VI.—A COMPARISON OF THE AVERAGE COMPOSITION OF SOME OF THE MORE IMPORTANT GRAINS AND SEEDS *

	Soft wheat	Hard wheat	Rye	Corn	Oats	Barley	Rough rice	Buck-wheat	Sorg-hum	Flax seed	Soy-beans	Polished rice
Moisture...........	12.0	12.0	11.0	11.7	11.6	11.3	11.1	11.7	9.5	5.3	7.0	11.7 –16.0
Ash...............	1.9	1.8	2.0	1.4	3.4	2.9	3.7	2.4	1.8	3.7	5.2	0.14– 0.46
Protein (N × 6.25)..........	9.0	12.4	11.6	10.0	11.5	10.8	7.2	11.7	13.3	27.0	38.0	5.50– 7.50
Ether extract........	1.7	1.7	1.7	4.8	4.7	2.0	1.7	2.5	3.5	35.1	18.5	0.15– 0.60
Crude Fiber........	2.5	2.5	2.0	2.2	11.0	5.7	8.5	11.7	1.8	5.1	4.0	0.18– 0.39
Pentosans........	7.0	7.0	8.5	6.0	12.0	8.6	5.5	6.5	4.8	......	4.5	......
Total sugar........	2.7	2.7	4.0	1.2	1.5	3.2	0.8	2.2	1.0	......	8.0	......
Other carbohydrates.......	63.2	59.9	59.2	62.7	44.3	55.5	61.5	51.3	64.3	23.8	14.8	75.8 –81.0
Weight per 1,000 grains, gms.	38.7	38.7	25.0	367.0	29.2	39.2	29.2	27.9	24.8	4.0	150.0	13.4 –37.2
Weight per bushel, lbs.......	60.0	60.0	56.0	56.0	32.0	48.0	45.0	50.0	56.0	56.0		

* G. Issoglio, Chim. degli Alimenti, Vol. 2, p. 311.

Since the United States is geographically so large and offers for grape culture at one spot or another every feature that can be found elsewhere, the following tabulation and discussion is confined to this country. The principles, of course, and possibly excepting some special niceties, the practice are of universal application.

The wine industry is somewhat less than 200 years old in the United States. It has successfully taken root in some of the Eastern and Middle States and principally in California and Texas. The first successful cultivation of American wine grapevines was carried out in California in 1770 at the San Diego Mission of the Franciscan Padres. They used a Spanish vine which had been successfully transplanted in Mexico. The first real attempt to cultivate the grapevine on a commercial scale in California was probably around 1861 when the state government backed Colonel Haraszthy in a trip to the principal grape growing districts of the world. He visited Europe, Asia Minor, Persia and Egypt and returned with 200,000 cuttings of vines which he believed likely to grow in California. A nursery was established at Sonoma and a study was made of the possibilities of the transplanted vines. Subsequently, the first State Board of Viticultural Commission, the United States Department of Agriculture and University of California, imported other clippings which were also tested. As a result vines were soon distributed and planted in all suitable localities of the state. Experienced European grape-growers and wine-makers were then induced to settle in the state and in course of time the well-known Californian industry was established successfully. As a result of this farseeing and well organized undertaking California, today, is able to imitate the wines of practically any district of the world.

Many attempts were also made in Eastern states to establish vineyards and to found an Eastern wine industry, procedure being much along the same lines as those tried successfully in California. Unfortunately, European grapes will not flourish in these sections of the United States and all such attempts ended in disaster. About the middle of the last century Nicholas Longworth, Sr., experimented with the native Catawba grape in Ohio and was so successful that the Ohio River was often referred to as the "Rhineland of America." This undertaking also came to a disastrous ending when disease attacked the vines. However, fresh plantings of American Vine Roots and Hybrids along the shores of Lake Erie and in Steuben County, New York, and parts of New Jersey have been more successful.

Geographical Considerations.—The choice of a site for grape production, as indicated previously, depends very largely on the nature of the soil and the climate. The following tabulation of grapes successfully grown in the United States will illustrate this statement:

TABLE VII.—AMERICAN GRAPE VARIETIES

Lake Shore District of Ohio
    Catawba, Delaware, Concord, Norton

Steuben County, New York
    Arranged in the order of merit:
        Delaware, Iona, Diana, Catawba, Concord, Isabellas, Norton

Southern Texas
    Devereux (Black July), Mustang, Herbemont, Lenoir (known as Burgundy in eastern Texas and Black Spanish in western Texas).

California
                Fresno County
Zinfandel, Malvoisie and Fahirzozos.
Zinfandel is considered best grape; color, excellent; flavor and acid, splendid.
                Sonoma Valley
Mission, Riesling, Gutedel (Chasselais), Muscatel, Burger, Zinfandel.
The Mission grape has spread over the whole state and is much used in the production of Hock, Claret, Port and Angelica.

                Napa County
Riesling, White, Pineau and Chasselais for dry white wines.
Black Burgundy, Zinfandel and Charboneau for Claret. The first makes a dark, full bodied and richly flavored wine. The second has a fine raspberry flavor but an excess of acid and is a little light in body and color.
Black Malvoisie is the best Port wine grape.

Bioletti (California Agr. Exp. Sta. Bull. 193. *"The Best Wine Grapes of California"*) has summarized the geographical and climatic factors which must be borne in mind. While his statement is with particular reference to California, the principles which he enunciates are of general validity. He says:

"For the good of the industry at large it is desirable that varieties should be planted which will produce as large a crop as is compatible with such quality as will maintain and extend the markets for our wine. These markets are varied in character. For some, cheapness is the essential factor; for others, quality. Cheap wines can be produced with profit only from heavy-bearing varieties grown in rich soil; wines of the highest quality only from fine varieties grown on hillsides or other locations where the crops are necessarily less. It is therefore unwise to plant poor-bearing varieties in the rich valleys where no variety can produce a fine wine. It is equally unwise to plant common varieties on the hill slopes of the Coast Ranges where no variety will produce heavy crops. The vineyards of the San Joaquin, Sacramento, and other valleys can not compete with the vineyards of the Coast Ranges in quality, and the latter can not compete with the former in cheapness.

Each region has its own special advantages which, if properly used, will make grape-growing profitable in all, and instead of competing each will be a help to the other. The danger to be feared by the grape-growers of the Coast Ranges from the production of dry wine in the interior is not competition, but lies in the bad reputation given to California wines by the production of spoiled and inferior wines. If the cheap wines of the valleys are uniformly good and sound the market for the high-priced fine wines of the hills will increase, and large quantities of the Coast Range wines will be used for blending with the valley wines to give them the acidity, flavor and freshness which they lack.

In order to obtain these results it is necessary that varieties suited to each region and to the kind of wine should be planted. No variety which is not capable of yielding from 5 to 8 tons per acre in the rich valley soils or from 1½ to 3 tons on the hill slopes should be considered. On the other hand, no variety which will not give a clean-tasting, agreeable wine in the valley or a wine of high quality on the hills should be planted, however heavily it may bear. To plant heavy-bearing inferior varieties such as Burger, Feher Szagos, Charbono, or Mataro on the hills of Napa or Santa Cruz is to throw away the chief advantage of the location. The same is true of planting poor-bearing varieties such as Verdelho, Chardonay, Pinot, or Cabernet Sauvignon in the plains of the San Joaquin.

With these considerations in view, the following suggestions are made for planting in the chief regions of California:

WINE GRAPES RECOMMENDED FOR CALIFORNIA

For Coast Counties

Red Wine Grapes	*White Wine Grapes*
1. Petite Sirah	1. Semillon
2. Cabernet Sauvignon	2. Colombar (Sauvignon vert)
3. Beclan	3. Sauvignon blanc
4. Tannat	4. Franken Riesling
5. Serine	5. Johannisberger
6. Mondeuse	6. Tsaminer
7. Blue Portuguese	7. Peverella
8. Verdot	

For Interior Valleys

Red Grapes	*White Grapes*
1. Valdepenas	1. Burger
2. St. Macaire	2. West's White Prolific
3. Lagrain	3. Vernaccia Sarda
4. Gros Mansene	4. Marsanne
5. Barbera	5. Folle blanche
6. Refosco	
7. Pagadebito	

For Sweet Wines

Red Grapes	*White Grapes*
1. Grenache	1. Palomino
2. Alicante Bouschet	2. Beba
3. Tinta Madeira	3. Boal
4. California Black Malvoisie	4. Perruno
5. Monica	5. Mantuo
6. Mission	6. Mourisco branco
7. Mourastel	7. Pedro Ximenez
8. Tinta Amarelia	

Finally, a few suggestions as to what "not to do."

Don't plant Mataro, Feher Szagos, Charbono, Lenoir, or any variety which makes a poor wine everywhere.

Don't plant Burger, Green Hungarian, Mourastel, Grenache, or any common heavy-bearing varieties on the hill slopes of the Coast Ranges.

Vineyards in such situations must produce fine wines, or they will not be profitable.

Don't plant Chardonay, Pinot, Cabernet Suavignon, Malbec, or any light-bearing varieties in rich valley soils. No variety will make fine, high-priced wine in such situations, and heavy bearers are essential to the production of cheap wines.

Don't plant Zinfandel, Alicante Bouschet, or any of the varieties which have already been planted in large quantities, unless one is sure that the conditions of his soil and locality are peculiarly favorable to these varieties and will allow him to compete successfully."

Liqueurs and Cordials.—The raw materials for the manufacture of liqueurs and cordials are necessarily classified in a different manner from those used in the distilled spirit or wine industries. In general they may be divided into three broad groups with subdivisions as indicated:

A. *Flavoring Agents*
  1. Herbs, spices, seeds, roots, and fruits.
  2. Aromatic spirits and tinctures.
  3. Aromatic waters.
  4. Essential Oils.

B. *Coloring Agents*
  1. Vegetable Coloring Matters.
  2. "Aniline" dyes. (Synthetic dyestuffs.)

C. *Bulk Ingredients*
  1. Sugar and glucose.
  2. Brandies or rectified spirits.
  3. Water.

The number of materials included in Group A1 above is very large; among others the following find principal use:

HERBS, SEEDS, ROOTS, SPICES, BARKS, FRUITS, ETC.

Chinese aniseed	Cocoa—Caraque
Bitter almonds	Cocoa—Maragnan
Green anis	Mace
Coriander	Vanilla
Fennel	Figs
Angelica—root	Cumin
Angelica—seed	Calamus, aromatic
Lemon peel	Peel of Dutch Curacao

Orange peel	Aloes
Anis de Tours	Saffron
Anis d'Albi	Curacao—reeds
Ceylon cinnamon	Wormwood
Orris	Hyssop Flowers
Cloves	Lemon balm
Marjoram	Cherries
Sweet almonds	Alpine mugwort
Nutmeg	Arnica Flowers
Sassafras	Peppermint
Muskseed	Balsamite
Apricot stones	Thyme
Cherry stones	Tonka beans
Dried peach leaves	Black currants
Dried peaches	Black currant leaves
Myrrh	Quinces
Quince juice	Red Sandalwood
Strawberries	Liquorice wood
Nuts	Ginger root
Pineapple	Galanga root
Angostura bark	

In Group A 2 will be found:

AROMATIC SPIRITS

Anis	Celery
Angelica—root	Aloes
Angelica—seeds	Myrrh
Curacao	Saffron
Peppermint	Chinese cinnamon
Apricot	Cloves
Lemons	Nutmeg
Coriander	Orange Flowers
Amberseed	Bitter Almonds
Dill	Cardamom—major
Caraway	Cardamom—minor
Daucus	Oranges
Fennel	Nuts
Strawberries	

Group A 3 includes:

AROMATIC WATERS

Orange Flower	Moka
Peppermint	Chinese aniseed
Rose	Clove

Group A4 includes the flavoring principles derived from any of the raw materials listed under the heading of Group A2. The term "essential oils" is a generic commercial phrase used to designate the volatile oils obtained by various processes such as steam distillation, expression, maceration, enfleurage and solvent extraction, from the specific plant or part of a plant which gives them their name. The term is derived from the word "essence" meaning that in these oils is contained the soul or strength of the plant. In fact, in some countries, the single word is used instead of the longer "essential oil." In the United States, however, the usage is that the term "essence" or "spirits of" applies rather to a solution in alcohol of the "essential oil." And a new phrase "soluble essence" has been invented to describe essences which are completely miscible with water or dilute alcohol without separation of the oil taking place.

The essential oils exist in all odorous vegetable tissues, sometimes pervading the plant, sometimes confined to a single part; in some instances, contained in distinct cells, and partially retained after desiccation, in others, formed upon the surface, as in many flowers, and evaporating as soon as formed. Occasionally two or more oils are found in different parts of the same plant. Thus, the orange tree produces one oil in its leaves, another in its flowers, and a third in the rind of its fruit.

The volatile oils are usually colorless when freshly distilled, or at most yellowish, but some few are colored brown, red, green or blue. There is reason, however, to believe that in all instances the color depends on foreign matter dissolved in the oils. They have strong odors, resembling that of the plants from which they were procured, though generally less agreeable. Their taste is hot and pungent, and, when they are diluted, is often gratefully aromatic. The greater number are lighter than water, though some are heavier; their specific gravity varies from 0.847 to 1.17. They vaporize at ordinary temperatures, diffusing their particular odor and are completely volatilized by heat.

The following is a partial list of coloring agents used in the production of liqueurs and cordials:

Caramel	Indigo (Sulphonated)
Cherry Extract	Orchil
Chlorophyll preparations	Saffron
Cochineal	Vanilla Extract
Cudbear	

SYNTHETIC DYESTUFFS

In France the artificial coloring agents used are subject to less limitation than in the United States, and since France is the largest producer of cordials the following list is appended:

Rose Bengal	Bleu de lumière
Rouge de Bordeaux	Bleu coupier
Fuchsine, acid	Malachite green
Bleu de Lyon	Violet de Paris (Gentian Violet 3B)

It should be noted that not one of these dyes is included in the United States list of colors certified for use in foods, which only are permitted to be used in this country in interstate commerce and in many states in intrastate commerce. This list which follows is varied enough to allow for the production of any desired shade.

CERTIFIED FOOD COLORS

RED SHADES:
- 80. Ponceau 3R.
- 184. Amaranth.
- 773. Erythrosine.
- Ponceau SX.

ORANGE SHADE:
- 150. Orange 1.

GREEN SHADES:
- 666. Guinea Green B.
- 670. Light Green SF Yellowish.
- Fast Green FCF.

BLUE SHADES:
- 1180. Indigotine.
- Brilliant Blue FCF.

YELLOW SHADES:
- 10. Naphthol Yellow S.
- 640. Tartrazine.
- 22. Yellow AB.
- 61. Yellow OB.
- Sunset Yellow FCF.

The numbers preceding the names refer to the colors as listed in the Colour Index published in 1924 by the Society of Dyers and Colourists of England, which gives the composition of these dyes. Names not preceded by numbers are not listed in the Colour Index.

No description is here necessary of the materials which we have classified as group C above. A description of their specific employment will be found in subsequent chapters devoted to manufacturing operations. The same applies to a number of minor ingredients or materials used for special purposes such as souring, fining, preserving, sterilizing, etc.

CHAPTER V

YEASTS AND OTHER ORGANISMS

General Statement.—In the preceding chapter on Fermentation it was stated that the production of alcohol is performed by enzymes which act on the sugars present in the fermenting liquor. It was also indicated previously that the enzymes which accomplish this transformation are representative of a very large group of such materials which function in every chemical change involved in the life process. The enzymes of value to the fermentation industry are produced by living plants, yeasts, which are allowed to grow in the liquor to be fermented and which, as an incident of their own life process, achieve the desired production of alcohol.

Yeasts belong to the second broad group of vegetable growths: those which do not contain chlorophyll and are, therefore, unable to manufacture their own food. This group is distinguished from the first group of plants which can extract inorganic materials from the soil and the air and from these manufacture their own food. The ability to perform this function is ascribed to the presence in the plant of a green coloring matter, chlorophyll.

The group name of the non-chlorophyll bearing plants is fungi. They are all dependent for their food supply upon the materials built up by living plants of group one, or upon animal matter. Among the fungi, which naturally vary in complexity of structure, are a group of simply constructed plants having but one cell and which are called yeasts. The cells vary in shape and size, being round, oval or elongated, but of the order of 0.003 inches in diameter.

Each such cell consists of a transparent elastic sac or membrane enclosing a more or less granular mass of jelly-like substance which is called protoplasm. The name was originated by

46

Purkinje (*ca.* 1840) to apply to the formative material of young animal embryos. Later, von Mohl (*ca.* 1846) used the term to distinguish the substance of the cell body from that of the nucleus which he called cytoplasm. In modern usage dating from Strasburger (*ca.* 1882) the name protoplasm has been applied to all of the essential living substance within the cell wall and means the form of matter in or by which the phenomena of life are manifested. Protoplasm can exist in many modifications varying from its usual one of a thick, viscous, semi-fluid, colorless, translucent mass containing a high proportion of water and holding in suspension fine granular material. Chemical examination of protoplasm, necessarily after death, has shown that it is composed largely of protein material. During its life, however, it appears probable that the chemical composition is both more complex and more unstable. For our purpose it is only necessary to state that the protoplasm is that portion of the plant which is alive and which carries on the vital processes of the plant.

Life Processes.—All living organisms exemplify the cycle of birth, growth, multiplication and death. Starting, for yeast, at the stage of growth we see the process as follows: When the yeast is supplied with an abundance of nutrient material, it grows vigorously and the cellular protoplasm is homogeneous. As the growth continues and the nutrient material becomes exhausted, clear, apparently empty, spaces called vacuoles appear in the protoplasm. Actually these spaces are filled with serum or sap. At a little later stage, granules appear, some of which are fat globules while others are more condensed portions of protoplasm. Finally, as the cell nears the end of its life, the protoplasm shrinks to a thin layer against the cell wall while the balance of the cell is occupied by a large vacuole. The nucleus of the cell is rarely visible and does not, for this reason, enter into our consideration.

While this growth of the single cell continues, multiplication also takes place at a rapid rate. According as the life conditions are favorable or not multiplication takes place in one of two ways:

1. By budding or germination.
2. By endogenous division or ascospore formation.

The first process occurs under favorable conditions when the yeast is growing rapidly. It consists in a bulging of part of the cell wall and the pressing of part of the cell contents into the bulge which is formed. This is the bud. As the bud grows the wall between it and the mother cell constricts and finally closes and there are then two distinct cells. Whether the bud-cell stays in connection with the mother cell depends somewhat on the conditions of growth but more on the variety of yeast. With some varieties the bud stays in connection with the mother cell and itself multiplies through many generations so that long chains or branching clusters are formed. In all cases the rapidity of multiplication depends on the vigor of the yeast and the suitability of the conditions under which it is growing.

In distinction to the method of reproduction just described, when a vigorous yeast growth is placed into adverse conditions it prepares to survive by a different mode. The protoplasm of the cells becomes granular, then divides so as to form separate masses. These round off and become invested with a wall so that the original cell wall acts merely as a sac to contain the new bodies. Because of this last fact the name ascospores is given them, meaning spores formed within a sac. The conditions required to bring about ascospore formation are not completely understood. Usually, however, a suitable temperature, plenty of moisture and lack of nutrient material will cause young, vigorously growing yeast cells to form from one to four ascospores each. The spores have remarkable ability to survive under conditions which would be fatal to ordinary cells; such as extremes of temperature, lack of food, drying out, etc.

When at some future time the spores are placed into favorable conditions for growth they germinate and start a new series of the ordinary type of yeast cells. In germinating they exert a pressure against the wall of the mother cell which finally breaks and permits the escape of the new cells. In some cases actual dissolution of the cell occurs during germination.

The changes described above are well illustrated in the accompanying figure, which shows stages in the life of a favorable pure culture yeast and also some of the mixed forms.

Classification of Yeasts.—Although the structure of yeasts is so exceedingly simple that it seems difficult for varieties to exist, nevertheless there are many different kinds or species. The distinctions are partly morphological, *i.e.*, in the physical form of the cells, and partly chemical, *i.e.*, in the enzymes elaborated by the cells which, of course, means that the products resulting from

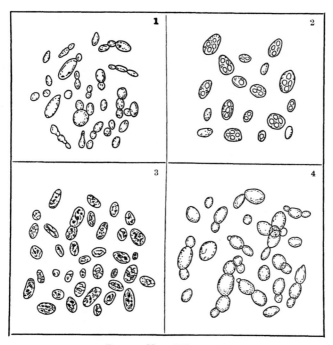

FIG. 2.—True Wine yeast.

1. Yeast cells in fermenting grape juice (*mixed forms*).
2. Saccharomyces ellipsoideus (*spores*).
3. Saccharomyces ellipsoideus (*old*).
4. Saccharomyces ellipsoideus (*young*).

the action of different yeasts on the same medium or substrate will be different. When a pure yeast is used in fermentation, an entirely different result is obtained than from the use of an impure. The flavor is different, as also the odor, and with a pure yeast the keeping properties are better. For the purposes of the fermentation industry, yeasts are divisible into two groups:

1. The wild yeasts—those which occur in nature; floating in the air, in the soil, and on the skins of fruits.

2. The cultivated yeasts—which have been selected from the wild yeasts for their favorable action and are carefully guarded in laboratory and factory to insure purity of strain.

There are two schools of thought in the fermentation industry and especially among wine makers as to whether it is preferable to use pure culture yeasts or to depend on wild yeasts. On the one hand it is claimed that pure culture yeast will result in a more accurately controlled fermentation, in a reduction in the time and labor required for racking and aging, a cleaner taste and flavor, and in the ability to select beforehand the flavor desired.

On the other hand, it is claimed that these advantages are offset by the necessity of sterilizing the must, pressed grapes and juice. In practice, however, the benefits derived from the use of pure culture yeast can often be had without resorting to the costly operation of sterilization. The pure culture yeasts are especially advantageous in the manufacture of white and sparkling wines and in the refermentation of the latter. They are used extensively abroad for these purposes and have secured some recognition and produced excellent results in some wineries in this country.

In whiskey distilleries, pure culture yeasts have been used extensively in recent years. The types are selected for high alcohol yields and propagated by the Hansen single-cell method. There can be no doubt at the present time but that pure culture yeasts are an absolute necessity for the manufacture of distilled spirits and very much preferable for the manufacture of wines.

The principal yeasts encountered in spirit and wine manufacture are:

Saccharomyces cerevisiae. The ordinary yeast of the brewer and the distiller. Two kinds are recognized: (a) top fermentation, and (b) bottom fermentation.

Top yeast, as its name implies, is a type which rises in a frothy mass to the top of the mash during fermentation. Bottom yeast sinks to the bottom of the vat during fermentation. Higher temperatures favor the formation of the former and lower temperatures favor the latter. Distillers' yeast is a high attenuating top variety.

Saccharomyces ellipsoideus. This is the yeast which converts must, or grape juice, into wine.

Saccharomyces pastorianus. This also occurs in wine making and when present during brewing gives a bitter taste to the beer.

Saccharomyces mycoderma. This yeast is the cause of "mother" which appears on the surface of wine or beer after exposure for some days to the air.

Bakers use either compressed yeast (compressed cakes of top yeast) or dried yeast (a mixture of yeast cells with starch). The former has high fermenting capacity and gives uniform results, but it will keep only a day or two; while the latter retains its capacity to produce fermentation for a long period. Brewers' yeast is not desirable for bread making because it is likely to give a bitter flavor and its activity is slow in a dough mixture.

Pure Cultures.—To determine the properties due to a particular yeast, it must be separated from all other organisms with which it is associated and when grown thus, free from all contamination, it is then known as a "pure culture." A commercially pure yeast is different, as this simply means "free from added non-yeast matter." This is the condition of most compressed yeasts as found on the market; they are commercially, but not bacteriologically, pure, since they have numbers of bacteria and molds associated usually with more than one yeast variety.

Micro-Organisms Found on Grapes.—The surfaces of grapes in the vineyard will hold any or all of the bacteria and fungi usually carried in the air and by insects. Many of these, especially the bacteria, cannot grow in grape juice on account of its acidity. These, of course, have a negligible effect on the wine. Others, such as most yeasts and molds and a few varieties of bacteria find grape juice to be a favorable medium for their development. Wine is a somewhat less suitable medium than unfermented grape juice (*must*), owing to its alcohol content, but still, a large number of forms are capable of growing in the wine. As the wine ages, the less suitable it becomes for the growth of micro-organisms, but it is never quite immune.

Among the great variety of organisms the only ones desired

are the wine yeasts. Many different types of wine yeast have been isolated and studied. It has been demonstrated that not only do slight morphological differences exist, but also that they vary in the flavor and quality of wine produced and also in the speed and completeness with which they split sugar and consume acids. The true yeasts occur much less abundantly on grapes than the molds. Until the grapes are ripe they are practically absent, as first shown by Pasteur. Later, they gradually increase in number; on very ripe grapes being often abundant. In all cases and at all seasons, however, their numbers are much inferior to those of the molds and pseudo-yeasts. The cause of this seems to be that, in the vineyard, the common molds find conditions favorable to their development at nearly all seasons of the year, but yeasts only during the vintage season.

Investigations of Hansen, Wortmann and others show that yeasts exist in the soil of the vineyard at all times, but in widely varying amounts. For a month or two following the vintage, a particle of soil added to nutritive solution contains so much yeast that it acts like a leaven. For the next few months the amount of yeast present decreases until a little before the vintage, when the soil must be carefully examined to find any yeast at all. As soon as the grapes are ripe, however, any rupture of the skin of the fruit will offer a favorable nidus for the development and increase of any yeast cells which reach it. Where these first cells come from has not been determined, but as there are still a few yeast cells in the soil, they may be brought by the wind, or bees and wasps may carry them from other fruits or from their hives and nests.

The increase of the amount of yeast present on the ripe grapes is often very rapid and seems to have (according to Wortmann) a direct relation to the abundance of wasps. These insects passing from vine to vine, crawling over the bunches to feed on the juice of ruptured berries, soon inoculate all exposed juice and pulp. New yeast colonies are thus produced and the resulting yeast cells quickly disseminated over the skins and other surfaces visited.

The more unsound or broken grapes present, the more honeydew or dust adhering to the skins, the larger the amount of yeast will be. The same is true, however, also of molds and other organisms.

True Wine Yeasts—*Saccharomyces ellipsoideus.*—In the older wine-making districts, much of the yeast present on the grapes consists of the true wine yeast, *S. ellipsoideus.* The race or variety of this yeast differs, however, in different districts. Usually several varieties occur in each district. The idea prevalent at one time, that each variety of grape has its own variety of yeast seems to have been disproved, though there seems to be some basis for the idea that grapes differing very much in composition, varying in acidity and tannin contents, may vary also in the kind of yeast present. Several varieties of ellipsoideus may occur on the same grapes. In new grape-growing districts, where wine has never been made, ellipsoideus may be completely absent.

Besides the true wine yeast, other yeasts usually occur. The commonest forms are cylindrical cells grouped as *S. pasteurianus.* These forms are particularly abundant in the newer districts, where they may take a notable part in the fermentation. Their presence in large numbers is always undesirable, and results in inferior wine. Many other yeasts may occur occasionally, and are all more or less harmful. Some have been noted as producing sliminess in the wine. Many of these yeasts produce little or no alcohol and will grow only in the presence of oxygen.

Pseudo-yeasts.—Yeast-like organisms producing no endospores always occur on grapes. Their annual life cycle and their distribution are similar to those of the true yeast but some of them are much more abundant than the latter. They live at the expense of the food materials of the must and, when allowed to develop, cause cloudiness and various defects in the wine.

The most important and abundant is the apiculate yeast, *S. apiculatus* (according to Lindner this is a true yeast producing endospores). The cells of this organism are much smaller than those of *S. ellipsoideus* and very distinct in form. In pure cultures these cells show various forms, ranging from ellipsoidal to pear-shaped (apiculate at one end) and lemon-shaped (apiculate at both ends). These forms represent simple stages of development. The apiculations are the first stage in the formation of daughter cells; the ellipsoidal cells, the newly separated daughter cells, which, later, produce apiculations and new cells in turn.

Many varieties of this yeast occur, similar in degree to those of *S. ellipsoideus*. They are widely distributed in nature, occurring on most fruits, and are particularly abundant on acid fruits such as grapes. Apiculate yeast appears on the partially ripe grapes before the true wine yeast and even on ripe grapes is more abundant than the latter. The rate of multiplication of this yeast

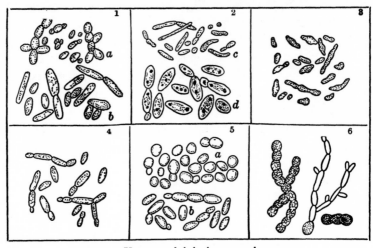

Fig. 3.—Yeasts and injurious pseudo-yeasts.
1. Torulæ and pseudo-yeasts.
2. Torulæ and pseudo-yeasts.
3. Saccharomyces apiculatus.
4. Saccharomyces pasteurianus.
5. Mycoderma vini (2 forms).
6. Dematium pullulans.

is very rapid under favoring conditions and much exceeds that of wine yeast. The first part of the fermentation, especially at the beginning of the vintage and with acid grapes, is, therefore, often almost entirely the work of the apiculate yeast.

The amount of alcohol produced by this yeast is about 4 per cent, varying with the variety from 2 to 6 per cent. When the fermentation has produced this amount of alcohol, the activity of the yeast slackens and finally stops, allowing the more resistant ellipsoideus to multiply and finish the destruction of the sugar.

The growth of the apiculatus, however, has a deterring effect on that of the true yeast so that where much of the former has been present, during the first stages of the fermentation, the latter often fails to eliminate all the sugar during the final stages.

Wines in which the apiculate yeast has had a large part in the fermentation are apt to retain some unfermented sugar and are very liable to the attacks of disease organisms. Their taste and color are defective, often suggestive of cider, and they are difficult to clear. This yeast attacks the fixed acids of the must, the amount of which is therefore diminished in the wine while, on the other hand, the volatile acids are increased.

Many other yeast-like organisms may occur on grapes; but, under ordinary conditions, fail to develop sufficiently in competition with apiculatus to have any appreciable effect on the wine. Most of them are small round cells, classed usually as Torulae. They destroy the sugar but produce little or no alcohol.

A group of similar forms, known collectively as *Mycoderma vini*, occurs constantly on the grapes but, all being strongly aerobic, they do not develop in the fermenting vat. Under favoring conditions, however, they may be harmful to the fermented wine.

Bacteria of many kinds occur on grapes as on all surfaces exposed to the air. Most of these are unable to develop in solutions as acid as grape juice or wine. Of the acid-resisting kinds, a number may cause serious defects and even completely destroy the wine. These, the "disease bacteria" of wine are mostly anaerobic and can develop only after the grapes are crushed and the oxygen of the must exhausted by other organisms. Practically all grape-must contains some of these bacteria, which, unless the work of the wine-maker is properly done, will seriously interfere with the work of the yeast, and may finally spoil the wine. The only bacteria which may injure the grapes before crushing are the aerobic, vinegar bacteria, which may develop on injured or carelessly handled grapes sufficiently to interfere with fermentation and seriously impair the quality of the wine.

Among the organisms which can infect wine and cause so-called "diseases" are the following:

Molds. The spores of the common saprophytic molds, *Penicillium, Aspergillus, Mucor, Dematium,* are always present on the grapes, boxes, and crushers, as on all surfaces exposed to dust laden air, and most of them find in grape must, excellent conditions for development. *Botrytis cinerea,* a facultative parasite of the leaves and fruit of the vine, is also nearly constantly present in larger or smaller quantities. All of these molds are harmful, in varying degrees, to the grapes and the wine. Some of them, such as *Penicillium,* may give a disagreeable moldy taste to the wine, sufficient to spoil its commercial value. Others, such as *Mucor* and *Aspergillus* may affect the taste of the wine but slightly and injure it only by destroying some of the sugar and thereby diminishing the final alcohol content. *Dematium pullulans* may produce a slimy condition in weak. white musts, and most of them injure the brightness and flavor to some extent and often render the wine more susceptible to the attacks of more destructive forms of micro-organisms.

On sound, ripe grapes, these molds occur in relatively small number, and, being in the spore or dormant condition, they are unable to develop sufficiently to injure the wine under the conditions of proper wine-making. On grapes which are injured by diseases, insects or rain, they may develop in sufficient quantities to spoil the crop before it is gathered. On sound grapes which are gathered and handled carelessly, they may develop sufficiently before fermentation to injure or spoil the wine.

The molds are recognized by their white or grayish cobwebby growth over the surface of the fruit. This consists of fine branching and interlacing filaments known as mycelium. This is the vegetative stage of the fungus and the active part in the destruction of the material attacked. When mature, it produces spores which differ for each mold in form, size and color. The spores are the chief means of multiplication and distribution. They are minute, single celled bodies which are easily distributed as dust through the air, and are capable, after remaining dormant for a longer or shorter period, of germinating, under favorable conditions and giving rise to a new growth of mycelium.

The commonest molds on grapes in California are the Blue Mold, the Black Mold and the Gray Mold. Usually only one of these occurs plentifully at the same time. Which this one will be depends principally upon the temperature and humidity. In the hotter regions the Black Mold is most common during the earlier part of the vintage, later the Blue Mold takes its place. In the cooler regions only Gray and Blue Molds occur commonly.

Blue Mold (Penicillium glaucum). This is the common mold which attacks all kinds of fruit and foods kept for a length of time in a damp place. It is distinguished by the greenish or bluish color of its spores which cover the grapes attacked, and by its strong disagreeable moldy smell. It sometimes attacks late grapes in the vineyard after autumn rains have caused some of them to split. Grapes lying on the ground are especially liable to attack. The principal damage of this mold occurs usually, after the grapes are gathered, while they lie in boxes or other containers. It will grow on almost any organic matter if supplied with sufficient moisture and at almost any ordinary temperature. It is almost the sole cause of all moldiness in boxes, hoses, and casks, and the most troublesome of all the molds with which the wine-maker has to deal.

The conditions most favorable to its development are an atmosphere saturated with moisture and the presence of oxygen.

Black Mold (*Aspergillus niger*). This is very common in the hotter and irrigated parts of California. It annually destroys many tons of grapes before they are gathered. It attacks the grapes just as they ripen and is distinguished by the black color of its spores, which sometimes fill the air with a black cloud at the wineries where the grapes are being crushed. It is especially harmful to varieties which have compact bunches and thin skins, such as Zinfandel. Its effect on the wine has not been well studied but it is much less harmful than Green Mold. Large quantities of grapes badly attacked are made every year into merchantable wine. The main damage done is in the destruction of crop and it is therefore a greater enemy to the grape-grower than to the wine-maker.

Gray Mold (*Botrytis cinerea*). This fungus in certain parts of Europe is a harmful parasite of the vine, injuring seriously leaves, shoots and growing fruit. The only injury of this kind noted in California is in the "callousing" beds of bench grafts.

As a saprophyte it may attack the ripe grapes in much the same manner as the Black Mold. It occurs apparently all over California but seldom does much damage. It attacks principally second crop and late table grapes.

Under certain circumstances this fungus may have a beneficial action. When the conditions of temperature and moisture are favorable, it will attack the skin of the grape, facilitating evaporation of water from the pulp. This results in a concentration of the juice. The mycelial threads of the fungus then penetrate the pulp, consuming both sugar and acid but principally the latter. The net result is a relative increase in the percentage of sugar and a decrease in that of acid. This, where grapes ripen with difficulty, is an advantage, as no moldy flavor is produced. Two harmful effects, however, follow: First, the growth of the mold results in the destruction of a certain amount of material and a consequent loss of quantity. This is, in certain circumstances, more than counterbalanced by an increase in quality, as is the case with the finest wines of the Rhine and Sauternes. For this reason, the fungus is called in those regions the "Noble Mold." Second, an oxydase is produced which tends to destroy the color brightness and flavor of the vine. This may be counteracted by the judicious use of sulfurous acid.

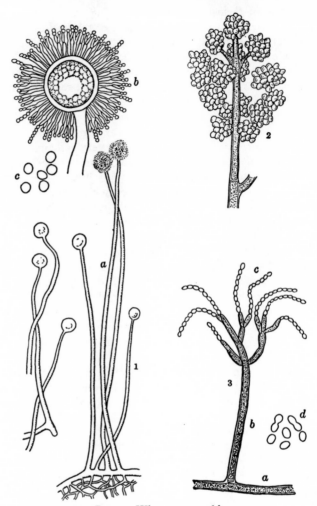

FIG. 4.—Wine grape molds.

1. Black mold (*Aspergillus niger*).   (After Duclaux.)
   a. Fruiting hyphae.
   b. Sporecarp showing formation of spores.
   c. Spores.

2. Gray mold (*Botrytis cinerea*).   (After Ravaz.)

3. Blue mold (*Penicillium glaucum*).   (From skin of moldy grape.)
   a. Mycelium.
   b. Fruiting hypha.
   c. Chains of spores.
   d. Spores.

This mold is not of great importance in California as its beneficial effects are not needed and there is seldom enough to do much harm.

The special organisms which cause diseases in wine include:
Anaerobic organisms such as *Dematium pullulans* induce slimy fermentation which results in "ropiness." These bacteria attack the sugar, but not glycerin nor alcohol and produce mannite, carbon dioxide, lactic and acetic acids and alcohol. Their

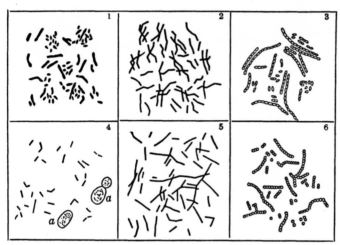

FIG. 5.—Disease Bacteria of wine.

1. Bacteria of mannitic wine.
2. Bacteria of bitter wine (*butyric*).
3. Bacteria of vinegar (*b. aceticum*).
4. Bacteria of lactic acid, *young*.
   (*a*) Cell of wine yeast.
5. Bacteria of lactic acid, *old*.
6. Bacteria of slimy wine.

growth is entirely prevented by the presence of alcohol above thirteen per cent, free tartaric, tannic or small amounts of sulphurous acid. The infection is ordinarily not very serious and disappears under ordinary cellar treatment.

Botrytis and *Penicillium* which when present cause oxidation of the tannin causing a bitter taste. This is more common in red wines.

Acetic acid bacteria which cause the further oxidation of alcohol to acetic acid and result in a "pricking" taste. This taste is noticeable even when there is only 0.1-0.15% of acetic acid. A dry wine becomes practically undrinkable at 0.25% of acetic acid.

Saccharomyces apiculatus will cause some production of alcohol but affects the flavor adversely.

Mycoderma vini attack the alcohol changing it to carbon dioxide and water and hence weaken the wine directly as well as rendering it more susceptible to infection by other disease organisms. It is sometimes the cause of film.

Control of Yeasts.—Control of the growth of these organisms and even to some extent selection of the variety which shall grow is largely possible by a consideration of the factors affecting their vigor.

Nutrition.—The preferred food of the yeasts is the sweet juice of more or less acid fruits. Most of them are active agents of alcoholic fermentation breaking up the sugar into alcohol and carbonic acid gas. Wine yeast may carry on the fermentation until the liquid contains 15 per cent or slightly more of alcohol. Other yeasts, such as ordinary beer yeast cease their activity when the alcoholic strength of the liquid reaches 8 to 10 per cent, while some wild yeasts are restrained by 2 to 3 per cent.

Relation to Oxygen.—They are aerobic, that is, they require the oxygen of the air for their development. Most of them are, however, capable of living and multiplying for a limited time in the anaerobic condition, that is, in the absence of atmospheric oxygen. It is in the latter condition that they exhibit their greatest power of alcoholic fermentation. They multiply most rapidly and attain their greatest vigor in the presence of a full supply of air. In fermentation, therefore, it is necessary, first, to promote their multiplication and vigor by growing in a nutritive solution containing a full supply of oxygen and, then, to make use of their numbers and vigor to produce alcoholic fermentation in a saccharine solution containing a limited supply of oxygen. These conditions are brought about automatically in the usual methods of wine-making. The stemming and crushing of the grapes thoroughly aerates the must. The yeast multiplies vigorously in this aerated nutritive solution until it has consumed most of the dissolved oxygen. It then exercises its fermentative power to break up the sugar, with the production of alcohol. With many musts it is able in this way to completely destroy all the sugar without further oxygen. In other musts, especially those containing a high percentage of sugar, the yeast becomes debilitated before the fermentation is complete. In such cases it is generally necessary to reinvigorate it by pumping over the wine or by some other method of aeration before it can complete its work.

Relation to Temperature.—Yeast cells can not be killed or appreciably injured by any low temperature. They do not become active, however, until the temperature exceeds 32° F. Wine yeast shows scarcely any activity below 50° F., and multiplies very slowly below 60° F. Above this temperature the activity of the yeast gradually increases. Between 70° F. and 80° F. it is very active and it attains its maximum degree of activity between 90° F. and 93° F. Above 93° F. it is weakened, and between 95° F. and 100° F. its activity ceases. At still higher temperatures the yeast cell dies. The exact death point depends on the condition of the yeast, the nature of the solution and the time of exposure. In must and wine a temperature of 140° F. to 145° F. continued for one minute is usually enough to destroy the yeast.

The best temperature in wine-making will depend on the kind of wine to be made and will lie between 70° F. and 90° F.

Relation to Acids.—The natural acids of the grapes, in the amounts in which they occur in must, have little direct effect on wine yeast. Indirectly they may be favorable by discouraging the growth of competing organisms more sensitive to acidity. Acetic acid has a strong retarding influence which commences at about 0.2 per cent and increases with larger amounts until at 0.5 per cent to 1.0 per cent, according to the variety of the yeast, all activity ceases.

Relation to Sulfurous Acid.—Sulfurous acid is an antiseptic, mild or strong, according to the quantities used. The fumes of burning sulfur are used in various ways and for various purposes in wine-making. The active principle of these fumes is sulfurous acid gas of which the chemical formula SO_2 shows that it is composed of one atom of sulfur combined with two atoms of oxygen. As sulfur has just twice the atomic weight of oxygen this means that one part by weight of sulfur combines with one part by weight of oxygen to produce two parts by weight of sulfurous acid gas. This combination takes place when sulfur is burned in free contact with air. The same substance can be obtained from certain salts, one of which is most suitable for use in wine-making. This is a potash salt known as potassium metabisulfite. This salt is composed of nearly equal weights of potash and sulfurous acid. In contact with the acids of the must, the sulfurous acid is set free and the potash combines with the tartaric acid of the must to form bi-tartrate of potash, some of which is already present as a natural constituent of the must.

Bacteria of all kinds are much more sensitive to the effects of sulfurous acid than are yeasts. If used, therefore, in properly regulated amounts it can be made a very efficient means of preventing bacterial action and thus indirectly of aiding the work of the yeast. It has also the very valuable property of preventing the injurious action of the oxydase produced by Botrytis and other molds. Finally, it is necessary in most cases to prevent the too rapid or overoxidation of the wine during aging.

CHAPTER VI

PRODUCTION OF YEAST

Commercial Yeast.—The application of the principles just developed is well illustrated in the manufacture of yeast for general use. The same niceties observed in this process must also be followed in the production of so-called "starters" for the fermentation of whiskey mashes or of wine must. Figure 6 is the flow sheet of such a process. The exact proportions of the various grains used are naturally varied according to the secret formula of the manufacturer.

It will be noted that the steps on this flow sheet may be divided by two horizontal lines into three broad divisions:

1. The first set of *mechanical* operations has for its object the conditioning of the raw materials for the next set. It includes very thorough cleaning and purification of all the materials, grinding the cereals to make them more reactive and steeping them in water to further ease the dissolution of the nutritive ingredients.

2. The next set of *biochemical and chemical* operations includes bringing the food for the yeast cells into the most readily assimilable form and then growing the yeast in the medium so produced under conditions which will result in the most vigorous and prolific production of yeast.

3. In the final set of *mechanical* operations the yeast cells are separated from the fermented liquor under the optimum conditions to ensure their survival and prepared for marketing. The actual operations involved in the process are somewhat as follows:

The cleaned, ground and steeped grains are cooked to pastify the starch. Usually the corn is cooked first at the highest temperature, then the rye is added and when the mash has cooled to the proper temperature for the most effective action of diastase (*ca.* 55° C., 130° F.) the malt is added.

63

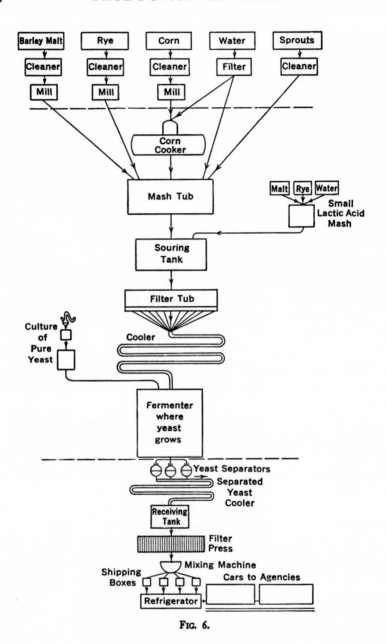

FIG. 6.

When the diastase of the malt has had time to act the mash is inoculated with a smaller special mash of rye and malt in which a pure culture of lactic acid bacteria (*Bacillus delbruckii*) is growing. The mash is now incubated for about sixteen hours at the proper temperature (*ca.* 50° C., 122° F.). During this time the proteins of the grains are partially hydrolyzed and some lactic acid is formed. The liquor now contains largely sugars, resulting from the action of malt diastase on the starch, lactic acid, amino acids, and other hydrolysis products of the proteins, all in a highly assimilable form for the yeasts, and the cellulose residues from the cereals.

This sour mash is filtered and the filtrate, now called "wort," is heated (Pasteurized) to kill off the lactic acid bacteria and any undesirable organisms. After cooling the wort is run into fermentation tanks and inoculated with a pure yeast culture of sufficient size. The wort is then aerated by passing in compressed air either through bottom or side inlets. The oxygen of the air bubbling through the wort stimulates the growth and reproduction of the yeast cells.

When the growth of yeast has reached the desired extent, the cells are separated from the fermenting wort in the third set of operations. The ordinary equipment for this purpose is identical in action with the familiar centrifugal cream separator. The heavy cream of separated yeast is cooled, further water is removed by means of a filter press. The press cake is churned, squeezed in hydraulic presses, packed, and stored in a refrigerator.

Distillers' Yeast.—Each manufacturer of distilled spirit prepares the yeast for carrying on the production of alcohol in a manner generally similar to that just described. The start is usually made by reserving a portion from each completed fermentation. The yeast in this reserved portion is propagated in a small special mash made often from equal amounts of barley malt and rye. In other distilleries only malt, either rye or barley, is used; or possibly a mixture of one of these malts with a ground, unmalted cereal such as wheat, barley or rye. In any case, the object is to produce yeast cells which are young, vigorous and so active that they will rapidly reproduce and will have a high sugar splitting or fermentive capacity, to the end that the highest possible yield of alcohol will result. A flow sheet of this process is shown in Figure 7. It will be noted that it corresponds quite closely to the main steps in the manufacture of ordinary yeast.

The ground rye is first scalded with water of about 170° F. temperature. Then it is stirred, the ground malt added and the whole mash kept for about two hours at a temperature of approximately 150° F. Its sugar content should be about 22 to 25 per cent as indicated on the Balling hydrometer.

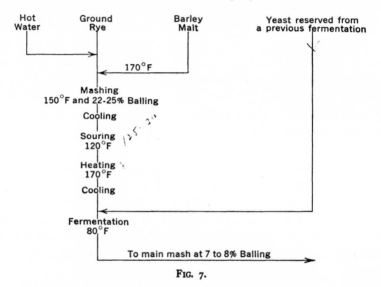

FIG. 7.

The mash is now cooled and soured. Sometimes a pure culture of lactic acid bacteria is added to speed up the souring process. Cooling reduced the temperature of the mash about thirty degrees, or to 120° F. When no lactic acid bacteria are added it is kept at this temperature for about forty-eight hours and souring is usually completed by that time. The addition of bacteria reduces the time for souring to about eighteen hours, or a little longer.

The souring process not only helps in the development of a nutritive medium for the yeast but also the lactic acid formed prevents, or retards, the development of unfavorable micro-organisms during fermentation, notably acetic acid bacteria.

As in commercial yeast manufacture, the lactic acid bacteria are killed by heating the mash again; this time to 170° F. After holding it at this point for about twenty minutes the temperature is brought down to about 85° F. and the seed yeast is added. Fermentation commences and the temperature is gradually lowered about five degrees. When the sugar content of the mash has dropped to about 8 per cent Balling it is added to the main mash where it represents about 5 per cent of the total volume. Heating and cooling in all cases is obtained by the use of coils for the circulation of hot or cold water in the tanks.

Wine Starters.—Grapes ordinarily will produce a must which contains sufficient yeast to carry on the fermentation. Unfortunately, the must is also almost certain to contain many varieties of unfavorable micro-organisms. Hence, especially in the manufacture of white wines some purging or sterilizing process is necessary. The process ordinarily used is called defecation. This consists of treating the must with sulfurous acid and is ordinarily accomplished by pumping it into sulfurized casks (as described in chapter on wine making). In from twelve to twenty-four hours, the must is purged, and all its gross impurities, including micro-organisms, dust and solid particles derived from the skins, stems, pulp and leaves have settled to the bottom. It may be slightly cloudy or nearly clear. It should then be drawn off into clean casks and fermentation started. Sometimes it is sterilized by Pasteurization following defecation, but this is not a very satisfactory operation from the flavor standpoint; it is costly and is generally dispensed with. In defecating must to eliminate unfavorable micro-organisms the wine maker, unfortunately also removes the true yeasts. The more perfect the process the more necessary it is to add wine yeast. It is, therefore, necessary to add a starter.

Natural Starters.—One method of producing such a starter is to gather a suitable quantity of the cleanest and soundest ripe grapes in the vineyard, crush them carefully and allow them to undergo spontaneous fermentation in a warm place. An addition of a quarter to a third of an ounce of potassium meta-bisulfite per hundred pounds of grapes is of great assistance in promoting a good yeast fermentation in the starter. Perfectly ripe grapes should be selected and the fermentation allowed to proceed until at least 10 per cent of alcohol is produced. If imperfectly ripe grapes are used or the starter used too soon, the principal yeast present may be *S. apiculatus*. Towards the end of the fermentation *S. ellipsoideus* predominates. From one to three gallons of this starter should be used for each hundred gallons of crushed grapes or must to be fermented. Too much should not be used in hot weather or with warm grapes, as it may become impossible to control the temperature.

This starter is used only for the first vat or cask. Those following are started from previous fermentations, care being taken always to use the must only from a vat at the proper stage of fermentation and to avoid all vats that show any defect.

Pure Yeast Starters.—An improvement on a natural starter of this kind is a pure culture of tested yeast. There are two ways of using these yeasts. One is to obtain, from a pure yeast laboratory, a separate starter for each fermenting vat or cask. All the wine-maker has to do is to distribute this starter in the grapes or must as they run into the vat. If the starter is used when in full vigor this method is simple and effective. Unfortunately, it is difficult to have it on hand in just the right condition at the right moment. If the starter is too young, it will not contain enough yeast cells; if too old, the cells will be inactive or dead. The usual starter is in full vigor for only a few days at the most. Recent improvements in the methods of preparing pure yeast starters are said to overcome this difficulty and to produce starters which maintain their full vigor for weeks or months.

The other method is for the wine-maker to obtain a small culture of pure yeast from a reliable source and from this to make his own starter.

To do this he prepares an innoculum of two or three gallons of must defecated with sulfurous acid and sterilized by boiling. This, on cooling, is placed in a large demijohn plugged with sterilized cotton and the pure culture of yeast added. The demijohn must be placed in a warm place (70° to 80° F.) and thoroughly shaken several times a day to aerate the must. In a few days a vigorous fermentation occurs.

When the fermentation is at its height in the demijohn, which will be when the must still contains 3 or 4 per cent of sugar, it is ready to use to prepare a bulk starter. This is best prepared in a small open vat or tub, varying in size according to the amount of starter needed daily. Into this tub are poured twenty to fifty gallons of well-defecated must extracted from clean, sound grapes. It is not necessary to boil it, as the few micro-organisms it may contain will be without effect in the presence of the vastly more numerous yeast cells introduced from the pure culture in the demijohn.

The whole of the pure culture is poured into the tub of must, the temperature of which should be between 80° and 90° F. This temperature is maintained either by warming the room or by occasionally placing a large can full of boiling water in the tub. This can should, of course, be tightly stoppered in order that none of the water may get into the must. The must should be well aerated several times a day to invigorate the yeast. This is done by dipping out some of the must with a bucket or ladle and pouring it back into the tub from a height of several feet or by the use of compressed air. The tub should be covered with a cloth to exclude dust, and everything with which the must comes in contact should be thoroughly cleaned with boiling water.

In a day or two the must is in full fermentation and may be used as a starter. From ten to thirty gallons of starter are used for every thousand gallons of must or crushed grapes. The cooler the grapes the more should be added. Too much added to warm grapes may make the fermentation so rapid that it will be difficult to control the temperature. Moldy or dirty grapes require more than clean, because there are more injurious germs to overcome.

Every twenty-four hours, nine tenths of the contents of the starter tub can be used and immediately replaced with fresh defecated must. The yeast in the remaining tenth is sufficient to start a vigorous fermentation and multiplication of yeast. Two things must be watched with special care if the starter is to maintain its vigor. The temperature must be kept above 80° F. and thorough and frequent aeration must be given.

With care, a starter of this kind will remain sufficiently pure to be used continuously throughout the vintage.

CHAPTER VII

MALT

In one very important respect the manufacture of spiritous liquors from a grain base differs from that commencing with a fruit juice base. This difference is that the fruit juices contain *preformed* sugar directly available for fermentation while the cereals contain starch and proteins in a relatively *insoluble* form. It was stated in previous chapters that by suitable processes this insoluble starch and proteins can be converted into soluble forms which are then fermentable. The processes by which this conversion is accomplished are the subject of this chapter.

There are two general means by which starch and proteins are solubilized (hydrolyzed) for the purpose of fermentation. These are by the action of suitable enzymes or by treatment with acids. Of these the former is much more common in the liquor industry. For the production of suitable enzymes the natural changes which occur in the sprouting of seeds are used. The employment of this natural chemical process is called malting, and the product, malt.

Malt.—Malt may be made from any cereal but is commonly made from barley and unless otherwise specified, the term "malt" is understood to refer to barley malt. The general preference for barley is due to its high enzyme productivity, its ability to retain its husk in threshing (husk subsequently serving as a filtering material in the mash-tun) and the responsiveness of its endosperm during growth to modifications and mellowing.

The following Figure 8 shows an enlarged cross-section of a barley grain.

A grain of barley, or other cereal, consists essentially of two parts, the main starchy portion, known as the endo-sperm and a smaller part at one end of the corn known as the embryo. The embryo is the rudimentary plant. From it rootlets eventually develop which extract nourishment from the soil for the development or growth of the plant. The rootlets do not appear in the first stages and it is necessary for nature to provide some means of feeding the embryonic plant. The nutrient media provided by nature are principally starch, and smaller amounts of proteins and other products. The nutrient materials are contained in the endo-sperm and are insoluble and therefore non-diffusible through the cellular structures to the germ where they are needed. Nature has adjusted for this situation by providing that the plant

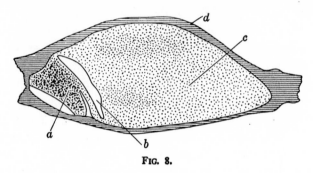

FIG. 8.

secretes certain substances, namely enzymes, very actively as soon as germination commences. The enzyme, cytase, attacks the cellular wall structures and its action makes it possible for the enzyme, diastase, to act upon the starch, changing it into a soluble form and therefore rendering it diffusible, assimilable and available to the germ as food. A portion of the insoluble proteins is also acted upon and changed into soluble, diffusible and chemically simpler substances, *e.g.*, peptones, proteoses, and amino acids. This change, commonly called proteolysis, is thought to be brought about by other enzymes in the plant but little is definitely known with respect to these agents, the action partly resembling that of trypsin and partly that of pepsin or peptase. The actual mechanism by means of which cytase enables the other enzymes to reach the starch and proteins is also not definitely known; *i.e.*, it is not known whether the cytase dissolves the cellulose or whether its action is merely a softening one whereby the cellulose is rendered permeable. Figure 9 is a diagram of the various changes which take place within the grain during the malting operation.

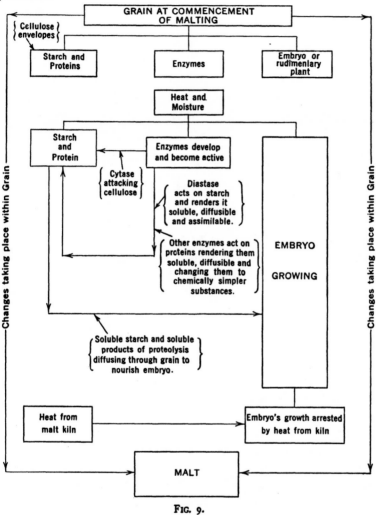

FIG. 9.

This is the process on which the maltster relies, but his endeavor is to keep the consumption of starch as low as possible and hence when the action has proceeded to a point where he judges the starch rendered soluble and the enzymes developed to the most useful point the process is arrested and the plant killed by drying the germinating grain in the malt kiln. This calls for considerable judgment and experience on the part of the maltster, but, in a general way, the end point may be said to have been reached when the plumule has grown almost to the length of the grain or corn. At this stage the starch should be in the proper degree of conversion into sugar and the enzymes developed to such a point as to be able to act upon and saccharify not only the starch of the malt but also the starch of the unmalted cereal with which it is often mixed in the mash tun. Since the yield of alcohol is dependent upon the yield of starch converted to sugar it is important to develop a malt of maximum diastatic properties. It is also economical because it eliminates the necessity of converting all of the cereal to malt.

Generally speaking, American malts have high diastatic (starch-converting) power and are mixed with other cereals in proportions ranging from 10 to 15 per cent of the whole for the production of lower grade whiskies and from 20 to 50 per cent for the higher grades. The United States Dispensatory requires that malt shall be capable of converting not less than five times its weight of starch into sugars. Distillers' malt, however, is usually capable of accomplishing the conversion of nearly fifteen times its weight.

Malt is used, either alone or in combination with unmalted cereals, as a raw material in the manufacture of whiskey, gin, vodka and kornbranntwein. It is also used as a raw material in the manufacture of beer. It is, therefore, one of the most important products used by the liquor industry.

The finished product may be light yellow, yellowish-brown or blackish-brown in color depending upon the intensity of the heat treatment received in processing. Caramel malts are yellowish-brown and are so called because they have been made from ordinary malt submitted to a secondary process consisting of steeping, drying and progressive heating until the sugar formed at the lower temperatures is finally caramelized. Black malts are those of the darkest color and are the product of comparatively high temperature drying.

The malting operation offers several distinct advantages. It assists in the development of the enzymes diastase, cytase and peptase which are of great importance. It influences the solubility of the albumen, starch and phosphates in the grains and affects the condition of the starch for subsequent conversion in the mashing operation.

Practical Malting.—The malting operation consists in: 1. cleaning the grain, 2. adding water and steeping, 3. germinating,

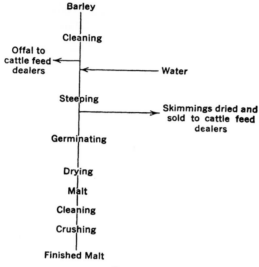

FIG. 10.

4. drying, 5. cleaning and crushing. The sequence of these steps is shown in the flow sheet, Figure 10.

Equipment and processes for cleaning, steeping and drying are more or less standardized but various methods are available for carrying out the germinating operation.

Steeping.—This consists of soaking the cereal in water for approximately 48 hours. Allowance must be made for variables such as kind and condition of barley, water temperature, hardness of water, humidity of air, etc. Float or ladle off skimmings. Change the water every 12 hours the first day and every 24 hours thereafter. When loading have tank half full of water and then add barley, allowing water to stand 1 to 2 feet above the barley when full. When cereal is properly steeped drain off water.

Germination.—There are three methods available for carrying out the germinating operation on a large scale. These are respectively known as the floor, compartment, and drum systems. The floor method was the first used, the other two representing the contributions of modern chemical engineering. They are commonly referred to as pneumatic malting and are based on mechanical as compared to hand turning over the grain and on subjecting it to a flow of conditioned air either intermittently or continuously during the raking or tumbling. Pneumatic malting is of great assistance to the patent still distilleries; it permits all-year-round malting.

Floor system.—In this method the damp cereal is spread upon the floor and is periodically shovelled over to aerate the germinating mass and keep its temperature below the scorching or burning point. The right air temperature is about 60° F. and the grain temperature should be about 75° F.

This part of the process must be watched closely and its success depends on the judgment and experience of the maltster who varies the depth of the grain at each working over. He will usually start with a depth of about 8 to 10 inches and then alternately and gradually increase and reduce it from a maximum of 14 inches to a minimum of about 5 inches.

Germination proceeds and in a few days reaches a point where further growth must be arrested. This is usually when the leaf or acrospire has grown to the length of the kernel. The germination is stopped by drying the green malt in the kiln. Manchurian barley reaches this point in about five days while the two-rowed and Bay Brewing types take about eight days.

Drum system.—These are large drums of the rotating, tumbling type found in many chemical and other industrial plants. They are arranged for continuous admission of conditioned air and their speed can be varied to suit operating requirements.

It is customary to revolve the drums very slowly at first but, as germination proceeds, the speed of rotation is gradually stepped up. No definite operating rules can be set up since germinating conditions are never absolutely standard, but the following procedure may be used as a guide: twelve complete revolutions every twenty-four hours for the first three days, then change to a full revolution every hour and a half for the next thirty-six hours and thereafter speed up to one revolution every forty minutes until germination reaches the arrestation point in about twelve hours more.

Compartment system.—This consists of an oblong tank with perforated bottom, usually of galvanized steel. The conditioned air can be either top or bottom delivered. A carriage fitted with revolving helices for stirring the grain travels across the top of the tank and a sprinkling device is usually provided for moistening the grain as it is turned over. Unloading is accomplished by means of a scraper which draws the grain to one end of the tank where an automatic device is located for feeding the germinating mass to a conveyor.

Kiln drying.—This is carried out in a three-storied building fitted with a suction fan on the top floor and a furnace on the ground floor. The furnace is fed with smokeless coal and air and the hot gases are sucked up through the floors and exhausted from the top of the building by the fan. The grain is spread on the top floor and given a preliminary drying. It is then dropped to the second floor where it is gradually subjected to higher temperatures. This is an operation of some delicacy, and considerable care and experience are required in order that the malt may be dried gradually and not spoiled by scorching. Exposing the green malt to too high temperatures at the beginning of the drying operation will reduce its diastatic strength. For the first 24 hours 90° F. is about right and it should then be dry to the touch; thereafter the heat is gradually increased until a temperature of 120° to 130° F. is reached in the 40th to 48th hours.

The maximum diastatic properties of the malt are obtained when the drying is stopped at about 123° F. and this is an important point for the distiller of mixed mashes. Malts dried at temperatures no higher than this point are usually referred to as green malts.

On the other hand, where high diastatic power is not so

essential, as at all-malt distilleries, it is better to continue drying to a higher temperature. This gives the following advantages: (1) a more friable product, easier to grind, (2) more suitable for storage and (3) better fermentations and superior flavoring properties. Offsetting these favorable characteristics is the fact that green malts are more nourishing to yeasts and have about ten times more diastatic power. Barley gives a malt of the highest diastatic power with rye, wheat, oats and corn following in the order named.

Yield.—After the malting and kilning operations the acreened malt produced will weigh approximately 20 per cent less than the green grain, but, taken by measure, the malt will exceed the original grain by 6 or 7 per cent. This is to say, the malt produced is bulkier but less dense than the original green grain.

Acid Conversion.—The theory of the acid conversion of starch into sugars was discussed in Chapter I. Practically, this method finds some use in the preparation of cereal raw materials prior to fermentation although probably less in this country than abroad. In the United States the hydrolysis of starch for fermentation is almost invariably accomplished by the diastatic action of malts. These malts are mixed with unmalted grain (the starch of which has been pastified by prior cooking) and the conversion to fermentable sugar carried to completion by a subsequent operation called "mashing."

In Great Britain there is more variation in the means employed to secure a completely fermentable mash. There is first of all the common process in which all of the diastatic action comes from malt. Then there are two processes for making mixed mashes. In the first of these, mixtures are made of malt and unmalted grain, the starch of the latter having been completely converted by the acid process. In the second process the action of the acid on the unmalted grain is halted before completion and a small proportion of malt added to finish the task of hydrolysis. The preparation of these mashes will be discussed again under the general subject of the whiskies prepared from them.

CHAPTER VIII

DISTILLATION

Definitions.—*Distillation* is defined as the separation of the constituents of a liquid mixture by partial vaporization of the mixture and separate recovery of the vapor and the residue. The more volatile constituents of the original mixture are obtained in increased concentration in the vapor; the less volatile remaining in greater concentration in the residue. The apparatus in which this process is carried on is called a *still*. Generally speaking, the essential parts of a still are: 1. The kettle in which vaporization is effected, 2. the connecting tube which conveys the vapors to 3. the condenser in which the vapors are reliquefied. Modifications involving the addition of other parts to the still are introduced for various purposes such as the conservation of heat and to effect rectification. *Rectification* is a distillation carried out in such a way that the vapor rising from a still comes into contact with a condensed portion of the vapor previously evolved from the same still. A transfer of material and an interchange of heat result from this contact, thereby securing a greater enrichment of the vapor in the more volatile components than could be secured with a single distillation operation using the same amount of heat. The condensed vapors, returning to accomplish this object, are called *"reflux."* (The above definitions are substantially based on "The Chemical Engineers' Handbook" by Perry, p. 1107-8.)

In less precise language, a *simple distillation* is a means of separating a volatile liquid from a non-volatile residue. A *fractional distillation* is a means of separating liquids of different volatility. This latter process rests on the fact that no two liquids of different chemical composition have the same vapor pressure at all temperatures, nor very often the same boiling point.

On the other hand, while its actual amount may be almost vanishingly small, every liquid or even solid substance has a definite vapor pressure at any given temperature. Furthermore, that vapor pressure is unchangeable at a fixed temperature by any external means, but only by a change in the composition of the liquid. From this it seems probable that the vapor pressure depends partly on the nature of the liquid molecules and partly on their mutual attraction. We have neither need nor space here to develop the proof of this theory. Its application is as follows: The molecules of water (B. P. 100° C.) and of alcohol (B. P. 78.3° C.) do possess a strong attraction for each other as shown by the contraction which is readily observed when the

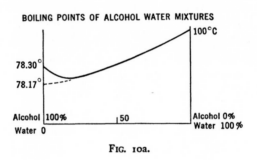

BOILING POINTS OF ALCOHOL WATER MIXTURES

Fig. 10a.

two liquids are mixed. The effect of this on the vapor pressure and hence on the boiling point is shown in Figure 10a. From this diagram, the proportions of which are exaggerated, it will be noted that a mixture containing approximately 95% of alcohol to 5% of water by volume has a lower boiling point (*i.e.*, higher vapor pressure) than either pure compound. From this it follows that alcohol higher than 95.57% cannot be produced by distillation and also that in a simple still, starting with a mixture of alcohol and water of relatively low alcoholic strength, the first distillate will be higher in alcohol content and as the distillation continues the alcohol content of each succeeding portion of distillate will be lower until finally pure water comes over. The relation between the alcohol content of the first vapors and distillate and that of the original boiling liquid as determined by

Sorel (Distillation et rectification industrielle, 1899) is shown in Figure 10b.

If the first portion of distillate were condensed and redistilled, the new distillate would be still richer in alcohol. For instance, if the liquid being distilled contained 10% of alcohol, the first distillate would contain 48.6% and this if condensed and redistilled would contain 69-70% alcohol. Obviously, a practical operation cannot be conducted in this manner. What is done, therefore, is to introduce into the head of the still a number of plates in each of which a portion of the vapor is condensed, yielding a liquid somewhat richer in alcohol than the original liquid, and this is again partly evaporated so that as we ascend the

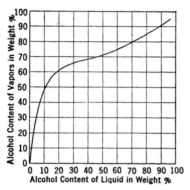

FIG. 10b.

column each plate is progressively of higher alcoholic strength. It is possible by the application of experimental results such as Sorel's (loc. cit.) to these considerations to calculate the number of plates required, and the proportion of condensate return required, to produce alcohol of any desired strength from a given dilute supply. In general it can be seen that there is an inverse ratio between the number of plates and the amount of reflux so that as a practical matter it is advisable to increase the number of plates as far as economically feasible in order to economize fuel.

The fact that it is difficult to secure alcohol concentrations in excess of 12-14% by fermentation alone, requires that for the production of stronger liquors the process of distillation be applied. The increase in alcohol concentration which can be achieved thereby depends on the effectiveness of the rectification and the completeness with which it is desired to recover all the alcohol. It can range up to a recovery in excess of 99% and alcohol of 95% strength by volume. The type and size of still

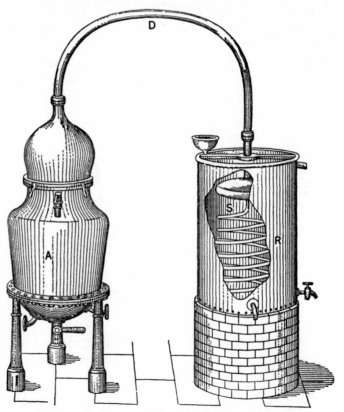

Fig. 11.—Simple pot still used in liqueur manufacture.

actually employed in the distilled liquor industry depends on the industrial development of the country in which the process is being applied, upon the beverage being made, the raw materials used, and the amount of material being processed at one time. The various types may be classified as pot stills, Coffey or patent stills, vat stills and continuous stills.

Pot Stills.—The simplest form of pot still is used in the manufacture of liqueurs both on account of the small lots which are worked and the method of manufacture. Such a still is shown in Figure 11. "A" is the kettle; "D" is the "swan's neck" for conveying the vapors to "R" the condenser; and "S" is the worm. The mode of operation of this apparatus is obvious from an inspection of the figure.

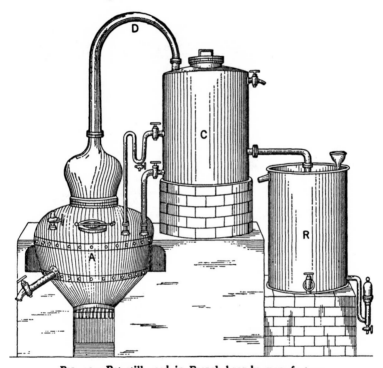

FIG. 12.—Pot still used in French brandy manufacture.
C is the chauffe-vin used for pre-heating the wine fed to the still.
R is the condenser.

Figures 12 and 13 illustrate the addition of another part to the simple pot still as used in France for the production of brandy. This is the device marked "C," called in French the *"chauffe-vin,"* from its function of pre-heating the wine which is fed to the kettle "A." This pre-heating is a mode of conserving some of the latent heat of the vapors by passing them through the feed to the kettle before leading them to the condenser. The types of still illustrated are specially designed for brandy manufacture and their peculiar adaptation for this purpose will be found in the chapter on Brandy.

Improved Pot Still.—In Great Britain the chief distilled liquor is whiskey. Some of this continues to be made in pot-stills

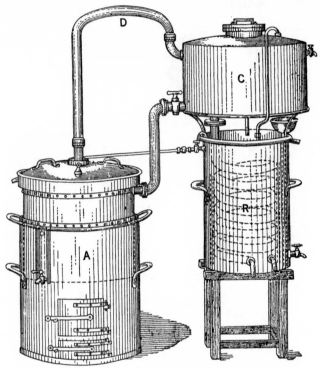

FIG. 13.—French brandy still fitted with chauffe-vin.

A is the kettle.
C the chauffe-vin.
R the condenser.

of somewhat improved design and considerably larger size as shown in Figure 14. The pot stills used for this purpose are divided into two classes, wash stills and low-wines stills. The mash in which fermentation is complete, now called "wash," is distilled in the former. The distilled product, low wines, is approximately a third of the volume of the original wash and

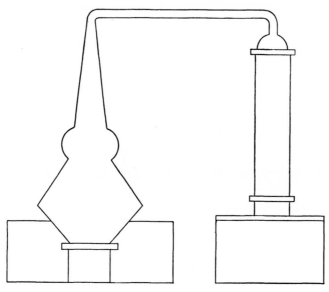

FIG. 14.

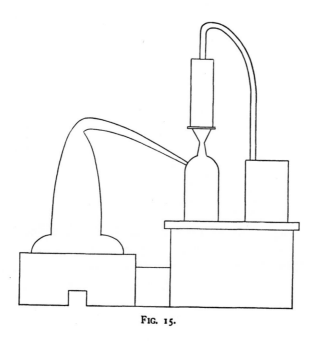

FIG. 15.

somewhere in the neighborhood of 25% alcoholic concentration. Since a large volume of wash must be handled to produce a much smaller volume of whiskey, the wash stills are very large, ranging

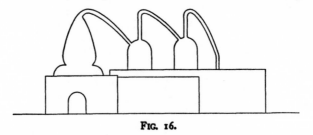

FIG. 16.

from 7,000 to 12,000 gallons capacity. It is not safe practice, however, to charge these stills beyond 50-75% of their capacity, to avoid foaming and priming, that is, the carrying over of some

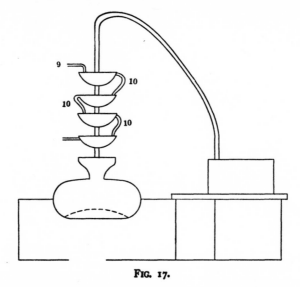

FIG. 17.

of the boiling wash into the condenser. A still of the size range indicated may be expected to distill about 600 gallons of wash to low wines per hour.

Figure 15 is an improved pot still, arranged for direct firing and equipped with a rectifier in addition. The vapors pass from the kettle into the rectifier, which is similar to an ordinary condenser. The least volatile portions are condensed and returned to the still. The more volatile portions of the vapor pass on to the regular condenser.

A doubling of this arrangement is shown in Figure 16. This kind of equipment is used in the British West Indies for the production of rum. Either Figure 15 or 16 may also be arranged for heating by steam instead of direct firing.

Another type of rectifying arrangement called a "Corty's head" is shown in Figure 17. Four traps are fitted into the neck of the still. Each trap contains a diaphragm by means of which the direction of the rising vapors is changed, forcing them to circulate around the trap. Cooling water enters the system by means of the pipe (9) and flows downwards through pipes (10) from trap to trap. A fractional condensation occurs in each trap causing progressive rectification of the ascending vapors in an effective manner and resulting in the vapors passing to the total condenser being much richer in alcohol than those evolving from the boiling liquid in the kettle.

Coffey Still.—The Blair, Campbell and McLean form of the "Coffey" or patent still, shown in Figure 18 in plan, and diagrammatically in Figure 19, is a much more effective rectifying and distilling equipment. Instead of applying direct heat to a large volume of wash in a kettle, the wash is spread in thin layers over a large surface and heat supplied by the introduction of steam from an external boiler. The wash enters the still at the top and trickles down over a series of perforated copper plates. Steam enters the still at the bottom and bubbles upward through the perforations, each of which is in effect a trap. By this means the wash is heated and the alcohol vaporized so that by the time the wash has reached the lowest plate it has lost all of its alcohol and can be discharged from the still with its dissolved and suspended solids. As the mixture of alcohol vapor, volatile impurities, and steam rises toward the top of the apparatus, the lower becomes its boiling point, more steam is condensed from it, and the richer it becomes in alcohol. The part of the still in which this operation takes place is called the "analyzer."

FIG. 18.—(From Martin *Industrial and Manufacturing Chemistry-Organic*, Crosby, Lockwood and Son, London.)

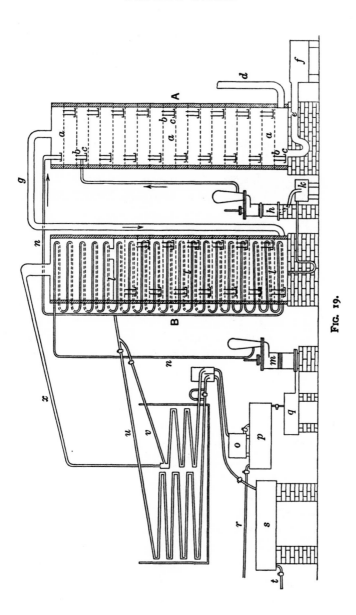

FIG. 19.

The vapors from the top of the "analyzer" are led into the bottom of a second column of perforated plates which is called the "rectifier." There is a zig-zag tube full of cooling liquid extending the full length of this column to serve as a condenser. Usually cold wash on the way to the analyzer is employed as cooling liquor, while it is thereby preheated, thus effecting economy of heat. The alcoholic vapors on their upward passage through the plates are fractionally condensed in each cooling chamber and lose their water until finally they are condensed on an unperforated sheet at the top of the column (the "spirit plate") as very strong alcohol, and are removed. It is stated that from 94-96% spirit can be continuously obtained from this type of still, whereas a simple still at best and even by repeated distillations will only yield a small quantity of strong alcohol not over 90-92% by volume from each operation.

A mixture of weak, impure alcohol and "fusel oil" ("hot feints") collects in the bottom of the "rectifier." This is returned to the "analyzer" to recover the alcohol. Towards the end of the distillation of a batch, however, instead of completing the purification of all the alcohol, it is found more economical to raise the temperature of the apparatus and distill off the whole residue of impure spirit. This is condensed and collected as "feints" in a separate receiver. Here the fusel oil which has accumulated throughout the process separates, to a large extent, from the weak spirit and is skimmed off. The remaining "feints" are redistilled with the wash of a succeeding operation to recover their ethyl alcohol.

In the United States alcohol is distilled and rectified from its wash by means of continuous stills. In smaller establishments all non-volatile materials and a substantial portion of the water are removed in a so-called "beer still," Figure 20. On account of the partial rectification in the preheater the distillate from a 6% beer will frequently run as high as 40 to 60 per cent alcohol. The crude alcohol is neutralized with some suitable alkali such as soda ash, and then purified and concentrated in an intermittent still, Figure 21. Assuming that the feed runs as high as 60 per cent alcohol, the feed is diluted so as to reduce the concentration to 50 per cent and the distillate will run as high as 95 per cent alcohol but is collected in a number of portions of which 70 to 75 per cent can be used as spirits. The neutralization of the distillate from the beer still must be very carefully done, because if the solution is boiled when alkaline, the nitrogenous

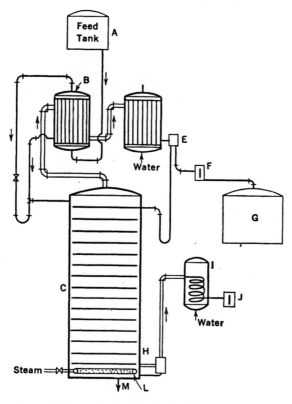

Fig. 20.—Beer Still. (Redrawn from Robinson's *Fractional Distillation.* Courtesy McGraw-Hill Book Co., Inc.)

bodies set free amines whose disagreeable odor is difficult to remove in the finished alcohol, and which also form blue compounds with copper which discolor the alcohol. Another drawback is that they tend to combine with the aldehydes, forming resins which may gum up the column, or impart a yellow color to the alcohol withdrawn from the column. On the other hand, if the solution is acid, esters will form, and any undecomposed ammonium acetate will react with strong alcohol forming ethyl acetate and setting free ammonia. (See Robinson, "The Elements of Fractional Distillation.")

The operation of the beer still shown in Figure 20 is performed as follows: The alcoholic feed is supplied by a constant level feed tank "A" containing a ball float which controls a steam

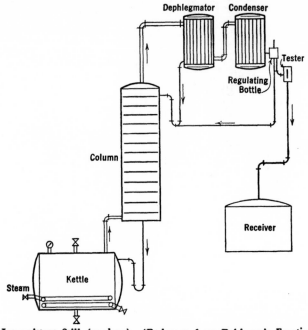

Fig. 21.—Intermittent Still (modern). (Redrawn from Robinson's *Fractional Distillation*. Courtesy McGraw-Hill Book Co., Inc.)

pump to pump the feed from the storage tank to A as rapidly as it is used. The feed then flows by gravity through the feed heater "B" where it is raised nearly to its boiling point. Continuous stills are often fitted with recuperators in which the incoming feed is heated by the outgoing hot waste from the bottom of the exhausting column. If the liquor contains solid materials that are likely to form deposits on the heating surfaces, recuperators are dispensed with, on account of the difficulty of cleaning the outside of the tubes. The vapor heater shown at "B" has the liquor only inside of the tubes, and the fouled surfaces can very easily be cleaned by removing the top and bottom heads of the heater, and passing a cleaning device through the tubes.

The hot feed is introduced into the top section of the exhausting column "C" where it flows downward from plate to plate; the volatile portions are gradually removed as the liquor comes into contact with the steam blown in at the bottom through the perforated sparger pipe "L." The exhausting column has usually from 12 to 15 plates, each plate being large and deep to give a long time of contact of the feed in the column in order to insure complete removal of the volatile substances. The complete removal of these substances is readily tested by what is known as the slop tester. Vapor is withdrawn from a plate near the bottom at "H," any entrained liquid removed by the separating bottle, and the vapor condensed in a suitable condenser "I," from which it flows to a tester "J" where it can be tested, or its specific gravity measured by means of a hydrometer. The exhausted liquor is then discharged from the bottom of the still through a suitable seal pipe "M." The rate of introduction of steam into the column is governed by means of a suitable pressure regulator.

The vapor leaving the exhausting column to pass to the heater is substantially in equilibrium with the liquid on the top plate of the column. It is partially condensed in the heater, enriched in its alcohol content, and then passes to the condenser where it is completely condensed. The portion of the vapor condensed in the heater is returned to the top plate of the column together with a controlled portion of the vapor condensed in the condenser, from the regulating bottle "E." The distillate flows, through the tester "F" where its quantity and specific gravity may be measured, to the storage tank "G." The water supply for the condenser is obtained from the constant level feed tank "N."

In larger establishments a continuous rectifying still is used in place of the intermittent still for the second operation. Figure 22 is a diagram of such a continuous rectifying still. The still consists essentially of a purifying column "C" and a concentrating and exhausting column "D." The function of the purifying column is to remove the volatile head products, which are separated from the alcohol by fractionation. The function of the concentrating column is to separate the alcohol from water, as well as from the less volatile impurities which are not removed in the heads. This rectifying unit will produce an alcohol of higher grade than the best produced by the intermittent still with

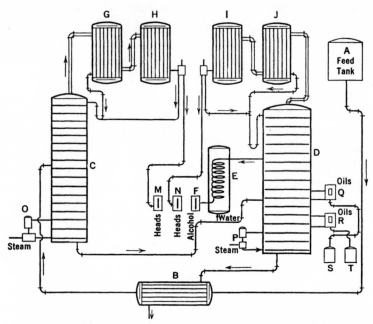

Fig. 22.—Ethyl Alcohol Still (continuous). (Redrawn from Robinson's *Fractional Distillation*. Courtesy McGraw-Hill Book Co., Inc.)

a recovery of perhaps 75 to 85 per cent. It will avoid the necessity of subsequent purifying treatments with charcoal, etc., and rehandling of intermediate fractions at a considerable saving in time, labor, and expense. The purifying column is fitted with a partial reflux condenser "G" and a total condenser "H" and is independent of the rest of the apparatus except that it receives the hot feed continuously from the recuperator "B" and the feed supply tank "A" and delivers the purified dilute alcohol continuously from its base to the other column "D." It has its own steam regulator "O" and cooling water supply and its rate of operation can be controlled according to the amount of the impurities to be removed.

A further improvement of this system of alcohol production is shown in the plan in Chapter IX on Whiskey. Here the beer still is connected directly to the rectifying unit so that the feed to the rectifier is in the form of a vapor instead of a liquid. This effects considerable saving of steam for heating purposes and the final product contains not less than 98 per cent of all the alcohol present in the beer and produces 90 per cent of this alcohol as high grade, pure spirits, the remainder as heads, and a washed fusel oil. Chemical analysis cannot determine whether the alcohol produced comes from molasses or grain. These units use about 40 pounds of steam per gallon of alcohol produced. Further improvements now pending will give an even higher yield.

The most modern American whiskey still consists of a column and an extra head which contains two rectifying plates and one washing plate. This still and its mode of operation are described in Chapter IX on Whiskey.

CHAPTER IX

WHISKEY MANUFACTURE

Historical.—Whiskey is essentially an English and American beverage. It was first developed in the United Kingdom and subsequently its manufacture was taken up in this country.

When whiskey was first made is not definitely known. But Usquebagh (from which the word "whiskey" derives) is said to have been made in Ireland in the twelfth century and it is reported that its manufacture there had assumed sizable proportions even before Queen Elizabeth's time. It is also said that distilled spirits were made by the monks prior to the fifteenth century and that they jealously guarded the secrets of their formulae and methods of manufacture. However, commencing with the fifteenth century the process became more widely known. There is a treatise on distillation which is one of the very first of printed books. At first the manufacture was carried on in a small way in the household, but a young whiskey distilling industry was gradually established. At that time spirits were made from malt in Scotland and from wort and sour beer in England. The industry was operated under government supervision in all three countries and its products were taxed. As a result, the history of spirit distilling can be followed fairly accurately by a study of the taxation legislation.

The industry has had a very stormy career and it is interesting to note that in the sixteenth, seventeenth and eighteenth centuries many of its troubled periods and their subsequent developments paralleled conditions under the prohibition era in the United States in the last thirteen or fourteen years. For example, restrictive legislation, high taxation, manufacture under strict government supervision and last, but not least, the bootlegger, or as he was then called, the illicit distiller, all existed long before our times. Smuggling was often rife, and England had its rum row two or three hundred years before America. In 1556 there was a death penalty in Ireland for illicit distillation.

In the reign of Charles I there was formed the Distillers' Company of London, which received a charter of incorporation and was empowered to regulate the manufacture of whiskey from the point of the quality of materials to be used. A little later, in the reign of Charles II, distilling materials included such varied substances as sugar, molasses, sour wines, sour ales, cider, and wort from grain and malt; and the products included whiskey, brandy, gin, and rum (although much of this was made in Jamaica). By 1694 annual production in England had risen to 900,000 gallons.

The seventeenth, eighteenth and nineteenth centuries were periods of experimental taxation and other governmental regulation in Scotland, England, and Ireland, and the industry may be said to have grown in spite of, rather than because of, these regulations. Prior to 1860 the taxation and regulations were different in all three countries; and in Scotland, during one period, there was even a wide difference in regulations for Lowland and Highland distilleries. At one time taxes were either collected by local authorities or were farmed out to private persons or business houses, who received a percentage of their collections for their services. It was quite common for such tax collecting contracts to be let and sub-let. Later, a fixed minimum of receipts was stipulated for each distilling region. Following this another form of taxation was tried, based on the capacity of the still and its rate of operation, but the legitimate distillers displayed considerable ingenuity in beating the law, often by faster distillation. Morewood (Inventions, etc., in Intoxicating Liquors, 1824) describes a still for this purpose which was built with the unique proportions of 48 inches in diameter and only three or four inches deep!

Thus, three centuries have passed and a satisfactory solution of the problem is still sought. When taxation was high and regulations very restrictive, legitimate production waned and illicit distillation and smuggling increased. When taxation was lowered and regulations made less restrictive, legitimate manufacture would prosper and smuggling and illicit manufacture diminish.

In 1730 the laws almost killed the industry, but illicit distilling became so profitable that the government was forced to revise the regulations in 1743. Legitimate production then jumped to 5,000,000 gallons of proof spirit. In 1751, and again in 1756, taxation was increased and legitimate production gradually dropped till it amounted only to 3,000,000 gallons in 1820. In 1760, 500,000 gallons of spirit were smuggled into England from Scotland. About the year 1800, 6,000 illicit stills were seized in Ireland in one year and illicit production exceeded legitimate production three or four times. Taxation was revised in Scotland in 1817, and in England in 1823, and finally in 1860 all legislation was consolidated and restrictions were somewhat relaxed. Since then further concessions have been made from time to time, the greatest being those since the war. At present the tax is about $2.00 per bottle.

The history of whiskey manufacture in the United States is not so easily traced. The date of the building of the first distillery is uncertain, as is the progress of the industry in the eighteenth and nineteenth centuries. However, no book on this subject can omit a reference to the "Whiskey Insurrection" which occured in Western Pennsylvania in 1792 to 1794, when President Washington was compelled to call out the militia to quell the insurrectionists, so strong was the reaction against the excise regulations put into force about that time.

The prohibition question, especially its legislation and developments in the past two decades in the United States are too well known to warrant reviewing. According to D. S. Bliss, U. S. Commissioner of Industrial Alcohol, only seven distilleries were legally operated during this period for the purpose of manufacturing whiskey for medicinal purposes. They were allowed to manufacture only limited quantities and they started operation during the fall of 1929. On the manufacture of whiskey during the last three decades see also Chapter XV on Statistics.

Definition and Types.—Whiskey may be defined as an alcoholic beverage produced from cereal grains by the following general series of operations:

1. Transformation of the starch of the grains, either malted, unmalted, or mixed, into fermentable sugar.
2. Fermentation of the sugar to produce alcohol.
3. Distillation to concentrate the alcohol.
4. Ripening by aging in oak barrels.

There are available on the market a number of types of whiskey, which as the result of variations in the details of processing or in the raw materials used, possess different flavors and other characteristics of importance to the consumer. In general these may be classified as:

American

Rye:
Made from a mash composed of unmalted rye and either rye or barley malt.

Bourbon:
Made from a mash composed of maize and either wheat or barley malt.

Low grade American whiskeys are made from mashes containing from 10 to 15 per cent malt.

High grade American whiskeys are made from mashes containing from 20 to 50 per cent malt.

Most American whiskeys are made in patent stills.

Scotch

Pot still:
Made from barley malt and having a smoky taste, obtained by using peat instead of coal as fuel in the kiln drying of the malt. Changes in the variety of peat used materially affect the flavor. This includes scotch whiskeys commonly classified in the British Isles as follows: (1) Highland malts, (2) Lowland malts, (3) Campbelltowns, (4) Islays.

Patent still:
Made from a mash composed of unmalted cereals and barley malt. The former may be either rye or oats but commonly is American maize (corn). These whiskeys do not have the smoky taste and are more American in character.

Irish

Pot still:

Made from an all-malt mash or from a mixed mash composed of barley malt and unmalted cereals. The latter may be barley, oats, wheat, rye or variously proportioned mixtures. Malt runs high, from 30 to 50 per cent of the whole.

Patent still:

Made from a mash composed of unmalted cereals and barley malt.

On a different basis all mixed mash whiskies may be classified as either:

Sour or sweet mash

Sour mash:

A whiskey made by cooking the ground, unmalted cereal with spent liquor of a previous mash which has been dealcoholized by distillation.

Sweet mash:

A whiskey made by cooking the ground unmalted cereal in the ordinary way with water.

Blends.—In addition to the straight whiskeys described above, both in this country and abroad various blends have come into public favor. Especially in Great Britain blending has become a very large trade as it is stated that the public taste demands a whiskey of less prominent but more uniform characteristics than formerly. To gratify this desire blends are made, in the United States presumably of straight whiskeys; but in Great Britain either by the mixture of various pot still whiskies of varying age, etc., with the possible addition of silent spirits from patent stills. In the latter case cheapness is often the purpose of the blend, but it is also stated that it unites the several whiskies in the mixture more completely and enables the blender to produce a whiskey of more uniform character. Blends, even when made from aged spirits of various kinds, are frequently stored in bond for considerable time. The addition of patent still spirits, even those containing very small amounts of secondary products, is viewed as dilution rather than as adulteration. Methods of blending are discussed under that heading later in this chapter.

MANUFACTURE OF WHISKEY

General Outline.—The manufacture of whiskey is essentially a chemical process based on changes in the composition of materials brought about by temperature alterations and the effect of the activity of ferments and other reagents. Very little depends on mechanical manipulation and there is a lack of spectacular features.

Successful operation depends on a complete understanding of the changes taking place in the composition of the materials and on accurate temperature control. Technical knowledge, experience and judgment are required to select and control conditions and materials so that a high yield of uniform product is obtained.

It has been the object of the preceding chapters to explain the theoretical bases on which the process of whiskey making rests. In the present chapter the sequence of the operations and some of the manners of control are discussed. In actual fact, it is very easy to make a sort of crude whiskey by simple performance in regular order of the first three or four basic operations listed in this chapter in the section on definitions and types. The commercial production of whiskey in quantities is very largely a magnification in scale of these operations with the introduction of refinements and modifications designed to facilitate the operation, secure a more uniform product, and obtain a maximum yield. It is to be expected, therefore, that the historical steps in the change from the simple "home still" of earliest times to the largest scale continuous operation of American practice have been preserved and can be seen in the manufacture of whiskey in various establishments in different countries. This is the case to such an extent that the common varieties of whiskey are each identified with a different degree of evolution in the whiskey making process. A number of distinct process sequences can be formulated on this basis, of which the following are outstanding. (See Table VIII).

Mashing.—In all types of whiskey, the process, by which all the starch of the grains used is brought into solution, is called mashing. It involves both extraction and conversion of the starch

TABLE VIII.—TABULAR COMPARISON OF WHISKEY PROCESSES

Whiskey type	Scotch or Irish	Scotch or Irish	Scotch or Irish	American small scale	American large scale
Materials	All malt	Malt and grain	Malt and grain	Malt and grain	Malt and grain
Pre-mashing	None	None	Partial acid conversion	Cooking at normal pressure	High pressure cooking of grain
Filtration of mash	Yes Only wort is fermented	Yes Only wort is fermented	Yes Only wort is fermented	No Whole mash is fermented	No Whole mash is fermented
Method of distillation	Pot still	Pot still	Patent still	Patent still	Patent still

into sugars. The process is carried out in an apparatus called a "mash tun" as illustrated in Figure 22a. The essential parts of a mash tun are a vat equipped either with steam coils or means

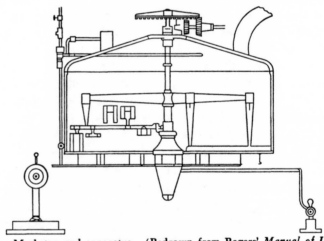

FIG. 22a.—Mash tun and apparatus. (Redrawn from Rogers' *Manual of Industrial Chemistry,* D. Van Nostrand Company, Inc.)

of heating by direct injection of steam, and an efficient agitator. The latter must have both scraper and stirrer arms to ensure that all the ground grain comes into contact with the water.

Water.—The quality of water used in mashing is very important both on account of its influence on the quality of the finished liquor and in its own right, since it is used by the distiller in many times greater volume than any other of his materials. As used in mashing it is possible for impurities in the water to cause irreparable damage.   It is also claimed that the water used influences the flavor of the finished whiskey.   There is even told a tale of a Scotch distillery being built on the banks of a stream and then abandoned and a new distillery built on another stream twenty miles away because the water from the latter resulted in a product of superior flavor.   As the Italians say *Se non è vero, è ben trovato.*   Certainly it is known that Scotch and Irish distillers emphasize greatly the purity of their water supply.   They select by preference, moss water, or some special location such as Loch Katrine or the river Bush, whose name is part of the trade name of "Old. Bush Mills."

Lacking such ideal locations, an effective water purification and softening plant may be necessary if the water supply is in the least questionable.   The magnitude of the problem is readily seen from the fact that a pot still distillery, on a basis of 1,000 bushels of malt mashed weekly, will require about 240,000 gallons of water, and a patent still distillery, producing 20,000 proof gallons of alcohol per week, will use about 700,000 gallons of water in its production.

Scotch or Irish Pot Still Whiskey.—*Preparation of Wort.*— As can be seen from Table VIII (p. 102), the manufacture of this type of whiskey involves the least introduction of modern improved processes.   The mashing procedure as shown diagrammatically in Figure 23 consists of three extractions of the ground grain, either all malted or a mixture of malted and unmalted, with separate portions of liquor.   Oat husks are added to assist in the drainage or filtration of the wort and the third or final liquor from one batch of grain is used as the first liquor on the succeeding batch. The liquor is heated to the proper temperature, poured over and mixed with the cereals in the mash tun, allowed to soak for a suitable time and drained off.   The first two liquors obtained in this manner are cooled to the proper temperature for fermentation and run to the fermenting vats. The third liquor or "weak wort" is returned for use on the next batch of malt.

Fermentation.—This stage of the process of whiskey making permits of only minor variations in methods of inoculation, time, temperature control, etc. The general practice is the same both in

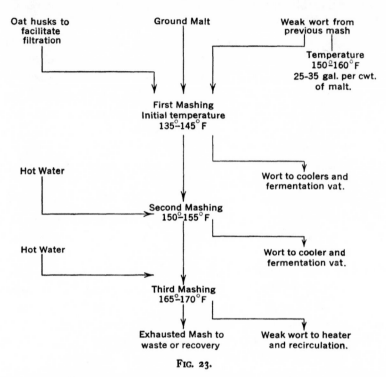

FIG. 23.

America and in the British Isles. The principles to be observed have been outlined in Chapters III and VI.

The customary procedure is as follows: The wort coming from the mash tuns, filtered abroad, unfiltered here, is cooled to between 68° and 70° F. Yeast in a vigorous state of activity is added and the fermentation proceeds. The temperature of the fermenting liquor increases and must be carefully controlled by passing cold water through coils in the fermentation vat. The amount of temperature rise permitted has a direct effect on the time of fermentation. In some distilleries the rise is kept small and the fermentation slow. In most, however, the temperature is allowed to advance about twenty degrees in the first twenty hours. The temperature is never permitted, however, to exceed 90° F. Since the distillers' yeast is very active, a sweet rye fermentation, for example, is usually complete in 72 hours.

It is observed that high and rapid fermentations on the one hand are more likely to exhaust the sugar in the wort, but on the other hand, it is claimed that they are responsible for the forma tion of larger amounts of the congeneric substances including esters, fusel oil and aldehydes.

Aeration.—Some form of aeration is necessary both before and during fermentation. It may vary from the simple raising of buckets full of wort and pouring them back, to elaborate perforated pipe and air pump assemblies. The results of aeration and the objects of the practice include: 1. thorough stirring and intermingling of wort and yeast, 2. maintenance of uniformity of temperature throughout the vat, 3. expulsion of carbon dioxide from the wort, 4. stimulation of yeast in vigor and multiplication, 5. flocculation of suspended matter.

Distillation.—On the conclusion of the fermentation, the liquor, now called "wash," is ready for distillation. In Scotch pot still practice two distillations are required for preparing whiskey from the wash. The first takes place in the wash still. The distillation is continued until all the alcohol has been driven off from the wash and collected in one distillate. The liquor remaining in the still is termed "pot ale" or "burnt ale"; and is either run to waste or dried for fertilizer. The distillate, which is technically termed "low wines," contains all the alcohol and secondary constituents from the wash, and considerable water. The low wines are transferred to a second and smaller still and are redistilled. Three fractions are obtained from this distillation. The first is termed "foreshots," the second constitutes the clean or finished whiskey, the third is called "feints." The foreshots and feints are collected together, while the residue in the still, called "spent lees" is run to waste like the pot ale.

The judgment and experience of the distiller determine the point at which the collection of foreshots is stopped, and that of whiskey commenced, and similarly that at which the latter is stopped and the collection of feints begun.

The strength at which the whiskey fraction is run is of great importance as regards the character of the spirit. In Scotland this is generally from about 11 to 25 degrees over proof (11–25° o.p.).

The foreshots and feints from one distillation are mixed and added to the charge of low wines for the next distillation and so throughout the distilling season. The feints collected from the last distillation of the season are kept to be added to the low wines from the first distillation of the succeeding season.

In some distilleries in Scotland the whiskey is produced in three distillations. This practice is very general in the Lowlands; the spirit being then run at 40 to 45 degrees over proof.

The volatile secondary constituents, which pass over with the alcohol into the low wines receiver, on the distillation of the wash, are thus incorporated as far as possible with the finished whiskey finally produced. There can be no doubt, however, that a portion escapes with the spent lees since it is known that partial decomposition is undergone during the process of distillation, *e.g.*, certain esters are easily decomposed when boiled with water under such conditions as those which obtain during distillation in the wash or low wines stills and the products of decomposition may wholly or partially remain in the spent lees and may consequently be absent from the whiskey.

Again, some of the constituents which boil at much higher temperatures than water, may not wholly pass over with the alcohol in the distillation of the low wines, but may remain in the spent lees, and so also be lost to the finished whiskey.

The extent to which the loss of secondary constituents may thus occur and affect the character of the whiskey depends largely upon the variety of pot still employed, and the manner of its operation; whether, for instance, the process of distillation be carried on slowly or rapidly; and it also depends on the strength at which the whiskey fraction is run.

In Irish pot still practice the stills employed differ somewhat from those used in Scotland in that they are generally much larger, the wash still occasionally being of a capacity of 20,000 gallons. The head of the still is shorter and in the still used for the distillation of low wines and feints the pipe connecting the head of the still with the worm is of considerable length and passes through a trough of water, the result being that a certain amount of rectification of the spirit vapor is effected on its way to the worm. This pipe is termed the "Lyne arm" and is connected with the body of the still by what is known as a "return pipe" through which is conveyed to the still for redistillation any liquid which has condensed in the pipe.

Three distillations appear to be universally practiced in Ireland for obtaining pot still whiskey and the method of collecting various fractions during a distillation is somewhat more complicated than with the Scotch process. Strong low wines and weak low wines, strong feints and weak feints, are collected and blended in various orders, and the practices in this connection probably differ in every Irish distillery. The whiskey fraction is usually run at a higher strength than in the Scotch process, viz; from 25–50 per cent o.p.

The addition of charcoal and also of soap in distillation is common both in Ireland and in Scotland, the soap being used to prevent frothing in the wash still and the charcoal in the low wines still to remove undesired constituents by absorption.

The differences between Scotch High and Lowland and Irish practices in pot distillation are readily seen from the flow diagrams (Fig. 24).

British Patent Still Whiskey.—*General Statement.*—It is claimed for patent still operation in preference to pot still operation that various economies are achieved as follows:

1. Economy of time:
   a. Operation is continuous and rate of distillation is greater. There is a continuous feed of wash and a continuous discharge of spent wash, as compared with shut downs to charge and discharge in pot still practice.
   b. Rectification and distillation are carried out as part of the one process, whereas in pot still practice rectification is only partially achieved in one distillation, and two or even three distillations are necessary for complete rectification.

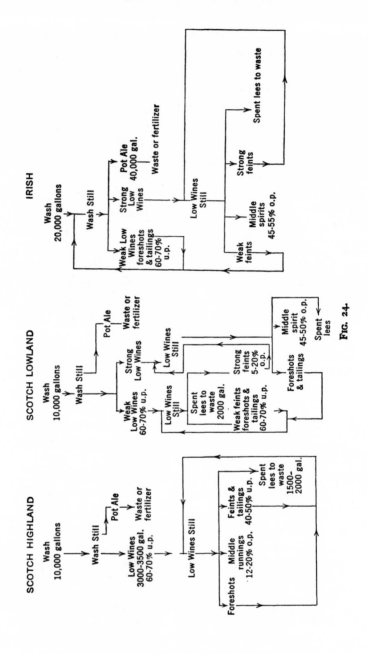

FIG. 24.

     c. Only one condensation of the distillates is necessary, with cooling from a maximum temperature of 150° F. down to 60° or 70° F., as compared with pot still practice in which two or three condensations are necessary.

     d. Operation can be carried on throughout the year.

2. Economy of operation:

     a. Less cooling water required for condensing vapors to distillates.

     b. Cold wash is used to condense vapors to distillate, and the latent and sensible heats of the vapors and of the distillate serve to preheat the wash and raise it almost to its boiling point. This results in large fuel savings for primary heating.

3. Efficiency of operation:

     a. Pure grain alcohol for use in the arts can be made as well as alcohol for denaturing.

     b. A highly rectified whiskey can be produced.

     c. Distillate is 148 to 154% proof.

     d. Strength of distillate can be practically constant despite wide variations in strength of wash.

4. Miscellaneous:

     a. A greater variety of materials can be used and from practically all sources of supply so that advantage can be taken of temporary low prices.

     b. Yeast can be made as a by-product.

On the other hand, the elimination of secondary products contained in the fermented wash is 95 per cent for the patent still as compared to only 90 per cent for the pot still. Hence many old time Irish and Scotch whiskey makers claim that it is the presence of some of these secondary products in the distillate which determines that their product is whiskey and not merely "neutral spirit"; *i.e.*, flavorless, pure grain alcohol.

The Scotch and Irish pot distillers claim that over a period of many years they built up a world-wide reputation for a beverage called "whiskey" to obtain which it is necessary to distill a wash, obtained from certain raw materials, in a pot still according to certain methods developed by long years of practical research and experimentation. They claim that no patent still spirit, however dosed with flavors, can duplicate the quality and flavor of their product.

Preparation of Wort.—In harmony with greater efficiency in the distillation, patent still operators have introduced modifications in the method of saccharification of their cereal grains, usually by some application of the "acid-conversion" process. This process has as its object partial conversion of the starch of the grains into fermentable sugars by the use of acid rather than the diastase of malt and depends on the latter only for the final completion of the conversion.

The general principles underlying this operation have been discussed in Chapter I on Starch. In practice small amounts of either sulphuric or hydrochloric (muriatic) acid are added to the mixture of ground grain and water, heat is applied until the action has proceeded to a sufficient extent and then the acid is neutralized. The procedure consists in mixing in a tun about 36 gallons of water per cwt. of ground cereal or grist. About 1–1.5 pounds of 60° B. sulphuric acid (oil of vitriol) suitably diluted (*by pouring the acid into the water—never the reverse*) is added for each cwt. of grist. Agitation is applied, and steam injected so that the temperature rises gradually. Care must be taken that the heating is neither too high nor too prolonged. When the starch has been gelatinized and the whole converted to a thin liquid the action is stopped by neutralizing the acid. Ordinarily milk of lime (a suspension of slaked lime in water) is used to accomplish most of the neutralization and the rest effected to a very faint acid reaction by the gradual addition of powdered chalk. At the optimum condition of acidity cold water is added to cool the batch to about 145° F.

The batch is then discharged into the mash tun in which some malt at a temperature of 125°–130° F. has been previously prepared. The temperature of the whole mash after mixing should be about 138° F.

There are various modifications of this acid conversion process in which small amounts of malt are added at different stages to supplement the action of the acid. Three such variations are shown diagrammatically in Figure 25.

It will be noted from Figure 25 that the first modification represents the acid conversion process exactly as described pre-

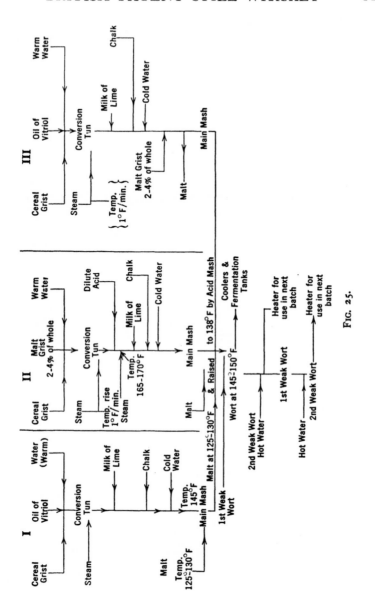

FIG. 25.

viously. In the second modification a small portion of malt is added immediately to the ground cereal which is steeped in tepid water. Heat by steam injection is carefully applied so that the rate of temperature rise is very closely 1° F. per minute. The acid is only added when the batch has heated up to 165°-170° F. and the cooking is then continued as previously. In the third modification it is customary to use more water (50 gal. per cwt.) and less acid (½ lb. per cwt.). The cooking follows the usual procedure except that a small portion of malt is added to the acid mash as a final step before it is poured into the main malt mash. In each case three extractions on a sort of counter-current system are used as in the all malt process.

Fermentation.—While equipment in a patent still distillery may be both larger and more elaborate, the process of fermentation is practically identical with that described under pot still whiskey (p. 104).

Distillation.—The operation of a patent still in the British Isles is in all respects as described in Chapter VIII and on page 126 of this Chapter under American Practice. The wash or beer is fed continuously from a storage tank or well through a heater, in which it absorbs heat from the ascending spirit vapors, into a distilling column consisting of perforated plates. Steam is fed at the bottom of the still causing evaporation of alcohol from the wash. By regulation of:

1. Rate of beer inflow
2. Point of beer entry
3. Steam input
4. Amount of condensed vapor returned to column (reflux)
5. Point of reflux entry

a dynamic equilibrium is established within the still. That is, the temperature and concentration of liquid at each point of the still remain constant even though there is continuous counter-current flow of liquid and vapor past the plates. Hence, it becomes possible to discharge from the still bottom a spent slop containing less than 1 per cent of its original alcohol and to recover over 90 per cent of the alcohol as whiskey of 50–75 per cent o.p. and of almost any desired content of the "congeneric substances" on which the flavor and odor depend.

Operation and Yield.—Nettleton (*"The Manufacture of Whiskey and Plain Spirit,"* Aberdeen, 1913) cites records of various distilling operations at pot and patent still distilleries from which the following have been abstracted for comparison:

LOWLAND POT STILL DISTILLERY—ALL MALT

Equipment:

  2 wash stills, each 6,200 gallons capacity
  2 low wines stills, each 3,800 gallons capacity

Batch:

  2,600 bushels of malt—835 cwt. malt

Wash:

  54,100 gallons
  4 consecutive mashings at 7 to 9 hour intervals

Yield:

  5,104 British proof gallons or 6 1/10 gal. per cwt.

SMALL LOWLAND POT STILL DISTILLERY—ALL MALT

Batch:

  1,000 bushels of malt

Wash:

  20,900 gallons

Yield:

  1,979 British proof gallons or 6.16 gallons per cwt. (with a high class malt, skilfully manipulated, yield should be 7 to 7¼ gallons per cwt.)

HIGHLAND POT STILL DISTILLERY—ALL MALT

Equipment:

  1 wash still, 8,500 gallons capacity
  1 low wines still, 4,500 gallons capacity

Batch:

  1,800 bushels of malt—659 cwt.

Wash:
37,730 gallons
4 successive mashings at 7 hour intervals

Yield:
4,370 British proof gallons or 6⅔ gallons per cwt.

PATENT STILL DISTILLERY

Batch:
6,080 bushels of corn and malt—3,052 cwt.
(5,168 bushels of corn)
(912 bushels of malt)

Wash:
216,500 gallons

Yield:
16,542 gallons of British proof spirit or 5.4 gallons per cwt.

Reduction in bulk, in one continuous operation, from 216,500 gallons of wash to 9,930 gallons of spirit, or of 100 gallons of wash to 4.5 gallons of spirit. The wash had an average spirit value of 7.8% of proof spirit and the spirit produced an average spirit value of 165.2% or 65.2% o.p. The relative quantities of spirit and feints in the collected distillates were in the ratio of 100 proof gallons of spirit to 2.7 proof gallons of feints (16,409.9 actual to 444.6 actual). In bulk, the ratio was 100 gallons of spirit to 5.7 gallons of feints.

PATENT STILL DISTILLERY

Batch:
10,500 bushels of corn and malt—5,089 cwt.
(3,000 bushels of malt)
(7,500 bushels of corn)

Wash:
500,000 gallons wort
50,000 gallons yeast washings

Yield:
31,068.1 proof gallons or 6.1 proof gallons per cwt.

Rate of distillation:
2 Coffey stills operating 32 hours; rate per hour—17,187 gallons

Reduction in bulk in one continuous distillation was 550,000 gallons (original wort, with yeast pressings, etc., recovered) reduced to 18,615 gallons of spirit and feints, or in the ratio of 100 wash to 3.38 spirit and feints. The wash had an average spirit value of 6.3 per cent proof spirit, and the spirit and feints one of 166.9 per cent proof (83.5 per cent by volume).

The relative quantities of spirit and feints collected as distillates were 31,045.3 and 826.2 proof gallons; or in a ratio of 100 spirit to 2.6 feints. The ratio of bulk was 100 spirit to 2.9 feints.

In considering yield figures it should always be borne in mind that yield does not wholly depend on efficiency of distillation. The quality of grain varies from year to year and though the distiller may pay more for his grain in one particular year it is quite possible that the yield per cwt. will be less than in the preceding year. This reduced yield is due to inferior grain and the rise in prices to economic factors.

American Whiskey.—*General Statement.*—While the practice of whiskey manufacture in the United States varies quite definitely from British practice, it is also divided within itself into two general schemes. The divergences from British practice are mainly in the method of preparing the grain before the actual mashing (saccharification of the starch); in the general custom of fermenting the whole mash rather than filtered wort; and in the almost universal employment of patent stills. Within itself American whiskey making may be classified as small scale and large scale. There are definite differences of procedure and not merely size of operation which distinguish the two groups.

Preparation of Wort.—Small Scale.—Not only is starch a highly resistant material, but in cereal grains it is present in such highly compressed masses that some variety of treatment is necessary to loosen and spread it thinner before it is possible to convert it to sugars with any degree of efficiency. In the British Isles this object is accomplished either by subjecting the entire batch of grain to the malting process or else by the use of one or another modification of the acid conversion process. In the United States, where mixed mash whiskey is made, the starch is "opened up" or "pastified" by boiling with water, with or without the addition of a small portion of malt. In "small scale" operation this cooking is done at atmospheric pressure. The grain is ground to grist, a portion of dry or green malt grist is added, hot water is poured on in the ratio of 20 gallons per 56 lb. bushel of grist, and heat (steam) and agitation applied until the starch is "pastified." While the primary object of the process is "pastification," the addition of malt undoubtedly causes a slight amount of saccharification to occur simultaneously. It is usual to start the cooking at about 140° F. Live steam is then run into the open tank until the batch boils. Boiling continues for about one hour and then the batch is cooled to about 150° F. and the mashing commenced by the addition of a suitable quantity of malt. Generally, in a small distillery of this character, an open tank equipped with agitator, live steam, and cooling coils is employed for the starch pastification and the mashing is carried in the same vessel. Usually the malt, when introduced into the mashing vat, is at a temperature of 125°-130° F. so that the batch after mixing will have a temperature of about 140° F. The mash is held at this temperature for a half hour or so and then warmed to about 150° F. After being held at this temperature for about one and a half hours it is cooled to about 66° F. in the summer or 72° F. in the winter.

Fermentation.—The batch is then ready to commence the fermentation. While it is possible to commence fermentation at a somewhat higher temperature than those stated, it is also dangerous as the reaction may overheat beyond control.

It has been suggested that the process just outlined may be modified in the interest of malt economy, as follows: The mixed mash is started at 135° F. and after thorough agitation, most of the wort is drained off and stored temporarily. In the while, the wet grain is again raised to boiling temperature and boiled for a short time. Cold liquor is then added to reduce the temperature to 140° F. and the stored wort pumped back and thoroughly mixed with the balance of the mash. The subsequent procedure is as above. This modified process was designed to permit a second pastification of such starch as was uncooked in the first operation and provides no advantage if the first cooking was sufficiently thorough.

Distillation.—Practically all American whiskey is distilled in patent stills by the process described in detail for large scale distillation.

LARGE SCALE OPERATION

The manufacture of whiskey on a large scale in the United States represents the application to this process of all the improvements in efficiency and economy of time and materials made available by modern Chemical Engineering knowledge. A distillery includes units for milling; yeast production; whiskey, spirits and gin manufacture; recovery of secondary constituents; blending; and reduction and recovery of slop. A diagrammatic layout is shown as a whole in Figure 26 and the separate parts in Figures 27-31.

Milling.—The first operation at the distillery is the preparation of the grain and malt. These are elevated to hoppers and passed over magnetic separators to remove tramp iron, etc., which might injure the crushers. The cereals are then fed to grinders of the type commonly used for flour milling, and reduced to meal. The separate meals are then elevated to receivers and hoppers by means of air conveyors, and fed to their respective storage bins.

Cooking.—Starch is completely pastified by cooking under pressure. In order to maintain semi-continuous operation this is accomplished in three cookers used cyclically at intervals of an hour. That is, cooking of each distinct batch consumes three hours, but each hour another cooker in rotation has completed its batch and commences with a fresh one. A scheme of this operation is shown in Figure 32, p. 125.

The charge is usually made up on the proportion of 15–20 gallons of water at 100° F. per bushel of grist. In the mash tun more water or "slop-back" is added until the ratio is about 40 gallons per bushel. It will be noted that economy of steam is obtained by using the high pressure steam from one cooker to preheat another. Similarly, the use of a barometric condenser serves to economize on cooling water. When the charge is properly

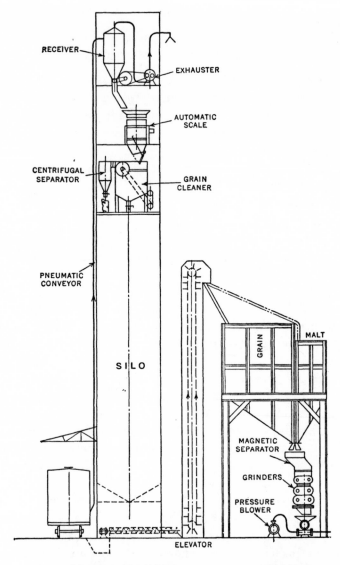

GRAIN STORAGE   MILLING

FIG. 27.—(Courtesy E. B. Badger & Sons Co.)

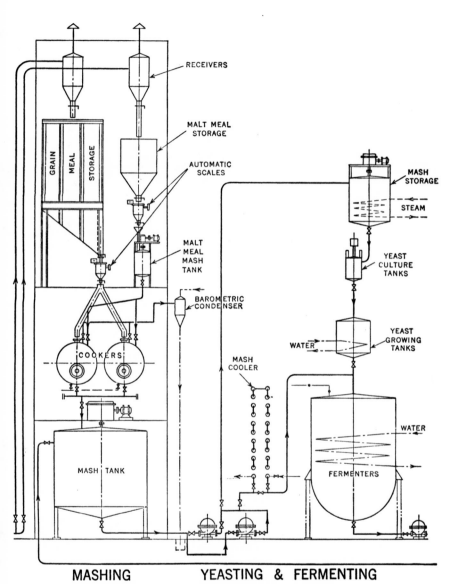

MASHING   YEASTING & FERMENTING

Fig. 28.—(Courtesy of E. B. Badger & Sons Co.)

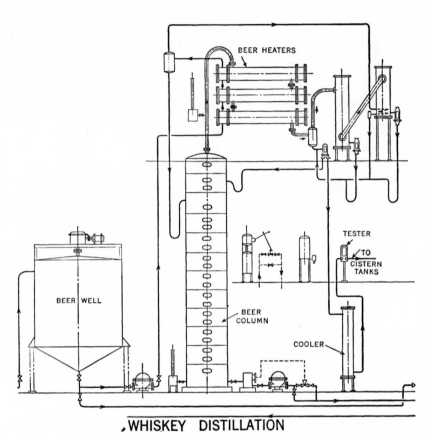

WHISKEY DISTILLATION

Fig. 29.—(Courtesy E. B. Badger & Sons Co.)

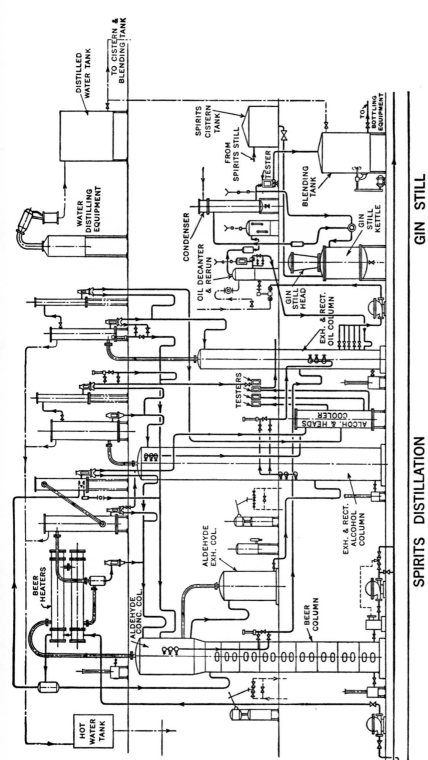

SPIRITS DISTILLATION   GIN STILL

Fig. 30.—(Courtesy E. B. Badger & Sons Co.)

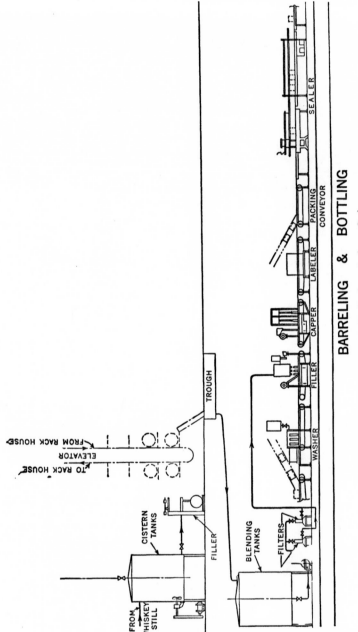

FIG. 31.—(Courtesy E. B. Badger & Sons Co.)

BARRELING & BOTTLING

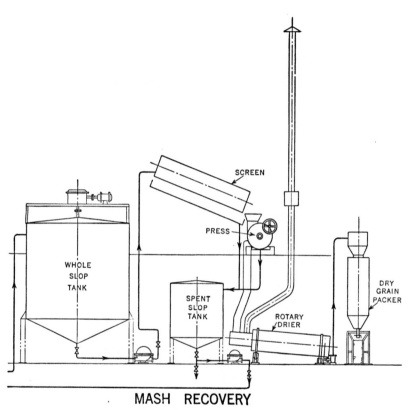

MASH RECOVERY

FIG. 32.—(Courtesy E. B. Badger & Sons Co.)

Time Scale	Cooker No.		
	1	2	3
0 Hour	Charge with grist and water. Steam bled from No. 2 cooker.	Steam bled to No. 1 cooker.	Cold malt mash added. Kept at 145°F. for 1st 15 min. 150°F. for 2nd 15 min.
	Live steam at 50 lb. pressure.	Steam bled to barometric condenser. Kept boiling until cooled to 150° F. 22″ vacuum.	Discharge to mash tun.
1 Hour	Steam bled to No. 3 cooker.	Cold malt mash added. Kept at 145° F. for 1st 15 min. 150° F. for 2nd 15 min.	Charge with grist and water. Steam bled from No. 1 Cooker.
	Steam bled to barometric condenser. Kept boiling until cooled to 150° F. 22″ vacuum.	Discharge to mash tun.	Live steam at 50 lb. pressure.
2 Hour	Cold malt mash added. Kept at 145° F. for 1st 15 min. 150° F. for 2nd 15 min.	Charge with grist and water. Steam bled to No. 3 cooker.	Steam bled to No. 2 cooker.
	Discharge to mash tun.	Live steam at 50 lb. pressure.	Steam bled to barometric condenser. Kept boiling until cooled to 150° F. 22″ vacuum.
3 Hour	Recharge as at 0 hour.	Steam bled to No. 1 cooker.	Cold malt mash added. Kept at 145°F. for 1st 15 min. 150° F. for 2nd 15 min.

FIGURE 33.—3-Hour Cooking Cycle in Large American Distillery.

cooked it has the consistency of a thick soup and all of its starch is quickly accessible to the diastatic action of the malt.

Mashing.—The mash is then pumped to the mash tun and water, or liquor from the spent slop tank if a sour mash is desired, is introduced in sufficient quantity to make the ratio of the total mash 40 gallons of liquid per bushel of grain and the batch is agitated while the mashing operation proceeds under temperatures and conditions approximately equivalent to those previously described under "mashing processes."

Fermentation.—At the conclusion of the mashing operation the whole contents of the tun are pumped, usually to an inclosed cooler. In this type of cooler the mash is forced through a double pipe system and is cooled by a counter current of water down to 70° to 80° F. This is a radical departure from Scotch and Irish practice in which only the wort is used for fermentation. A small portion of the mash is bypassed to an auxiliary mash tank which forms part of the yeast propagating system. The main mash is pumped from the coolers to the fermentation tanks and a certain amount of yeast is added from the yeast growing system. Fermentation soon commences and the temperature is controlled by means of cooling coils. It will be observed from the diagrams that the heating and cooling operations are effected by means of coils within the tanks. After completion of the fermentation the "beer" is pumped to the "beer well," a tank which is fitted with an agitating device thoroughly to stir the beer and prevent any settling.

Distillation.—At this stage of the process there are several interesting variations from Scotch and Irish methods of distillation. Note, on the diagrams, that by means of a bypass system the beer well may be connected to either the whiskey or spirit stills; these are both termed beer columns, although they are really distinct kinds of stills.

Various economies and several different methods of operations are effected by the aid of an elaborate piping system. Both the whiskey and the spirit stills are equipped with heat exchanging apparatus so that the still vapors passing through the heat exchangers give up their latent heat to the beer as it is pumped through the apparatus on its way to the stills. In other words, pre-heating of the beer is effected by means of the latent heat of the still vapors.

The whiskey still consists of a column equipped in the head with two rectifying plates and one washing plate. In this very efficient arrangement ascending alcoholic vapors are rectified so that their alcoholic content is increased from approximately 6 per cent to 72 per cent. This effects in one continuous operation, and more efficiently, the same doubling of concentration achieved by the European pot still equipped with a doubler as shown in Figure 15 in Chapter VIII. The rectified vapors then pass through the heat exchanger and then into the condensing system. As they emerge from the condensing system they are tested for strength and content of secondary products.

Another advance over European practice at this point is the fact that by means of the control system it is possible either to pass the condensates back to the column for reflux purposes or to withdraw them to a test box. By these means, a whiskey of almost any desired secondary product content can be produced at the will of the distiller, depending on the rate of reflux employed. The distillate when diluted with distilled water to about 100 proof (50% by volume) is ready to barrel as whiskey.

The spirit still is a typical continuous alcohol rectifying still equipped with heat exchanging device for pre-heating the beer fed from the beer well and with various rectifying units for removing both lower and higher boiling impurities, especially aldehydes and higher alcohols. By means of control and concentration devices a practically pure, 96 per cent spirit can be produced.

These spirit units recover more than 98 per cent of all the alcohol present in the beer and produce 90 per cent of this alcohol as high grade U. S. P. spirits and the remainder as heads and a washed fusel oil. Such a unit will use about 40 pounds of steam per gallon of alcohol produced. The largest of these units now in use produces 18,000 gallons of spirits per day of 24 hours.

Figure 31 shows a mash recovery system. Here the spent

still liquor, or slop, is received, agitated, and passed over a screen to filter out the solids. The solids are fed to a press to squeeze out residual moisture. The liquor is discharged to the spent slop tank, while the damp solids are then fed to a rotary drier of standard design, which drives off any remaining moisture. A blower delivers the dried solids to a receiver fitted with a hopper which feeds the material to a packaging device. Recovery of solids is approximately 12 to 15 pounds per bushel.

Unless legislation or peculiar local conditions require total evaporation the thin slop is discharged to the sewer or part is returned if process calls for "slopping back."

Aging.—Whiskey as first produced by any of the processes described is raw and unpleasant to the taste and disagreeable in odor. It has been known for very many years that by storage for a period of time, changes in the odor and taste are produced which result finally in the ripe smoothness of taste and pleasant odor associated with good whiskey. Despite much work the whole chemical nature of these changes is still incompletely known. As the late William Howard Taft concluded from an investigation made during his presidency: "It was supposed for a long time that by the aging of straight whiskey in the charred wood a chemical change took place which rid the liquor of fusel oil and this destroyed the unpleasant taste and odor. It now appears by chemical analysis that this is untrue—that the effect of the aging is only to dissipate the odor and modify the raw, unpleasant flavor, but to leave the fusel oil still in the straight whiskey."

Actually, comparative analysis of old and new whiskey shows a somewhat greater content of secondary constituents in the old matured whiskies, especially in the relative amounts of volatile acids and aldehydes. The esters also increase, but to a lesser extent, while the furfural and higher alcohol contents remain practically unaltered. Of course, whiskey stored in wooden barrels increases in proof due to a relatively more rapid diffusion of water through the pores of the wood. In obtaining the analytical results noted above, this change is compensated by calculation to a uniform base alcoholic strength. The solids content and color of the whiskey increase markedly on aging due to the extraction of tannin, resins and other materials from the wood. The density of color is directly proportional to the solids content in an aged whiskey.

Aging practices differ somewhat. British custom is to store the whiskey in uncharred oak barrels while American whiskies, both Rye and Bourbon, are stored in charred barrels. The color and solids of whiskey aged in uncharred packages are much smaller in amount and more water soluble than those of whiskey stored in charred packages. The charring also results in a "bead" of oilier consistency and greater permanence than the uncharred barrel imparts.

Rye whiskies are stored in heated warehouses, while Bourbons are matured in unheated buildings. As a result the former are stronger in color than the latter. In general, whiskey does not improve at all after about ten years of storage, although there still continue slight changes in composition; nor is there any very marked improvement in desirable character after the first four to six years of storage. The high price of very old whiskies is largely to compensate for evaporation losses which become very marked, and the carrying charges on investment tied up for long terms of years. Storing has its limitations. A fifteen year old whiskey may be a bad whiskey because, as President Taft pointed out, its fusel oil content has increased too much. There are whiskies only two years old far better in flavor than hoary distillates that have been kept in barrels for two decades.

Artificial Aging.—The chemist distinguishes between aging and maturation; that is, between the mere passage of time and the effects thereby produced. If the latter can be duplicated within a short period, the results, from the chemist's and from an economic point of view are much preferable. Hence much study has been given to the subject of the artificial aging of spirits. Many of the more scientific suggestions are admirably summarized by Snell and Fain in an article which appeared in Ind. & Eng. Chem. News Ed. XII, 7, p. 120. They state:

The Legalization of Traffic in spirituous liquors in this country has created a situation which puts a premium on naturally aged alcoholic beverages. To satisfy the demand for liquor at a popular price, the available stocks on hand have to be increased either by blending or accelerated aging.

During the aging process the constituents of alcoholic spirits undergo chemical change. A study of the changes taking place in whisky stored in wood over a period of eight years revealed (*5*) important relations between the acid, ester, color, and solid contents of a properly aged whisky which will differentiate it from artificial mixtures and from young spirits. High color, high solid content, and high alcohol concentration are generally accompanied by high acid and ester content; low color and solid content go with a small amount of acids and esters. In the aging process the acids are at first formed more rapidly than the esters. Later the esters form more rapidly so that by the end of the fourth year they are present in the same amounts. The equilibrium reached at this period is maintained. The amounts of higher alcohols increase in the matured whisky only in proportion to the alcohol concentration. The oily appearance of a matured whisky is due to material extracted from the charred container; this appearance is almost lacking in whiskies aged in uncharred wood. The improvement in flavor of whiskies in charred containers after the fourth year is due largely to concentration. The higher content of solids, acids, esters, etc., of rye over Bourbon whiskies is explained by the fact that heated warehouses were used for maturing rye whiskies and unheated warehouses for maturing Bourbon.

The aging of brandy, similar to that of whisky, takes place in oak casks. The conjoint action of the oxygen of the atmosphere and the resins, gums, and tannins extracted from the wood are responsible for the improvement of the liquor. These compounds, being capable of easy oxidation, pass through a series of reactions. Aromatic compounds particularly agreeable in taste and odor are formed.

Aging of spirits involves oxidation. It is this reaction which one attempts to hasten by the processes devised for accelerated aging. Methods for aging spirits artificially fall into four main classes as follows: (1) treatment with air, oxygen, or ozone; (2) exposure to actinic rays; (3) electrolytic treatment; and (4) use of catalysts. Combinations of these methods are likewise employed.

TREATMENT WITH GASEOUS OXIDANTS

A recent example of the first type provides for treatment (*17*) of the liquor with oxygen while exposed on large wooden surfaces which have been impregnated with a solution obtained by extracting seaweed ash. Brandy (*28*) is artificially aged by bringing it into contact with activated charcoal which may first be treated with a current of air or oxygen.

Oxidation may be accelerated by the use of compressed air (*6*). The liquid to be treated is run into a tank which can stand a pressure of several atmospheres. Compressed air enters from the bottom of the tank. The length of time required for treating by this process depends mainly on the pressure of the air, the nature of the liquid, and the extent of aging desired. In a special apparatus (*21*) for this purpose, the liquid is sprayed or atomized in a chamber containing air under pressure by delivering from two oppositely disposed nozzles with a double cone between. Intimate mixture is obtained. As a modification of methods for treatment with oxygen in high concentration, beverages such as brandy, cognac, and liqueurs (*13*) are cooled to a temperature below −18° C., saturated repeatedly with air while at this temperature, and afterward stored in a warm room until the acids combine with the alcohols forming esters. According to a Canadian process (*26*) oxygenating gas is bubbled through new spirits in a vat. The gas, after passing through the liquid, rises through a mass of shavings or cuttings of charred or desiccated wood (preferably oak) over which a counterflow of the spirits is maintained by withdrawing liquid from the bottom of the vat and discharging it into or over the wood.

Alcoholic beverages, such as whisky, cognac, etc., are also treated with air (*2*) which has been subjected to a high-tension electric arc. This air contains oxides of nitrogen. The claim is made that the flavor is improved.

Aging has been accomplished (*4*) by bringing the liquid in contact with bodies such as oak chips or shavings which have been treated with ozonized air or oxygen. This prevents local excess concentration of the ozone.

Apparatus has been developed (*1*) for the production of ozone for use in the accelerated aging of liquors. Sizes up to 300 kw. with capacities up to 10 to 12 kg. of ozone per hour are available. Concentrations of 2 to 4 grams of ozone per cubic meter can be obtained in air or oxygen, with an energy consumption of 25 to 35 kw-hr. per kg. of ozone. Treatment of liquor with ozone gives a mellowing effect in a short time which can be obtained otherwise only after months or years of storage. Analysis shows a decreased aldehyde content and an increased ester content. A suitable ozonizer for liquor treatment is a special 50-watt size with a capacity of 100 liters per hour.

USE OF ACTINIC RAYS

For artificial aging of wines and liquors by the action of ultra-violet light (*7*), a vapor electric arc having a quartz container is used. The liquor is passed over this lamp in a thin film. In several processes actinic rays are used in combination with oxygen or an oxidizing agent.

One process (16) subjects them to light from a neon lamp ranging from yellow to orange in color, in the presence of oxygen. Another (20) subjects them to ultra-violet rays after addition of a small amount of hydrogen peroxide, inorganic and organic peroxide, or ozonide.

As another variation (19) wine, cognac, etc., are aged and improved by pretreating with ultra-violet light the water used in their preparation. The water may also be aërated or treated with oxidizing agents. The product after addition of the water is sometimes irradiated.

ELECTROLYTIC TREATMENT

Beverages are artificially aged by an electrolytic treatment (14) producing hydrogen and oxygen in the liquid. The electrodes and the diaphragm between them are impregnated with insoluble inorganic salts or oxides capable of producing oxidation and reduction effects in the presence of the oxygen and hydrogen produced.

In aging and maturing alcoholic liquors by electrolysis (10, 11), a depolarizing cathode and a current of low intensity are used. The cathode is formed of a carbon electrode surrounded by manganese dioxide and carbon contained in a porous pot. The anode is formed of a nonoxidizable metal such as gold. The electrolysis is effected in the presence of the substances which the spirits will extract from oak wood. For this purpose the spirits are allowed to remain for some time before treatment in oak casks. Sometimes oak shavings are added to the spirits during treatment. The electrolysis will, owing to cataphoresis, assist in the extraction. The electrolysis may be effected in oak casks between anodes on the outside of the cask and a cathode inserted through the bunghole. Pads of moist linen or cotton are placed between the anode and the surface of the cask. The vats and casks are supported on insulators which may be bowls containing a liquid such as vaseline oil. The passage of current is maintained continuously for eight to ten days, according to the conditions adapted for each application.

Apparatus (8) for treating liquids such as wines, spirits, etc., with electrical currents of high voltage and low amperage, consists of two point-and-disk separators placed oppositely in parallel circuits connected to the terminals of a transformer, so that one alternation of the transformer current will pass through one circuit and the other alternation through the other. Barrels containing the liquid to be treated are inserted in each circuit. By the use of this method there is no heating of the liquid, and loss by volatilization of the aromatic compounds contained in the liquor is minimized. Another process (9) ages wine, cognac, and arrack by passing a high-tension electric discharge through them. The combination of the electrolytic treatment with the use of air, oxygen, or ozone has likewise proved effective.

Suitable apparatus (23, 24, 25) combines treating liquor in barrels with a gas, such as air, oxygen, or ozone, and the use of an electric current to accelerate aging. An electrode is inserted through the bunghole of the liquor container, and the liquid either alone or together with a fine wire of high resistance connecting the electrodes serves as conductor. Heating of the liquid occurs in either case.

An ingenious process (18) includes saturation of the liquid with oxygen, followed by the transformation of this oxygen into ozone by means of discharges of electricity through the liquid. The oxygen is introduced into the liquid under pressure and the electricity is discharged at short intervals through the liquid. Impurities from distillation are oxidized and the flavor is improved.

USE OF CATALYSTS

Artificial aging of spirits is aided by the use of catalysts. The vapors may be passed over finely dispersed metal oxides such as those of copper, nickel, and titanium (27) at 150° to 180° C.

Suitable catalysts for oxidation (22) are oxides of cobalt, cerium, vanadium, and uranium. Catalysts for ester formation are oxides of lead, molybdenum, silicon, uranium, and cerium. The best flavors are produced by the use of oxides of lead, copper, nickel, molybdenum, cobalt, titanium, and silicon.

Charcoal and charred sawdust have likewise been found to catalyze the maturing of spirits. The rising vapors, inside or outside the cask, may contact catalytically acting charred sawdust or charcoal (29) without the catalyst, however, coming in contact with the liquid. Other catalysts may be employed in this way, alone or together with the charcoal or charred sawdust.

A similar method (3) for maturing potable alcoholic liquors is to mix the vapors from a pot still with heated air, subdivide the mixture into narrow streams, and pass this through a narrow conduit heated to about 150° C. The streams are joined and the treated vapors condensed. The heated metal walls are supposed to act catalytically to produce the desired result.

MISCELLANEOUS PROCESSES

Spirits are also aged (15) by separating alcohol and water, and removing the fusel oil from the concentrated extract of oils, etc., by treatment with petroleum ether. The concentrated extracts are subjected to accelerated aging by one of the methods described above, and again mixed with alcohol and water free from fusel oil.

According to another process (*12*) an extract prepared from oak wood, such as is used in making the usual storage vats, is added. The wood, which may be the heart of the larger branches of the trees or the waste obtained in making casks, is disintegrated and submitted to two successive extractions with aqueous alcohol and a final extraction with water. The alcoholic extracts are distilled *in vacuo* at a low temperature, the residue is added to the aqueous extract, and the mixture is evaporated *in vacuo* to obtain the extract in the form of a dry solid.

The old principle of acceleration of a chemical reaction by heat is applied to wines and spirits by storing them in closed vessels and agitating (*30*) for some months at 43° C. A rocking, effected by oscillating a platform on which the cask rests, rather than by a tremulous vibration, gives the desired results.

The lines of attack on the problem are sound and can be expected to give results when properly applied. Details as to application vary in different processes. Some are in commercial use today in our newest large industry.

LITERATURE CITED

(1) Becker, J., *Chem. Fabrik,* 1929, 49.
(2) Brabender Elektromaschinen G. m. b. H., German Patent 500,-708 (1928).
(3) Carroll, J. E., U. S. Patent 968,832 (1910).
(4) Coffre, R., British Patent 340,647 (1928).
(5) Crampton, C. A., and Tolman, L. M., *J. Am. Chem. Soc.,* 30, 98 (1908).
(6) Deriques, J. L., *Z. Spiritusind.,* 31, 141 (1908).
(7) Henri, V., Helbronner A., and von Recklinghausen, M., U. S. Patent 1,130,400 (1910).
(8) Henry, C., British Patent 17,400 (1914).
(9) Hirschmann, W. A., German Patent 199,265 (1907).
(10) Jarraud, A., British Patent 141,687 (1920).
(11) Jarraud, A., German Patent 239,300 (1910).
(12) Jarraud, A., and Roussel, J., British Patent 148,829 (1920).
(13) Monti, E., U. S. Patent 1,108,777 (1905).
(14) Nottelli, L. E., French Patent 711,300 (1931).
(15) Philipsky, J. H., German Patent 549,524 (1929).
(16) *Ibid.,* 557,806 (1930).
(17) *Ibid.,* 572,351 (1932).

(18) Plotti, A., *Progrès agr. vit.*, 24, 674 (1908).

(19) Reinisch, E., Austrian Patent 112,976 (1928).

(20) *Ibid.*, 115,902 (1929).

(21) Saint-Martin, W., German Patent 237,280 (1910).

(22) Sándor, Z. de, *Mezögazdasági Kutatások*, 4, 468 (1931).

(23) Seitz, J., U. S. Patent 961,167 (1908).

(24) *Ibid.*, 967,574 (1908).

(25) *Ibid.*, 967,575 (1909).

(26) Sunderman, F. R., and Gaut, R. E., Canadian Patent 303,644 (1930).

(27) Toth, G., *Magyar Chem. Folyóirat*, 38, 129 (1932).

(28) Verdeaux, F., French Patent 716,829 (1931).

(29) Verein der Spiritus Fabrikanten, German Patent 291,349 (1915).

(30) Vianna, J. da V., British Patent 23, 548 (1913).

Dosing.—It is also claimed and has probably been practiced that the addition of small proportions of the materials listed below either singly, or in combination, will improve the flavor and appearance of whiskey.

Acetic Acid
Allspice
Almond shell extract
* Beechwood creosote
Caramel
Caraway seed
Cedar wood extract
Cherries, dried
Cherry juice
Cinnamon
Cloves
Glycerin
Glycerite of tannin
Oak extract
Peach juice
Peaches, dried
Plum juice
Plums, dried
Prune juice

* Beechwood creosote is said to have been the material used during prohibition for imparting the smoky taste to bootleg Scotch whiskey.

Prunes, dried
Sherry wine
Spirit of nitrous ether
Tannic acid
Vanilla
Walnut shell extract
Catechu tincture
Coumarin
Kino tincture
Orris root
Pekoe tea

One of the less scientific methods for accelerating the maturation of whiskey was aging for comparatively short periods in old sherry wine casks. This method was claimed not only to be very effective but also is probably the least objectionable. As a variation the casks could be subjected, before filling with whiskey, to a forced seasoning. The process consisted of placing the casks bung down and drying them thoroughly by forcing a current of warm air through the bunghole. Then enough wine to wet all the inner surface was poured into the cask, the cask revolved to coat all the wood and the wood impregnated by forcing in warm air under pressure.

Blending.—On account of the inherent variability of a product made in relatively small batches like pot still whiskey; and the natural fluctuations in the qualities of the raw materials available for patent still whiskey, the practice of blending whiskey of different distillations and different years arose early in the life of the industry, to enable the distiller to market a more uniform product. Later, the custom extended to the blending of the products of different distilleries and of distillates from both pot and patent stills. Still later, and in the United States possibly even more after the repeal of prohibition, the practice of spreading the flavor of an old whiskey over three to five times as much diluted "silent spirits" was exceedingly common.

In Great Britain the business of blending has assumed great importance within the industry as can be shown by reference to the Directory of Whiskey Brands and Blends which lists 3428

Scotch, 487 Irish, and 128 Scotch and Irish blends founded upon the output of 122 Scotch and Irish distilleries.

The practice has both good and bad features. For example, the blender can take the distillers' output and by skillful blending produce a product of uniform characteristics year after year with little variation. Again, qualities desired of whiskey vary according to the locality. For example, Scotland, Canada and Scandinavian countries favor stronger whiskies than those drunk in England, France, Belgium, Holland, Australia and India. It is hard to say what the United States now favors, probably a strong whiskey. The blender can meet these geographical differences in taste by skillful blending. On the other hand, cheap and inferior blends have often been foisted upon the public under misleading names.

Blending formulae are secret, and the practice as carried on by some of the oldest and most conservative blenders approaches an art. At its crudest it consists of pouring the various whiskies into a blending tank according to formula, dosing, coloring and stirring the mixture and then allowing it to rest for 24 hours. The blend is aged in a cask for a short time and then bottled.

This gives a raw whiskey, imperfectly blended, and fit only for a cheap trade. A good blender proceeds more as follows: First, he selects various fine malt whiskies. He blends these carefully, marrying one whiskey with another every three months until the desired body and flavor are obtained, and then ages them in an uncharred oak cask for about two years. When he deems the blending and aging to be complete, he mixes the product with patent still spirit and Lowland or equivalent malt, stirs them up and allows them to age again in an uncharred oak cask for a year or more.

Scotch pot distillers have admitted the necessity of blending both pot and patent still products, except when the pot still spirit has matured for a considerable number of years in wood, in which case they consider it unnecessary. They claim that it is the pot still product which imparts character to the blend and that consequently it must always be employed in preponderating proportion in the blend if the reputation, which the best classes of Scotch whiskey have gained, is to be maintained.

The production of cheap and palatable Scotch whiskies involves a different set of considerations. It is necessary for pot still spirits to mature in wood in order that they should acquire a pleasant flavor. Patent still whiskies, on the other hand, although they are improved by aging in wood, change to a less extent and mature much more quickly. It is stated that by blending immature pot still with patent still whiskey the pungent, unpleasant taste of the former is attenuated or toned down and that the mixture then becomes "a palatable and not unwholesome spirit." Such a mixture, if stored in wood, would mature in a shorter time than would the pot still whiskey alone.

The proportion of pot still to patent still whiskey in these cheap blends is varied chiefly in accordance with the price at which they are planned to sell. The cheapest blends may contain as little as 10 per cent of the former and even less.

Irish distillers contend that it is unnecessary to blend aged Irish pot still whiskey with patent still spirits, but admit that such blends are made for the cheaper trades. By British law the age of the whiskey in a bottle is determined by the age of the youngest whiskey in it, irrespective of the amount of that whiskey. Let us suppose that there are fifteen whiskies in a bottle, averaging fourteen years and making up 99 per cent of the contents, while 1 per cent is a whiskey aged five years. The legal declarable age of that whiskey is five years. White lies have been told on labels bearing such inscriptions as "Whiskey in this bottle is fifteen years old." It is true that there may be whiskey in that bottle aged fifteen years, but the real story is not there.

Post repeal American blending practice is still, at the time of this writing, in a chaotic state. The same general principles of desirable and undesirable blending are applicable as in British practice. The details of blending in this country as well as the tremendously complicated system of combined federal and separate state legislation are in an almost continuous state of flux so that there is little to gain by recording them. With the advent of repeal in this country, stocks of aged whiskey were much below the anticipated demand. Hence, most such stocks were "blended" or "cut" with very new whiskey or even with diluted alcohol and other materials, colored with caramel, dosed with "prune juice" and "bead oil" and sold quickly. Very shortly the flood of federal and state regulations appeared. These range from requirements that a "blended whiskey" shall contain not less than 20 per cent of four year old "U. S. P. Whiskey" to requirements similar to the British, that blended whiskey shall bear on its label the age of the youngest whiskey it contains. The reader is referred to Chapter XIII on interpretation of analysis for further details on this topic.

CHAPTER X

BRANDY, RUM, GIN

AND

OTHER DISTILLED LIQUORS

BRANDY

Brandy is the product prepared by distilling wine, wine lees and/or grape pomace and often by blending the results of these operations.

It is a yellowish-brown liquor of sweet, smooth ethereal flavor and of fine bouquet. Alcoholic content is usually from 45 to 55 per cent by volume.

As first made it is normally colorless, and the familiar yellowish-brown hue is obtained either naturally by aging in oak casks or artificially by addition of a solution of caramel.

The fine flavor and bouquet result from the secondary constituents of the brandy and are dependent upon a number of factors, principally raw materials, operating methods, aging, etc. The secondary constituents consist of various esters (acetic, butyric, oenanthic, valerianic), acetic acid, volatile oils, tannin, fixed acid and coloring matter. Ethyl pelargonate (oenanthic ester) and other volatile constituents are thought to be mainly responsible for the flavor.

Because of the fine quality of its products France is commonly thought of as the home of brandy. However, other countries are also large producers: e.g., Spain, Egypt, South Africa, Australia, Algeria, Germany and the United States (California). Spanish and Algerian brandies are of very high quality. Egyptian brandies are made from imported grapes (Asia Minor and Southern Turkey) and have a strong flavor. They are not so fine and compete with the cheaper brandies. Australian and

South African brandies are of fair quality. South African "dop" brandy is an *"eaux de vie de marc."*

Brandies are produced in various parts of France. The best are produced in the Cognac district which is located in the two departments, Charente and Charente Inférieure. The region is also divided, according to the fineness of the wine, into the Grande (or fine) Champagne, the Petite Champagne, the Borderies and the Bois. Next in order of commercial merit are those made in the Armagnac, including the Marmande district.

Other parts of France in which brandies are produced are: le Midi, Aude, Gard, Hérault and Pyrénées Orientales. Brandies from these districts are commonly known as the "Trois-Six de Montpellier."

Eau-de-vie is the French name for brandy. It is used there in a rather broad sense and may embrace spirit distilled from wine, cider, perry, marc, cherries, plums or other fruit and also to mixtures of such spirits, or to a blend of any such eau-de-vie with any *"alcool d'industrie"* which is a name for either grain or beet alcohol. In view of this all-embracing nature of the term it is customary for a Frenchman to qualify his order for an eau-de-vie by specifying *"un fine,"* or *"fine champagne,"* or *"un cognac."*

True brandies may be classified into the following grades:

First.—Distilled from high quality light white wine not less than a year old.

Second.—Distilled from second grade wines or spoiled and soured wines which have been specially treated before distillation.

Third.—Distilled from grape pomace which may have been refermented with sugar and water. The term "grape pomace" includes the skins, pulp and possibly the stems of the fruit. These brandies are naturally of very inferior quality. They are known as Marc Brandies or *"eau de vie de marc"* from the French term for pomace. During prohibition a similar product was supplied in the United States by bootleggers under the name "grappo."

Fourth.—Certain incrustations are left on the sides and bottoms of fermentation tanks and aging barrels. They are called "wine lees" and usually contain from 20-35% of potassium acid tartrate (Cream of Tartar) and up to 20% calcium tartrate. They also contain yeast cells, and protein and solid matter which had settled out from the grape juice. By acidifying the lees with sulphuric acid and distilling, a product of exceedingly strong flavor and odor is obtained which is used to give character to diluted silent spirits, and the product is marketed as brandy.

Distillation.—In the Cognac district the brandy is made either by the large distilleries or by the farmer himself right at the vineyards. It receives very little rectification, when distilled, in order to conserve the secondary constituents which produce the bouquet and flavor.

For this reason, the simplest pot stills or slight modifications thereof are generally used. The only usual modifications are the pot still with *"chauffe-vin"* and the *"à premier-jet"* still. Capacity of the stills as a rule is about 150 to 200 gallons.

A *"chauffe-vin"* is a heat exchanging device for preheating the wine before it reaches the still kettle. It consists of an arrangement whereby the neck of the pot still passes through a wine container so that the vapors, prior to condensation, give up part of their heat to the wine in the tank. See Figs. 12, 13.

The *"à premier jet"* is a device for returning the distillate to a heat exchanging attachment at the head of the still. By this arrangement the newly rising still vapors give up part of their heat to the first distillate which is thus vaporized. Some rectification of the still vapors takes place. The *à premier jet* gives some effect of continuous rectification as compared to discontinuous operation with the ordinary pot still and the resulting product is stronger but not of such fine quality. These brandies are usually considered more suitable for liqueur manufacture than for direct consumption.

In the simple pot still process, two distillations are used, which may be compared with the process of whiskey making in the Scotch pot still distilleries; the two distillates are respectively termed *"brouillis"* and *"bonne chauffe,"* the terms being directly equivalent to the "low wines" and "spirits" of the whiskey distiller. The stills are worked very slowly and regularly, ten hours are usually allowed to complete the distillation of a batch. The quality of the resulting brandy, still depends to a great degree on the judgment and skill of the operator.

In other districts where the wines have a strong, earthy flavor somewhat more elaborate apparatus is used. The La Rochelle district uses the *Alembic des Iles* which is a pot still with rectifying equipment. The Midi uses a continuous distilling column of the kind in favor in this country, excepting that it is equipped with a faucet or tap at each plate. This arrangement enables the operator to distil at higher or lower strengths at will.

The wine used in the manufacture of Cognac contains from 6 to 11 per cent of alcohol, or from 12 to 22 per cent of proof spirit; the average strength is from 7½ to 8½ per cent of alcohol or from 15 to 17 per cent of proof spirit. The final distillate as run from the still is about 25% over proof or 60 to 65% in alcoholic content.

Aging.—Following distillation the brandy is aged in oak casks. Considerable care must be taken in the ripening process if the distiller wishes to market a good product. Four or five years at least are required to develop the right bouquet, flavor and mellowness. The finest brandies are sometimes aged for twenty years or even longer.

Before filling, the casks are thoroughly sterilized, either by steaming, or by scalding with several changes of boiling water. Following this, the cask is filled with white wine to dissolve any objectionable coloring matters or substances which might affect the flavor of the brandy and drained.

Blending.—Aged brandies are very often blended, since they may vary in characteristics according to source of raw materials, district of production, and year of vintage. Blending has been found necessary to produce a product of uniform characteristics year after year. As in whiskey blending, cheapening may also be a desideratum.

Formulae are, of course, secret and are based on the experience of the blender. They are generally varied each year, to some extent, to compensate for the variation in characteristics of the brandies available. The methods of procedure outlined under

Blending in Chapter IX on Whiskey apply, on the whole, to brandy blending.

Many imitation brandies are on the market and it is very doubtful how much of the brandy consumed is genuine. Imitation brandies are made as a rule by cutting strongly flavored brandy with diluted, rectified grain alcohol, coloring and sweetening with caramel and cane-sugar syrup, adding small amounts of aromatic substances, and dosing with either "lees oil" or an extract of oak wood chips.

Various extracts are used to give to the brandy aged and other characteristics. For example a wine distillate extract of cedar wood chips, 1 to 10 (about 500 c.c. per 100 liters finished product), gives wine and brandy a herb-like, typically aged character. A wine distillate extract of bitter almond shells (100 to 300 c.c. per 100 liters finished product) gives in addition to the herb-like flavor a pleasant aroma resembling vanilla. The same quantities of extract of either dried, green walnut shells or dried, stoned plums round out the flavor nicely. Orris root, coumarin, cinnamon, Pekoe tea and vanilla are also used although the wine laws of some countries prohibit their employment. Many of the products listed as dosing agents in Chapter IX on Whiskey have also been used.

British Brandy.—This is a compounded spirit prepared by re-distilling duty paid grain alcohol with flavoring ingredients or by adding flavoring materials to such spirits. The flavoring materials used in any one case are a trade secret, but in general are to be found in the lists mentioned.

Hamburg Brandy.—This is an imitation grape brandy made by adding flavoring to potato or beet alcohol.

RUM

Rum is a spiritous beverage prepared by fermentation, distillation and aging, from molasses and the scum and foam which form on the top of sugar cane juice when it is boiled. Fresh sugar cane juice may also be used when the cost of sugar production makes it profitable. High quality rums are made from mashes containing comparatively small amounts of skimmings (scum).

So-called "Nigger rum" is made from mashes consisting principally of the skimmings and other waste products of the defecators of sugar cane factories.

Rum is a yellowish-brown liquor of fine bouquet and sweet, smooth, alcoholic taste and flavor which cannot be successfully imitated artificially. The alcohol content of the genuine product should not be less than 78 per cent by volume. "Nigger rum" has a raw, sour, burnt taste and flavor.

When first made, rum is normally colorless and the familiar yellowish hue is obtained by aging in casks. If an exceptionally dark color is required it is dosed with caramel.

Because of the fine quality of its products Jamaica is commonly thought of as the home of rum. However, rum is produced in all countries where sugar cane is abundantly grown; e.g., British Guiana, West Indies, Brazil, southern United States, Madagascar and the East Indies.

Jamaica rum is graded into three classes, namely: 1. "local trade" quality for home consumption, 2. "home trade" quality for consumption in the British Isles, and 3. "export trade" quality for export. Local trade rum, the lowest quality, is distilled with particular emphasis on its alcoholic strength to the neglect of the other substances, chiefly esters, from which the flavor is derived. The flavor of this grade is, therefore, decidedly inferior.

The "home trade" quality constitutes the bulk of the exported rum. It has a full flavor, and chemically is characterized by a higher proportion of esters of higher fatty acids. It is generally accepted that these acids result from bacterial decomposition of the dead yeasts found in the distilling materials. As compared with "local trade" goods the "home trade" have a fuller and more fruity aroma and a marked spicy residual flavor is noted on dilution. Sometimes, even, an excess of the higher alcohols and esters which produce this result will also cause an objectionable cloudiness on dilution with water. On occasion "home trade" rums will have a noticeable burnt flavor resulting from over-distillation by direct fire.

"Export trade" Jamaica Rum is manufactured principally for European, especially German, consumption. This class of goods is so high in flavoring ingredients that it is unsuitable for beverage use, as such. The chief uses are for blending with lighter rums or neutral spirits and for the fortification of hock and similar wines.

Rum Manufacture.—*Jamaica Rum.*—Rum is distilled from the by-products of sucrose recovery from sugar cane juice. The process of sugar recovery is here stated in brief outline to explain the origin of the raw materials for rum. The cane, within a few hours of cutting, is brought to the crushing mill. The lapse between cutting and crushing must be short to avoid losses of sucrose from various sources, especially inversion. Hence mills are usually rather small and serve only a limited territory. The sucrose content of the sugar cane varies from 10 to 18% of the total weight.

At the mill the canes are cut into short bits by the rapidly revolving knives of the cutter and then pass to a series of three-roll crushers which press out the juice. If no water is added the process is called "dry crushing." In "wet crushing" water is played on the cane at the second or third set of rollers. The pressed cane or "bagasse" may be treated for further extraction. The juice drops into troughs under the rollers and is strained, warmed to about 200° F. and left for a time in settling tanks.

Some sugar is still retained in the bagasse. Hence in some plants it is passed on an endless belt through a shallow trough containing water and then pressed once more in crushing rolls. This extract is added to the first juice.

The amount of juice in the cane varies according to district of origin and degree of maturity. About 60 to 80 per cent of the juice is extracted by the methods described and the juice conforms approximately to the following analysis:

	per cent
Sucrose	14.1
Reducing sugars	0.6
Water	83.6
Undetermined solids	1.7
	100.0

After settling, the juice, which is still turbid and has an acid reaction, is drawn into mixing tanks and treated with enough lime to make it slightly alkaline. This treatment results in the precipitation of a number of impurities. The limed juice is heated, and in about an hour albuminous material coagulates on the lime precipitate forming a crust, and the whole produces a thick scum.

After the juice has settled, the scum is removed and sent to the fermenting tank in the still house. The clear juice runs to evaporators, its sucrose content being about 14 per cent. In the first evaporation, it is concentrated to about 50 per cent sucrose. Further concentration is carried on until the desired point for proper crystal growth has been reached. The mass in the pan, then called "massecuite," contains a total of 82 per cent sucrose and perhaps 8 per cent water. Of the total sugar in the hot massecuite 56 per cent is in crystals and 44 per cent is in solution; after cooling 65 per cent has crystallized and 35 per cent remains in solution.

The thick semi-solid mass is placed in centrifugal baskets and "whizzed." The adhering solution which whirls off to the outside, is collected and stored as molasses. The crystals are first washed in the basket, then removed for shipment. They constitute the raw or centrifugal sugar which refineries buy.

The molasses may be concentrated again until all of its crystallizable sugar has been removed.

Fermentation.—In the meantime the scum which was sent to the fermenting pan in the still house was allowed to remain a few days until it soured, a certain amount of bagasse having been added to assist souring.

A mash is made up of diluted molasses containing about 25 to 30 per cent sugar and skimmings (sometimes juice is added). "Dunder" is also added. This is the name given to the spent liquor from the stills and has the color and consistency of pea-soup. It contains mineral salts, coagulated albuminoids and soluble nitrogenous substances; and not only stimulates fermentation but increases yield and has a distinct influence on taste and flavor.

The mash as mixed contains about 12 per cent fermentable

sugar. A vigorous fermentation soon sets in as a result of the sub-tropical climate and the composition of the mash. Fermentation is completed in about 6 to 12 days, sometimes longer, various organic acids being formed along with the alcohol.

Distillation.—Distillation is carried on in pot stills. The whole process must be carried out with a great deal of care. The first distillate has a nauseating odor and a raw burning taste so that it must be rectified to eliminate objectionable ethers, aldehydes and acids. It is also customary to trap off a portion of the total rectified distillate so that it may be used for blending with succeeding distillates.

Demerara Rum.—We are fortunate in having the following description of rum manufacture which is quoted from a "Communication of the British Guianas Planters Assn. to the British Royal Commission on Whiskey and The Potable Spirits, 1909."

"In British Guiana the wort is prepared by diluting molasses with water to a density of 1,060 and it is rendered slightly acid by the addition of sulfuric acid in quantity sufficient to set free more or less of the combined organic acids, but so as not to have uncombined sulfuric acid present in the wash; whilst in some of the distilleries additions of sulfate of ammonia in small proportions are made to the wash, in order to supply readily available nitrogenous food for the yeasts and to thus enable them to multiply with rapidity and to retain a healthy active condition. The reason for rendering the wash slightly acid is to guard against the excessive propagation of the butyric and lactic organisms, and to render it more suitable for active alcoholic fermentation. Within a very short time from the molasses being diluted it enters into vigorous fermentation and rapidly proceeds to more or less complete attenuation in 30 to 48 hours.

In British Guiana the distilleries are of three kinds:

1. Those using pot stills or vat stills which are practically only modified stills.
2. Those using both pot stills or vat stills and Coffey or other continuous rectifying stills.
3. Those using only Coffey or other continuous rectifying stills.

Vat stills consist of cylindrical wooden vessels built of staves strongly hooped with wrought iron. They have high copper domes covering openings in the heads of the vessels which communicate with a retort or retorts of the Jamaica pattern, but, as a rule, the retort acts as the lowest vessel of a rectifying column. As in Winter's still a spiral pipe or a series of small perpendicular pipes descend down the interior of the column through which cold water is run whenever distillation is in progress, and by which the spirit vapor undergoes a process of rectification as it ascends the column before passing into the condenser. Vat stills are heated by injection of steam."

Aging.—The aging of rum does not differ markedly from the aging of whiskey (*q.v.*). The temperatures are possibly a little higher and the time somewhat shorter. Either charred or uncharred casks are used and a deficiency of color in the finished product is made up by caramel.

Imitation Rum.—The practice of "stretching" rum is quite common, either of two general methods being used. In the first method rectified grain alcohol is diluted, "cut," with distilled water until it is of the same alcoholic strength as a previously selected rum of strong bouquet. This diluted alcohol is then used to mix with the rum in any ratio from one-to-one to one-to-four or five parts of alcohol to one of rum. The mixture is aged in casks for several months at a temperature of about 75° F. The product of this treatment might possibly be better called a "cut" rum than an imitation. In the second method, a mixture prepared as just described but before aging is further "cut" with distilled water and redistilled. The new distillate is treated with "rum essence" and then aged in casks. Rum essence, or the so-called "pelargonic ether" is a mixture of esters, alcohol etc, prepared in various ways. One favored method is said to consist in distilling a mixture of alcohol, crude acetic acid, starch, manganese dioxide and sulphuric acid. Rum essence is quite generally used in the preparation of imitation rum and also as a cooking flavor. An experienced taster, however, would have no difficulty in distinguishing it from the genuine article.

GIN

There are two essential differences between gin and the liquors which have previously been under consideration. The major difference is that gin derives the bulk of its flavor from pre-existing natural essential oils rather than from the products of fermentation. Secondly, gin is somewhat more of an international product, being made in the Continent especially Holland, in the British Isles and in the United States. In each country there are minor qualities which are distinctive. In general, however, gin is a colorless beverage containing from 40-55% of alcohol and having a perfume-like odor. It was originally made in Holland.

Holland Gin.—Since the production of alcohol for gin is a separate step from the introduction of the flavor it might seem that any sufficiently pure alcohol could be used in the manufacture of gin. As far as the American public is concerned, this is probably true. However, abroad, and especially in Holland, the congeneric substances of the pot-still distillate from a properly fermented mash of barley malt, rye and corn are required to round out the taste of the product. This distillate, called *moutwjn* or maltwine, is bought from distilleries by the gin manufacturers and redistilled by the latter through a "gin head" containing juniper berries and other flavoring materials. This is the material which under the various names "Geneva," "Hollands," "Hollands Geneva" or "Hollands Gin" has spread widely over the surface of the earth.

English Gin.—In England the same raw materials are used as in Holland. However, since the distillation of alcohol for use in gin is almost always done in patent stills, the flavor of the British gin is decidedly different from the Dutch. The English gin manufacturer usually requires a clean spirit which has been rectified until only a slight grain flavoring remains, as decided by the judgment of the operator. Molasses spirit is objected to both in England and the Netherlands on the ground that it gives a coarse flavor to the finished gin. The selected spirit is then made into gin in a number of ways. The more approved process is to re-distill the spirit, after dilution with water, in a pot-still equipped with a gin-head containing juniper berries and other flavoring materials as required. Some manufacturers, however, distill the flavoring materials separately and then add them to the diluted alcohol. Others distill before dilution, etc. The addition of from 2-4 or even 6% of sugar, or of $\frac{1}{2}$-1% of glycerin to gin is common practice to sweeten and "smoothen" the product. Gin is usually bottled as made and is unaffected by aging.

American Gin.—In the United States gin is made from the usual grain mash with juniper berries as the principal flavoring agent. Sloe gin has in addition the flavor and color extracted from "Black-haw" or Sloe berries. Among the flavoring agents used in gin are the following:

Angelica	Fennel
Anise	Grains of Paradise
Bitter Almonds	Juniper Berries
Caraway Seed	Orris Root
Coriander	Liquorice
Calamus	Turpentine
Cardamoms	Bitter Orange Peel
Cassia Bark	

Turpentine is only occasionally used as a substitute for the essential oil of juniper. A small addition of sulphuric acid to the spirit before rectification is sometimes made to produce an ethereal bouquet and flavor.

Gin manufacture in the United States may be carried on along with the manufacture of whiskey and spirits. Its place in this unitized operation is shown in Fig. 26. A specialized plant for the manufacture of gin is shown in Fig. 34. The process is as follows: Pure spirit from the charge tank is drawn as needed to the gin still. Sufficient good-quality water is added to dilute the alcohol to about 125% proof. The juniper berries and other flavors required for a batch are placed in the gin head. High pressure steam is run through a coil in the still to cause distillation. The heads and tails are discarded, and the middle run, after dilution with distilled water in the blending tank to 80 or 90% proof, is drawn off to bottles.

Bath-tub Gin.—This term was applied during the prohibition era to so-called "synthetic gin" made by adding mixtures of essential oils or essences to a suitably diluted alcohol. Smootheners were sometimes added and the product was then ready for the market. Actually, while the term might have some bearing as applied to the questionably sanitary methods of small bootleggers, gin has largely been made in this way in all countries and at most times. Nor is there any very cogent reason why gin thus made

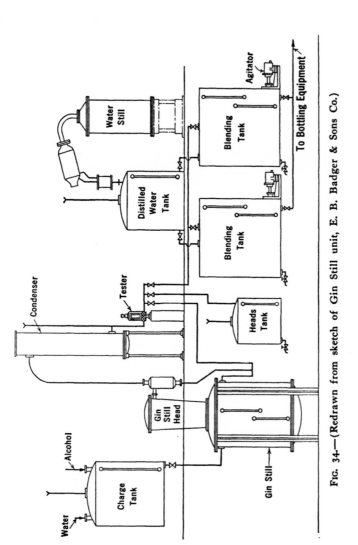

FIG. 34.—(Redrawn from sketch of Gin Still unit, E. B. Badger & Sons Co.)

"synthetically" should be inferior *qua* gin to the distilled product. The question seems to be largely one of taste and we have previously stated that the taste for gin varies in different countries. Certainly, the probability is in favor of a synthetic gin being consistently uniform, batch after batch, and the balance of oils used as flavors can be so selected that any desired aroma is achieved.

APPLEJACK

Applejack bears the same relation to cider that grape brandy bears to wine, but whereas fermented cider is more dealt in abroad than here; the reverse is true of applejack. It has been stated that applejack was first made in this country as early as 1698, and it is still a staple article of commerce.

Applejack is the product resulting from the distillation of fermented apple must. Hence, in general, the same considerations apply to the manufacture of applejack as to that of grape brandy, and the same processes are used. It should be noted in particular that the definition just stated differs from the popular impression of applejack as the unfrozen liquid removed from the core of a barrel of frozen cider. The product of this latter process would not only contain the alcohol of the cider, but also all its other dissolved substances and would be quite unpalatable. As far as any passable applejack is concerned, only that made by distillation need be considered, although it is possible that isolated farmers may in some few cases make their "hard liquor" in the more primitive fashion.

For the purpose of making applejack it is important that the cider be made from suitable apples both as regards variety and quality and that the fermentation be conducted in a manner to avoid as far as possible the formation of volatile acids especially acetic acid. The reader is referred to pp. 187-9 on Cider for further details regarding the fermentation operation. The next step after the fermentation and aging of the cider is the distillation of the brandy. Here particularly, the manner of procedure is similar to that in making grape brandy (*q.v.* pp. 141-2). Pot stills are generally preferred to continuous stills and two or three distillations are used to secure the desired alcoholic strength coupled with the proper removal of foreshots and tailings, which contain respectively aldehydes and fusel oil. After distillation the spirit is aged in oak barrels. Uncharred barrels of well seasoned white oak are used, and those which have previously held wine or other spirits are preferred to new barrels.

It is claimed, that on account of the lower starch and protein content of an apple must as compared with a whiskey mash, applejack can be aged in a much shorter time than whiskey. Hence it is usually considered potable after as little as three to six months' aging. It is, indeed, further claimed that apple brandy begins to lose its special character on aging in the wood for more than four to five years.

ARRACK

Arrack is prepared by distillation from toddy or a mixture of toddy with either fermented rice or rice and molasses mash. Toddy is palm wine which is obtained by fermenting the juice of the cocoanut palm. Arrack ranges from yellow to light brown in color. Its flavor and aroma resemble those of rum, if much molasses has been used in its preparation, but not to such an extent that the one could be substituted for the other without instant discovery. Normally arrack has a sourish aroma and taste which are claimed to derive from the toddy. The alcoholic content of arrack ranges from 70 to 80 per cent by volume.

It is used either as such or in the preparation of hot drinks (grog, punch), particularly in making Swedish Punch, and also for strengthening and improving the aroma of ginger liqueurs and bitters such as Angostura, Boonekamp, etc. It is also used in the preparation of sweetmeats.

Arrack is made in Siam, the Malay Archipelago, East India and Jamaica.

Manufacturing Process.—When rice or rice and molasses are used it is customary to employ only rice of the highest quality.

Germination of the grain is started by the usual moistening with water and spreading in heaps or layers. As soon as the kernels start to sprout the grain is crushed between rollers and hot water is added until the temperature of the mash reaches the neighborhood of 140° F. Around this temperature the enzymes convert the starch to sugar. The wort is strained and cooled to approximately 70° F. Fermentation is started by adding either toddy or a toddy and molasses mixture according to the formula employed. The fermented wash is subjected to three or more distillations after fermentation is completed. The unusual number of distillations is required by the crude and inefficient stills employed.

VODKA

Genuine vodka is made from a mash of unmalted rye and either barley or rye malt. Potatoes and corn have been used as substitutes for rye in the cheaper grades. The alcoholic content of the better grades ranges from 40 to 60 per cent.

SCHNAPPS AND KORNBRANNTWEIN

These distilled liquors are consumed in considerable quantities in Germany, Holland and elsewhere on the continent of Europe.

Kornbranntwein is prepared from a mixed mash of malted and unmalted cereals, generally rye. Corn may also be used. Methods of manufacture are similar to whiskey processes.

Schnapps is diluted, rectified potato alcohol prepared by (1) heating a mash of potatoes under a pressure ranging from 30 to 60 pounds to achieve pastification of the starch; (2) converting the starch to sugar by mashing with malt; (3) fermenting according to standard methods; and (4) distilling so that the final product is well rectified.

CHAPTER XI

WINE

Definition.—The term wine is a very broad one. It includes, with proper qualifications, the product resulting from the fermentation, with or without the addition of sugar and other substances, of such diverse materials as dandelion blossoms, elderberries, etc. More particularly, however, it refers to the result of the alcoholic fermentation and other suitable treatments of grape juice. It is in this sense that we shall use the term here. The processes by which raw grapes are converted into wine include crushing, pressing, defecation, fermentation, fining, racking, fortification, etc. Historically, the preparation of wine is of immemorial age. As an industry it dates back for more than two millennia, which of course gives it rank among the oldest of human occupations.

Classification.—Even in the centuries before the Christian era, qualities of wine were distinguished and different grades were known and demanded by consumers. At the present time and for commercial use almost innumerable distinctions are made in the grades and qualities of wine, including naming by types, combined with geographical distinctions down to the name of a particular vineyard and further distinction by the year of the vintage. The brands and names recognized in commerce are to be reckoned by thousands.

A few citations are given to illustrate this point. Among the French red wines are many which are highly valued and of whose excellence there can be but one opinion. These include the red wines of Burgundy and especially those of Musigny, Richebourg, Romanee, Chambertin, Corton, Beaune des Hospices, Pommard, Volnay, Allos du Roy, and Clos Vougeot. The Clos Vougeot is one of the most highly prized of the products of the beautiful Burgundian vineyards; its origin can be traced back to A.D. 1110 when the monks of Cipeaux received the vineyards from Hugues le Blanc, lord of Vergy, and cultivating with infinite care, succeeded in producing a wine which has maintained its reputation for centuries. The wines of Beaujolais such as Macon, Thomis, Fleuric, and Moulin-a-vent are also known, and the pride of the banks of the Rhone are l'Hermitage, Cote-Rotie, and Châteauneuf-de-Pape. But the French wines, however, which enjoy perhaps the greatest popularity in the land which produces them are the world-famous red wines of Bordeaux. Some of the principal varieties of these are Haut Brion, Château-Margaux, Château-Leoville, Château-Lafite, Château-Lagrange, Château-Larose, Château-Millet, Mouton-Rothschild, Château-Latour, Branaire, Montrose-Dolfus, Ducru-Beaucailloux, Clos d'Issan, St. Estephe, St. Emilion, and Medoc. Although the wines of Bordeaux have been famous for centuries, it was not until towards the end of the eighteenth century that they became really fashionable, a state of affairs which was largely brought about by the influence of Marshal de Richelieu, who introduced them to the notice of the Parisians.

There are a great many varieties of white wine, and perhaps the most famous of all is the Rhenish wine known as "Johannisberger." This variety has a reputation which is world-wide, and is said to fetch the highest price among white wines. Enormous casks of Johannisberger which were casked and stored in their present position over three centuries ago are lying in the municipal wine cellars of the township of Bremen. This wine, known as "the Rose," is as one might suppose, the subject of many legends, and is offered in hospitality to royalties and other persons of distinguished rank who partake in the festivities of the town; it is also graciously given to the sick. Other Rhenish wines of great repute are Rauenthaler, Liebfraumilch, Marcobrunner, Rudesheimer, Hoheheimer, Kottenlocher, Zeitlinger and Riesling.

The white wines of Burgundy are also highly appreciated and Montrachet is regarded by some as the king of white wines. Meursault-Goutte-d'Or, Chablis Moutonne, Pouilly-Tuisse are also excellent. Among the white wines of Bordeaux, Château

Yquem is considered the best, and Château-Myrat, Latour-Blanche, Clos St. Marc, and the wines of Sautome, Barsac, and Graves also enjoy a high reputation.

There are many varieties of champagne, but some of the most famous are Pommery-Greno, St. Marceaux, G. H. Mumm, Moet et Chandon, Montebello, Heidsieck, Roederer, Mercier, Veuve Cliquot, and Lanson.

Most of these names correspond to real differences. Names taken from regions, such as Rhine wine or Sauterne, represent large differences in character easily distinguishable by taste and usually by chemical analysis. Names representing vineyards or vintage years represent differences of quality, which may be equally marked to the practiced taster, but are difficult to indicate by chemical means.

Names drawn from particular vineyards are properly considered proprietary and should not be used, nor the wines imitated elsewhere. Names drawn from localities or regions are of the same nature. They represent qualities due to special features of soil, climate, grape variety, and manufacturing methods which can not be identically duplicated in any other place. An exception should probably be made of certain names which, while originally derived from particular localities have come to represent, through long usage, characters due principally to methods of manufacture. Such names are Port, Sherry and Champagne.

The name Burgundy should properly be given only to wine made in Burgundy from Pinot grapes; the name Medoc only to wine made in Medoc from Cabernet and also to the three or four other varieties recognized there as capable of producing the wine to which the region owes its reputation.

There seems to be no sufficient reason, however, why a wine should not be called Port if it is made of suitable grapes in the recognized way and resembles those wines of the banks of the Douro, which first received this name. "Port," is no longer synonymous with "wine of Oporto." All the wines made in the region of Oporto are not port, nor does all port come from that region.

With these possible exceptions, locality names belong only to the wines produced in that locality. Not only is this fair to the

consumer, but it is sound policy in the selfish interest of the producer.   Wines are produced most profitably by those localities
which have an established reputation.   They have a sure market
whatever the abundance of crops in other localities.   It should
be the aim of each locality to obtain and maintain a reputation
which will make it independent of general competition.   This can
be done only by marketing consistently good wines under the
name of the locality.

The listing of wines in the manner indicated, while exceedingly important to the consumer of wines and especially to the
connoisseur, is of little aid to the student or technician.   Fortunately, it is possible to classify wines also in a more general
manner on a basis of their gross composition, disregarding the
fine distinctions made by the specialist.   There are four dicho-
tomous bases on which wines can be divided, namely:

1.   Dry or sweet
2.   Fortified or unfortified
3.   Sparkling or still
4.   Red or white

These classes are defined as follows:

a.   Dry wines are those in which practically all of the sugar
has been converted by fermentation into alcohol.   Usually they
are of comparatively low alcoholic content (*ca.* 8-12%).   This
class includes such wines as Chablis, Riesling, Hock, Moselle,
Claret, Burgundy, Gregnolino, and Chianti.

b.   Sweet wines contain some unfermented sugar and have
an alcoholic content usually between 13-15% by weight, all of
which has been produced by fermentation.   Auslese Rhine wine,
Sauterne, and Tokay are wines of this class.

c.   Unfortified wines are those whose alcoholic content is entirely derived from fermentation.   All of the wines mentioned in
classes a. and b. fall within this group.

d.   Fortified wines derive some of their alcoholic content from
fermentation and some from the addition of distilled spirits, usually grape brandy.   They contain usually from 18-22% of alcohol.
Madeira, malaga, muscatel, port and sherry are wines of this

class. Champagnes also fall into this group. Angelicas are sometimes considered to belong to this group although strictly speaking they are not wines at all, since they are made by adding sufficient grape brandy to fresh grape juice entirely to prevent any fermentation.

e. Still wines, which include most of those mentioned above, are those whose fermentation has been completed before bottling so that they contain only such proportion of the carbon-dioxide produced in the fermentation as can remain dissolved in the liquid in equilibrium with the air under the conditions of manipulation.

f. Sparkling wines are bottled before the fermentation has ceased so that they contain carbon-dioxide gas in solution at greater than atmospheric pressure. When they are served, the carbon-dioxide is liberated with effervescence. Their gas and alcoholic content vary according to the market for which they are intended. They may be dry or sweet, light or strong. Champagne, sparkling Burgundy, and Asti-Spumanti are examples of sparkling wines.

g. Red wines are those in which the skins, stems, etc., of the grapes are present during the fermentation so that the grape pigment is extracted and colors the fermented juice. This group includes the majority of wines as, for example, Claret, Burgundy, Port, Chianti, etc.

It should be particularly noted that the distinction between red wines and white is based on a difference in manufacturing process. This is of significance on the one hand because the variation in color of so-called red wines covers every possible tint from inky purple to pale pink and tan, and on the other hand because the inclusion of skins, stems etc., in the fermenting liquor leads to somewhat different composition of the wine and requires different handling from white wines.

h. White wines are produced by fermentation of the grape juice only, with removal of the marc (skins, stems, etc.) before the fermentation has proceeded to a point when the pigment becomes soluble. Riesling, sauterne, and champagne may be cited as examples.

It will be seen that by combination of the classes just given, sixteen possible categories are obtained into which wines may be classified. For instance, claret is a dry, unfortified, still, red wine. However, champagne may be a fortified, sparkling, white, either sweet or dry wine and many other wines can be placed in more than one category. Despite this ambiguity, these categories are sufficient for scientific purposes, being based partly on the nature of the raw material, partly on the composition of the wine and partly on the methods of manufacture.

A further subdivision of wines may be made within any of the groups mentioned, on the basis of quality, into three grades: fine, ordinary, and blending wines. A fine wine is one, all the components of which are in proper and harmonious proportion, and which has sufficient quality to repay aging and bottling. These constitute, in most regions, only a small part of the product. They are, however, the ideal toward which the efforts of every winemaker tend. Ordinary wines are those which are sufficiently harmonious in their composition for direct consumption, but which exhibit no great delicacy of flavor or bouquet. These are usually destined for bulk shipments and cheap markets. Blending wines are of various degrees of quality and character, but agree in having a deficiency or excess of some one or more essential components. There are blending wines with an excess of alcohol, or extract, or color which make them unsuitable for direct consumption. They serve, by blending, to correct other wines which are deficient in these components. Where the wine handlers have perfected their business, the bulk of wines are used for blending, for it is only the exceptional wines which cannot be improved by additions to correct their deficiencies and faults.

Functions of Wine.—The experience of many centuries has taught mankind that wines, when used in proper combination with foods, not only enhance the flavors of the food and the enjoyment in partaking of them, but aid in the digestion. There follows an abridged tabulation of the principal classes of wine together with the foods they should accompany:

Dry white wines
  Oysters, fish, fowl, turkey, vegetarian dinners, omelettes, etc.

Dry red wines

Roast meats such as beef, pork, lamb, steaks and chops, duck, goose, turkey, pheasant, venison, etc.; Italian dishes such as spaghetti, ravioli, macaroni, etc.

Sparkling wines

These are the proper accompaniment of the end of the meal, sweets and cheese.

Fortified wines

Sherry is preferred to any cocktail by almost all peoples but the American. It is also the proper wine to serve with soups and with hors d'oeuvre. Port and other heavy wines like Malaga, etc., are sipping rather than drinking wines, and should be used with circumspection during the evening, when the appetite no longer clamors and the excellencies of the wine can be savored slowly.

In fine cooking, wines play an important part which largely forms the basis of the reputation of the French cuisine.

MANUFACTURE OF WINE

Introductory.—The manufacture of wine is, in principle, a matter of the greatest simplicity. The grapes are crushed, the juice fermented, the sludge of exhausted yeast and precipitated matter is removed by decantation, and one has wine. Unfortunately, there is an equal probability that if no more than the above is done, the product will be vinegar or something equally unpotable. Much more must be done if the wine is to be of a high quality. The ultimate in quality, of course, the wines that elderly connoisseurs sip with tears of thankfulness, are dependent not only on full and thorough care in their processing but also on a combination of favorable weather, soil, etc. in a given year in a given locality. These accidental factors are beyond human control. The control of the stated steps in production of wine is, however, readily feasible and will be the subject of the immediately following pages.

Components.—The fine points of wine making are necessitated by the original composition of the fermenting mass and the nature of changes which may occur during the fermentation and after-processes. The must (fresh pressed juice) contains sugar, organic acids, tannin, flavoring substances, proteins, mineral salts, and pectin and mucilaginous substances which it derives from the grapes. It also contains a large variety of yeasts, bacteria, and fungi, some of which are favorable and some the reverse. An average composition is indicated in the following Table IX reported by Koenig:

TABLE IX.—ANALYSIS OF WINE MUSTS

	Sp. gr.	Water %	Nitro-genous matter %	Sugar %	Acid %	Other non-nitro-genous matter %	Ash %
Minimum...	1.0690	51.53	0.11	12.89	0.20	1.68	0.20
Maximum...	1.2075	82.10	0.57	35.45	1.18	11.62	0.63
Average.....	1.1024	74.49	0.28	19.71	0.64	4.48	0.40

These components and their changes are very largely interrelated.

When the sugar content is high enough the activity of the first fermentation prevents much action by harmful organisms. Later, enough alcohol has been produced to prevent the growth of these. A proper sugar content lies between 18 and 28%.

Organic acids, especially tartaric, serve to produce sound and healthy wine in a number of ways. Sufficient acidity encourages sound fermentation and inhibits the growth of the disease bacteria. Sufficient acid ensures a full tasting wine which will store well, while insufficient acid means a flat taste and short life. Acid also ensures a better extraction of color from the skins.

Tannin, which is derived by extraction from the skins, seeds and stems of the grape is an essential constituent of the wine. It serves to confer disease resistance on the wine, aids remarkably in the clarification and produces a more brilliant color. On the other hand, an excess of tannin confers an astringence on the wine which delays its final maturity, although in the end the wine is more mellow for it.

The flavoring substances present in the raw grapes undergo

many changes during the life of the wine. To some extent they control the flavor of the end product. However, the extent of the changes has never been followed completely by chemical research so that little can be said on this topic. The characteristic bouquet of the finished wine is only slightly due to methyl-anthranilate, which has the distinguishing character of fresh grapes. Various aliphatic ethyl esters are formed as the wine lives, and these, the

Fig. 35.—Corner of the still room in a large applejack distillery. Showing one of the pot stills. (Courtesy of the American Wine and Liquor Journal, New York.)

action of special varieties of wine yeast (*S. Ellipsoideus*), and even of frost (as in Reislings) each contribute their share to the final celestial bouquet of good wine.

The relation of the constituents of the raw must to the finished wine is shown in Figures 35, 35a.

Red Wines.—The production of wine falls naturally into two broad divisions, red and white wines respectively. The production of champagne and of other fortified wines may follow in

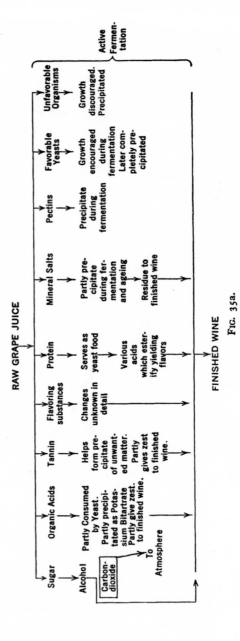

FIG. 35a.

their earlier stages either of the broad divisions stated and diverges only at the later stages. The apparent divergence between red and white wine processes consists only in that for the former, fermentation precedes drawing off and pressing, while for white wines the order is reversed. Actually, other differences are entailed which will be discussed under the topic of white wine manufacture. Red wine manufacture is somewhat simpler and will be considered here.

The sequence of operations in the manufacture of red wines is indicated in the flow sheet, Fig. 36.

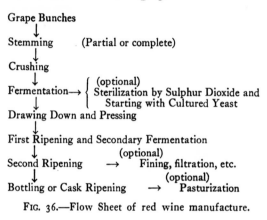

FIG. 36.—Flow Sheet of red wine manufacture.

When the grapes are at the proper stage of ripeness the bunches are plucked and brought to the winery. This may vary in size from the home of the French peasant to the large commercial wineries with capacities for thousands of gallons of wine. There can be, at this stage, no delay, since the grapes may become infected if left piled in bunches. The next stages to the starting of the fermentation must follow in rapid order.

Stemming.—If stemming is practiced, it may be done by hand, by the use of a screen which will pass the grapes, but not the stems, and which may be operated manually with the aid of a rake to spread the grapes over it, or may be mechanically shaken or vibrated. Another means of stemming, is a machine comprising a stationary, horizontal, perforated metal cylinder fitted internally with a revolving shaft having arms which cause the grapes to travel along the cylinder and drop through the holes or slots in its wall. Stemming may also be done after crushing. A strainer of suitable size with proper openings and fitted with revolving blades serves this purpose excellently.

The stems can furnish tannin to the juice if the skin and seeds are deficient in this respect. However, they also contain substances which "brown" the color and spoil the flavor of the finished wine. They may also introduce difficulties in handling the crushed mass.

Crushing.—In order to liberate the juice of the grapes and inoculate it with yeast, it is necessary that the grapes be crushed. Probably a sort of pestle or stamper operated in a container is even now used to crush grapes in small lots. By far the greatest amount, however, are crushed in a roll machine. This consists of a hopper into which the grapes are placed and from which they feed between two grooved rolls turning toward each other at different rates of rotation. The grooves catch the grapes from the hopper, and one roll passes over grapes held in the grooves of the other, crushing them. The crushed grapes may fall directly into the fermenting vat, or into a tank whence they may be pumped to the proper vat. In adjusting the machine it is only necessary that the rolls be so spaced that the seeds pass through uncrushed. Roll crushers are available in any size desired, from the domestic "one lug" to the large machines found in California. In any case their capacity is usually great compared with vat and other facilities required in the winery. There are possibly still existing, and there certainly were in the past, rural districts in which the grapes were placed directly in the fermenting vat, and the crushing done by the bare feet of men and women who walked around on the mass.

Fermentation.—The general considerations involved in the fermentation of grapes to produce wine have been discussed in Chapters V and VI. The fermentation of the crushed grapes is started as desired, either naturally or by means of a starter, and means to control the temperature must be available if very large batches are being fermented. These may include water cooling coils or as in parts of California, the construction of the vats with thin cement concrete walls so the evaporation serves to effect some cooling. Even chloroforming the yeasts, temporarily to arrest the fermentation and stop heat production has been attempted. This control of temperature, it is repeated for emphasis, is necessary because the desired yeasts function best

Fig. 37.—Redwood tanks. 17,000-25,000 gallons capacity. (Courtesy of the American Wine and Liquor Journal, New York.)

between 70° and 80° F. while the disease bacteria are favored by temperatures over 90°. It should be noted, however, that yeasts can be "trained" to work at unusually low or high temperatures when the climate, as in Southern California, requires it. The first visible evidence of active fermentation is the formation of carbon dioxide. This is liberated in bubbles which become entrapped in the skins and cause them to rise and mat at the top of the vat forming a "cap." When this cap is exposed to the air, the upper surface of the cap quickly ferments to completion and offers an excellent start to vinegar bacilli. At the same time the grape pigment becomes oxidized and bleached. To avoid this there are two alternative practices. The cap may be broken up and pushed down manually, or even as in some districts in France trodden down by men who enter the vats stripped, for this purpose. Alternatively a grid is installed in the vat about six inches below the level of the must, to keep the cap submerged. In this case it is necessary, to obtain aeration and uniform fermentation, that the liquor be circulated by pumping it from the bottom of the vat and allowing it to pour back at the top. A combination of these systems is often favored. The fermentation is kept open until the yeasts have multiplied and developed strongly and then the cap is submerged to avoid the dangers of open fermentation.

Changes During Fermentation.—While the most marked change during fermentation is the conversion of grape sugar into alcohol and carbon dioxide, other changes of smaller magnitude but equal importance to the quality of the finished wine also occur.

As indicated previously, some of the acid present in the fresh must is consumed by the yeast. The drop in acidity is about 40%, or from an original acidity of 0.5-1.5% expressed as tartaric, to an acidity of 0.3-0.7% in the finished wine. This includes volatile acids formed which should not exceed 0.15%.

Protein matter present in the grapes is also partly consumed during fermentation. The manner of this consumption was discussed in Chapter V. The end products, organic acids, esterify with the alcohol of the wine and contribute largely to its bouquet.

The liquid extracts tannin and coloring matter from the skins and seeds of the grapes. The former, as previously stated, is of great importance to the soundness and clarity of the wine. The coloring matter of the grapes is, of course, the obvious distinguishing feature of red wine. This color includes a group of substances of obscure composition, which are called enolic acids. They probably do not go into complete solution in the liquid, but rather into colloidal solution. Hence, the cell walls of the grapes must be broken down either by fermentation or heating (as in making bottled grape juice) before the color dissolves. It is possible that the alcohol may also have something to do with dissolving the color. Over-ripe grapes, in which it is possible that the pigment is over-oxidized, yield a paler wine than those in which the grapes were gathered at the peak of their ripeness.

Many other known and unknown changes of minor importance occur during the fermentation even when it is running its proper course. Coagulation of a portion of the albuminous and pectinous substance present may be mentioned as an example. However, the main changes have been indicated and the key to all is the conversion of sugar to alcohol.

Completion of First Fermentation.—The rate of this conversion varies, as might readily be foretold, with the temperature, the vigor and numbers, and the strain of yeast. At the best, about 4 per cent of sugar per day is converted, so that a must containing originally 20% of sugar, will be fermented dry in about five days. Actually the time required may be anything between three days and three weeks.

At some time during the active fermentation or very shortly after its completion the fermented juice or new wine must be separated from the marc, the pressed residue of grape skins and pulp. The exact stage at which this is done is very largely a matter of pure choice although to some slight extent climatic factors serve as a guide. In the French Bordeaux district pressing is not done until two or three weeks after the end of the violent fermentation on the theory that thereby strength is added to the wine. In the Burgundy district a reverse theory is held that the shorter the fermentation the better the wine. Hence no time is lost between the completion of the fermentation and the separation of new wine from marc. Here in the United States it is the general custom to draw down and press even before the completion of fermentation. It is stated that in the Sandusky wineries there was often as much as ten per cent of sugar in the must at the time of pressing. All three practices will produce good wines. Like the choice between open, submerged or combined fermentation, the election probably depends on the judgment and skill of the wine maker.

Pressing.—Once the choice of time has been made, the mode of separation between new wine and marc remains essentially the same. By opening a tap at the bottom of the fermenting vat, a very considerable portion of the new wine will drain off without pressing. This portion of the yield is usually less harsh and matures more rapidly than the press juice. Hence it is usually kept separate from succeeding portions of juice.

When the drainage is essentially complete, the remaining saturated mass is transferred to a press. These range from small hand-screw affairs to large hydraulic presses. Their principle of operation is nevertheless identical with that of the old home jelly-bag. The mass is placed in a press and the solid matter confined within a strong porous cloth, the filter cloth. When pressure is applied the new wine is squeezed out and the solids remain in the cloth. In order to avoid clogging of the pores of the cloth by the finer portions of albuminous and other matter in the mass, it is essential that pressure be applied gradually although it may finally be exerted to the limit of the machine used. Even with this precaution, it is usual to open the press, loosen the mass and repeat the pressing once or twice. Some of the wineries which are more interested in quantity than quality of product, moisten the residue in the press with water, before the third pressing to wash out as much extractable matter as possible. This last pressing is called *pinette* in France and may not be sold.

Aging and Racking.—The new wine produced in the manner described is still far from the finished product which is marketed. It is low in alcohol and it still contains unfermented sugar, excess tartaric acid and tannin in solution. It is cloudy due to the presence of suspended yeast cells, albuminoids, pectinous materials, etc. Its flavor is harsh and the aroma quite grapey rather than winey. Time is required for the completion of fermentation, the settling of suspended solids, and the ripening of the flavor. Usually these processes are accomplished by storing the new wine in oak casks at a cool temperature (50° F.) during the winter. A great deal of settling takes place, aided by the formation

of cream of tartar (potassium acid tartrate) crystals. With the
spring and approach of warm weather, the wine is "racked." That

FIG. 38.—Modern wine storage room. (Courtesy of the American Wine and Liquor
Journal, New York.)

is, it is carefully poured or siphoned off from the sediment into
new clean casks. The importance of this process is much greater

than appears from its simplicity. During the racking, some aeration takes place resulting in the solution of oxygen in the wine. This oxygen acts on the remaining albuminoids in the wine to precipitate them. It also improves the color and flavor. If the wine is to be fortified, the alcohol is added in portions at each racking. Artificial aeration is often employed to accelerate the processes desired. The sequence of aging and racking is repeated at intervals of a month to six months until the wine is ready for bottling.

Modifications in the temperature of storage, etc., all have their effect on the wine. Madeiras and sherries, for example, owe their special flavors to aging at a higher temperature than is usual for other wines. Some special wines may have sugar or condensed must added to them at the racking. For the simple red wines two or three rackings at intervals of six months are generally sufficient to produce a satisfactory wine.

Once the wine has ripened satisfactorily, efforts are made to arrest any further changes. Usually this requires that the wines be bottled and thereafter stored in a cool place with a minimum of disturbance. Sometimes the wine is pasteurized before bottling. This consists in heating it briefly to kill off as many bacteria as possible. The difficulty is always that more heat or time of exposure are required for complete sterilization than the flavor of the wine will permit without harm. Hence a compromise is made. Pasteurization temperatures range from 120°-150° F. and times from a very few seconds at the higher temperatures to a quarter hour at the lower heats. The entire operation is one which can only be performed on a large scale with the best possible equipment and control. Fortunately the dairy industry has furnished various machine designs and the technique of their operation which can be transferred unchanged to the wine industry.

WHITE WINES

In general the manufacture of white wines is very similar to that of red. The basic difference as will be noted by comparison of the flow sheet for red wines, Fig. 36, and a similar flow sheet for white wines, Fig. 39, is that red wines are fermented on the whole crushed mass, while white wines are pressed before fermentation so that only the juice is fermented.

This basic difference necessitates other variations between the manufacture of red wine and that of white. The plucking and pressing operations are the same. Since there has been no fermentation to break down the walls of the grape cells, higher pressures are required to ensure a good extraction of juice.

Sterilization.—Most of the wine yeasts are found on the skins of the grapes and remain there during the pressing. Hence the fresh grape juice is deficient in yeast and would ferment slowly

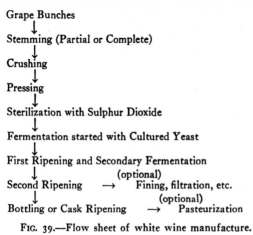

Grape Bunches
↓
Stemming (Partial or Complete)
↓
Crushing
↓
Pressing
↓
Sterilization with Sulphur Dioxide
↓
Fermentation started with Cultured Yeast
↓
First Ripening and Secondary Fermentation
↓   (optional)
Second Ripening   →   Fining, filtration, etc.
↓   (optional)
Bottling or Cask Ripening   →   Pasteurization

FIG. 39.—Flow sheet of white wine manufacture.

and poorly if only the natural yeast were relied on to cause fermentation. During this long slow process, disease bacteria would have ample opportunity to flourish and spoil the wine, especially since little or no alcohol is present, the acid is low and the tannin which comes from the skins and seeds is also very low. Hence it is preferable and indeed almost essential that the must or press juice be sterilized and then re-started with fresh vigorous yeast culture. The usual manner of sterilization is by means of sulphur dioxide which is introduced generally by burning sulphur in the cask into which the must will be placed.

The usual way of burning sulphur in a cask is to use sulphur matches or tapes, which are strips of thin cotton cloth which have been dipped several times in melted sulphur. These tapes are hung on an iron wire about eighteen inches long, bent up at one end to hold the tape and fastened to a bung at the other. This method has the defect that some of the sulphur melts and drips on to the bottom of the vat, where it is incompletely burned. This incompletely burned sulphur may communicate a bad taste to the wine. The same is true of the burning of the cloth to which the sulphur is attached. A better method is to use thin paper instead of cloth for the tapes and to burn them in a sulphur cage. A sulphur cage is simply a hollow cylinder of iron, or better, of porcelain, open on top and closed below, sufficiently narrow to enter the bunghole and sufficiently long to hold the required amount of sulphur tape. The cylinder is pierced with numerous holes in all parts except the bottom inch, which acts as a cup to catch all the melted sulphur. The cylinder is suspended at eighteen inches below the head of the vat by means of a piece of iron wire attached to a bung. An alternative method is to add to the must a suitable amount (about one one-hundredth of 1%, 0.01%) of potassium metabisulphite ($K_2S_2O_5$). This compound contains 56% of sulphur dioxide in an available form and therefore furnishes a more controllable means of dosing the wine than does burning sulphur. The effect of this addition is to kill off harmful bacteria and temporarily to inactivate the yeast. Hence the must is inactive for a time and may be clarified by filtration, centrifuging in a cream separator, or even by settling and decantation. The latter process, which is generally used only in the smaller wineries, must be performed within 24 hours of the addition of sulphur dioxide so that the commencement of fermentation by the yeasts still present in the residue does not stir up the sediment again.

The sterilized and clarified must is now ready for its fermentation. An active culture of yeast is added in about a proportion of 2-10% by volume of the must. The temperature must be raised if necessary to 80° F. so that the yeast multiplies rapidly and continues its activity until enough alcohol is formed to protect the wine against disease. In small wineries heating is done by suspending a milk can or bucket of hot water in the must. The large wineries are equipped with special coils in the vats which may be used. In any case, the manufacture of white wine, on account of its lower acid and tannin content, and on account of its slower fermentation requires greater care than does the manufacture of red wine.

Aging and Racking.—In these processes again, greater precautions are necessary for white wine than for red, and the racking must be performed a greater number of times and at shorter intervals to ensure health in the delicate constitution of the white wines. Similarly pasteurization is almost obligatory for the lighter white wines prior to bottling.

Correction of Wines.—From the time the grapes are crushed until the wine is bottled any number of inherent deficiencies may appear or new troubles and diseases may develop. It is the object of the correction of wines to so alter the deficient or diseased character that a normal wine results. Very many treatments have naturally been developed as part of the wine-makers' technique, for this purpose. Most of them are of limited or occasional application, but there are four or five which are very generally required. These include the correction of deficient sugar, acidity, and tannin, and the fining of wine.

When the grapes are crushed a portion of juice is tested for its sugar and acid content. A proper sugar content should be between 20 and 28% and a deficiency requires only a simple calculation and the addition of the calculated amount of ordinary cane sugar.

The normal acidity of the juice should be 0.5-1.5% expressed as tartaric. It was formerly the practice to correct low acidity by the addition of calcium sulphate, Plaster of Paris. Hence the operation is called "plastering the wine." This treatment has the advantage of ease since an excess of plaster may be added and comparatively small amounts as required will react. The reaction takes place as follows:

<table>
<tr><td>Calcium Sulphate
(Insoluble)</td><td>Cream of Tartar
(Insoluble)</td><td>Calcium Tartrate
(Insoluble)</td><td>Potassium
Acid Sulphate
(Soluble)</td></tr>
<tr><td>$CaSO_4$ +</td><td>$KH(C_4H_4O_6)$ →</td><td>$Ca(C_4H_4O_6)$ +</td><td>$KHSO_4$</td></tr>
</table>

The effect, therefore, is to increase the amount of acid in solution. On the other hand, the practice is undesirable since the dosage cannot be measured readily and also because it may result in the presence of sulphates in excess of the legal limits in the wine (corresponding to 2 grams of potassium sulphate per liter 0.2%). It is easier to correct deficient acidity by blending with a juice of excess acidity or by the addition of a properly determined and calculated dose of tartaric acid.

Excess acidity within reasonable limits is not important since on aging the excess will precipitate as cream of tartar. However, a new wine with excess acidity is harsh and matures slowly. There are two procedures which may be employed to correct excess acidity. These are called respectively *"gallizing"* and *"chaptalizing."* Gallizing consists in diluting the must with water and adding either grape or cane sugar to correct the deficiency in sugar which results. Within narrow limits it is a harmless practice, but naturally, the other essentials of the wine are equally diluted and a more watery wine results. Chaptalizing consists in partly neutralizing the acid with chalk and adding sugar. This is usually done with grapes that are insufficiently mature. Like gallizing it is not objectionable if practiced in great moderation.

Few red wines need treatment for insufficient tannin since they ferment over the seeds and skins and possibly some stems so that they have ample opportunity to acquire the tannin they need. White wines, on the other hand, are almost invariably deficient in tannin as the pressing immediately after the crushing offers no opportunity for extraction. The amount of tannin required in the finished wine is very slight, most of the excess being consumed in precipitating the albuminoids of the freshly fermented wine. However, the presence of tannin helps ensure a sound fermentation and to clarify the wine afterward, so that a slight addition of tannin, say one part to 20,000 of white wine must is unobjectionable and almost invariably beneficial.

Fining.—One of the qualities especially desired of wine is clarity in the highest degree. The various suspended solids which interfere with this clarity may settle out during aging and be removed in racking. Indeed, it has been stated that frequent racking of wine is practically the equivalent of both filtration and sterilization. However, it often happens that the solids are so finely divided that they do not flocculate or clump together in sufficient mass to settle out. When this happens the wine remains cloudy unless an agent is added which will assist the flocculation and settling. This operation is called "fining." There are a number of materials which are adapted for this purpose all of them being gelatinous in nature. That is, they first dissolve in the wine, then gradually they combine with the tannin, to form insoluble tannates which entrap the other solids dispersed in the liquid and cause the whole to settle to the bottom. Milk is used for this purpose, especially for wines deficient in tannin as the milk casein requires only acidity to cause it to change from a dissolved to an insoluble state. More commonly gelatin, egg white, and isinglass are used. Mechanical filtration, refrigeration, and centrifuging are all coming into use to effect clarification of the wine.

Bogue, The Chemistry and Technology of Gelatin and Glue, (1922, p. 355) discusses as follows the use of isinglass for fining:

"The efficacy of the isinglass for this service lies in the purely mechanical property it possesses of maintaining a fibrous structure in the solution, and as this settles slowly to the bottom it entangles in its netlike meshes the colloidal bodies that produce the undesirable turbidity. For clarifying wine the isinglass is first swollen in water and then in the wine until it is completely swollen and transparent. It is then thoroughly beaten into a small amount of the wine, strained through a linen cloth, and stirred into the rest of the wine. The temperature is kept low and the isinglass does not go into solution, but only into a very finely divided suspension. Thus the original fibrous structure of the sounds has at no time since it came from the fish been lost. In this lies the difference in the action of isinglass and gelatin for fining. If isinglass were heated and made into a true gelatin it would then have lost the properties which make it so valuable for this service."

A single ounce of isinglass will clarify, under the optimum conditions, 500 gallons of wine in 10 days. One ounce of gelatin will clarify 50-120 gallons of red wine. The white of an egg will fine about 10 gallons of red wine. This last material is chiefly used with only the highest quality of wine.

Finings may be prepared as follows:

Gelatin

Cover with wine and soak a few hours or overnight. Dissolve by gentle heating, cool and dilute with more wine. Mix thoroughly and add with stirring to the wine in the cask.

Egg White

Beat to a foam. Allow to settle and filter through heavy linen. Stir up with a small amount of wine and add with stirring to the wine in the cask.

CHAMPAGNE

General Statement.—Champagne is a sparkling wine of fine flavor and fragrant bouquet. Its effect upon the human system is the production of rapid, but transient, intoxication. Medical authorities have stated that fine, dry champagnes are among the safest wines that can be consumed. Champagne is said to have valuable medicinal properties and to be of definite benefit in the treatment of neuralgia, influenza and a run-down condition.

About the time of the Civil War pink or rose-colored champagnes were fashionable; the color being obtained by tinting with a small amount of a dark red wine. Today, a straw color is favored and that is the color of all current commercial champagnes. Occasionally, a pinkish wine is met, which owes its color to partial extraction from the grape skins and is the result of accident rather than design.

Champagnes are made dry or sweet, light or strong according to the markets for which they are designed. A dry champagne, of good quality and fragrant bouquet, free from added spirit, is made from the best vinbrut, to which a very small amount of liqueur has been added. Sweet champagne receives a heavier dosage of liqueur, which hides its original character and flavor, and therefore can be made from wine of less delicate flavor.

The dosage of champagne with syrup (liqueur) materially contributes to its sparkle, effervescence and explosiveness. It is not true, however, that the heavier the dosage the better the wine. Too heavy dosage causes an accumulation of carbon dioxide in the space between the wine and the cork and such a champagne explodes loudly and effervesces turbulently when the cork is withdrawn, but soon becomes flat and loses the characteristics one looks for in a good wine. On the other hand, a fine dry wine does not explode so violently, nor effervesce so turbulently, because it acquires its sparkle to a large extent from the natural sugar of the grape. This holds the carbon dioxide somewhat more firmly within the wine and so it continues to sparkle for a much longer time. While this helps it to hold the characteristics of a good wine for a longer period, it is only fair to say that a good champagne should retain its fine flavor even after the carbon dioxide is exhausted.

Russia and Germany prefer sweet champagnes and twenty or more per cent of liqueur in the wine is not unusual for the latter country. England buys very dry, sparkling wines, having about one-fourth the amount of dosage given wines intended for Germany. France, herself, prefers light and moderately sweet wines. The United States used to buy a wine of intermediate character before prohibition. Australia and South Africa like their champagne strong, while India and China and all hot countries favor light dry wines.

Champagne is made in Germany and the United States, but France is commonly considered the home of this king among wines. The heart of the industry is in the Department of the Marne and centers around the cities of Reims, Epernay, Ay, Mareuil, Pierry, Avise, and to a lesser extent, Chalons.

The best American sparkling wines come from the Finger Lake district of New York, and the Ohio region around Cincinnati. Very little sparkling wine was made in California before prohibition, but the industry is now being developed there.

Apart from incessant labor, skill, care and precaution the fine quality of French champagnes is attributed to the climate (which imparts a delicate sweetness and aroma, combined with finesse and lightness to the wine) and to careful selection of the vines, of which four types are cultivated, three of them yielding black and one white grapes. The soil is also said to impart a special quality which it has been found impossible to imitate in any other part of the earth. Claims are made that to the wine of Ay it imparts a peach flavor, to that of Avenay a strawberry flavor, to that of

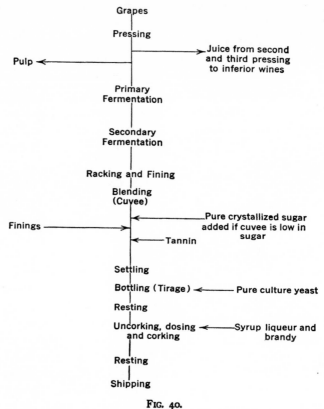

FIG. 40.

Hautvillers a nutty flavor, and to that of Pierry a flint taste known as the *"pierre à fusil"* flavor.

Manufacturing Process.—Figure 40 is a graphical presentation of the process of champagne manufacture. It will be noted that the early stages of the process are similar to those of white wine manufacture excepting that the juice of the first pressing is kept apart for first quality wines. Second and third pressings are given, but the wine made from second or third juices is inferior.

Following pressing the must is drawn into large vats and allowed to rest for 24 hours so that some settling of the sediment can take place.

It is then transferred to sterilized casks of about 40 gallons capacity. The cask is filled to about nine-tenths of its capacity, and the bunghole is generally covered with a vine leaf held in place by a small stone.

The must is then taken to one of the large underground caverns or cellars where a temperature of 60° to 70° F. usually prevails. The cask is bunged up, primary fermentation sets in and is almost completed in about two weeks to a month depending on whether the wine is high or low in sugar. At the proper point, as explained under red wine manufacture, primary fermentation is arrested by filling the cask up to the top, bunging it, and transferring it to a cooler cellar. Here a secondary and slower fermentation sets in. The object of this treatment is to preserve some of the sugar unsplit in order to insure to the wine its future effervescent properties.

About the third week in December the wine is racked and fined and then blended (cuvée) in large vats of about 12,000 gallons capacity. An agitating device worked by hand insures proper mixing of the wines. The proportions of the blend are never irrevocably standardized but about 80 per cent by volume of black grape wine to 20 per cent of white grape wine is average practice. The vintages of Bouzy and Verzenay are supposed to impart body and vinosity; those of Ay and Dizy softness and roundness; and those of Avize and Cramont lightness, delicacy and effervescence. Some blenders prefer a one-third mixture of vintages of Sillery, Verzenay and Bouzy, one-third of Mareuil, Ay and Dizy, and one-third Pierry, Cramont and Avize. Others advocate an equal mixture of Ay, Pierry and Cramont.

At this stage the important question arises of how much present and potential carbonic acid gas the wine contains. If it is too high in sugar and gas, there will be trouble because many will be lost after the bottling by explosive shattering of the

Fig. 41.—The vintage in the champagne: a wine press at work. (From Vizetelly, *The History of Champagne*, Henry Sotheran & Co., London.)

bottles and the cellars will be flooded. On the other hand, if sugar and gas are low the wine will not sparkle properly and the corks will refuse to pop.

The cuvée is, therefore, tested by means of a glucometer for sugar and if it registers low the addition is made up by sugar-candy. If it is high it is necessary to re-ferment it in a cask until it reaches the right condition.

The cuvée is now fined with isinglass and some tannin is added to offset ropiness and other defects. It is casked and allowed to rest for a month. If by that time it has not become clear and limpid, it is racked off, re-fined and allowed another month to settle. In some of the largest establishments, mechanical fining or filtration is now used.

The cuvée is put into new bottles (tirage). These are usually new, very strong, and weigh about 2 lbs. apiece. The pressure developed by the wine is such that the bottle is always weakened. It is, therefore, made of special glass which cannot liberate any alkali to act upon the wine and spoil it.

The tirage is accomplished by running the wine into vats from which it flows into oblong tanks provided with a row of syphon taps, at which the bottles are filled. The taps automatically close and stop the flow as soon as the bottle is full.

Next, a culture of selected pure yeast is often added and the bottles are corked, the corks secured by an iron clip (agrafe) and the pressure within the bottle determined by an instrument consisting of a sort of pressure gauge fitted with a hollow screw at the base. The screw is driven through the cork and the pressure in atmospheres registers on the gauge. A "grand mousseaux" represents a satisfactory wine. It has a pressure of 5 3/4 atmospheres and can safely be stored away in one of the cold subterranean caverns. If the pressure is about 4 atmospheres it is advisable to store the wine above ground until fermentation raises its pressure. If the pressure is less than 4 atmospheres it is advisable to put the wine back in a cask, add cane sugar and ferment further.

The temperature of the cavern is about 50° F., and must be kept as constant as possible to avoid too great changes in

Fig. 42.—The tuage or bottling of champagne. (From Vizetelly, *The History of Champagne*, Henry Sotheran & Co., London.)

pressure and the bottles bursting in consequence. This is always a most anxious time. The bottles are placed in preliminary stacks until this danger is passed and then they are placed in secondary stacks. Both stackings are arranged horizontally so that the sediment can work towards the neck of the bottle. The bottles are even marked with chalk so that the one side remains upwards. The duration of the bottle fermentation ranges from six months to two years.

FIG. 43.—Turning champagne bottles. (From Vizetelly, *The History of Champagne*, Henry Sotheran & Co., London.)

At the end of that time the bottles are placed in a slanting position on special stands, with neck tilted slightly down. The purpose is to induce the sediment to collect on the lower side of the bottle and to travel towards the neck. A short shaking and turning movement is given each bottle once each day for six weeks while the neck is gradually inclined further downwards until finally the bottle is vertical with the neck pointing straight down. The shaking and turning of the bottles require great skill but in spite of this the men develop tremendous speed and handle incredible numbers per day. The use of a proper variety of yeast aids tremendously in the settling which is the object of this part of the process.

FIG. 44.—The disgorging, liqueuring, corking, stringing, and wiring of champagne. (From Vizetelly, *The History of Champagne*, Henry Sotheran & Co., London.)

The bottles are now taken to the uncorking room and chilled to fix the carbon dioxide more firmly. Uncorking is an art. The workman holds the bottle in a slanting position and gradually loosens the cork until it together with the sediment is blown out by the pressure. A finger is inserted to scoop out any remaining sediment and to stop the wine flowing out. Large establishments freeze the neck of the bottle and disgorge the sediment in a semi-solid form. At the same time the bottle is turned upright and temporarily closed. The champagne is then dosed by adding a wine solution of sugar and possibly also some brandy and the bottle is permanently corked. A skilled operator now takes the bottles and swings them above his head Indian club fashion, which operation thoroughly distributes the sugar solution throughout the wine. The bottles are then aged for some time in order to develop and blend the taste further and to bind the carbon dioxide.

Imitation Champagne.—This is made by carbonating white wine, or even cider, under pressure in the same manner that soda water is made. The gas which is forced into solution by this process never is as firmly absorbed as is the natural carbon dioxide in true champagne. Hence the wine very quickly loses its sparkle and becomes flat and lifeless. Nevertheless these imitations are sold in large volume at considerably lower prices than the genuine goods since they can be made without either the losses by explosion, etc., the high labor cost, or the long storage of true champagne. The processing, however, is rather a branch of the carbonated beverage industry than of the wine industry. A stricter interpretation of the term "Imitation Champagne" is sometimes used which confines it to sparkling wines made by the champagne process but outside the French Champagne District.

CIDER

This term, in the United States, has generally been applied to the beverage produced from the unfermented juice of apples. As such, it is conceded that its consumption exceeds that of any other beverage juice. However, most farmers permit the cider which they make for their own use to ferment or become "hard."

Abroad the word "cider" is applied directly to the fermented product. The juice of the ripe apples from which cider is made contain from 7-15% of sugar, the average being around 11%. Hence the cider, if completely fermented can contain from 3.5-7.5% of alcohol, and the average product has about 5.5%.

The factors which affect the quality of the finished cider are, in general, the same as those which affect wine, namely: variety of fruit, quality of fruit, degree of maturity of fruit, and the organisms and temperature of fermentation. To make good cider, first quality, clean apples of a suitable variety, grown especially for cider making, must be selected. There are possibly as many varieties of apples as there are of grapes, if not more. The most important American varieties from their cider making possibilities are:

Sweet, sub-acid:
Baldwin, Esopus, Hubbardston, Fameuse, Mackintosh, Northwestern, Rome Beauty, and Stark.

Acid:
Winesap, Jonathan, Yellow Newton, Stayman, Northern Spy, and York Imperial.

Aromatic:
Delicious, Golden Delicious, Lady, Black Gilliflower, White Pearman, and Bana Bonum.

Astringent:
Florence, Hibernal, Soulard, Red Siberian, Hyslop, Transcendent, Launette, Martha, and Yellow Siberian.

Neutral:
Ben Davis, Black Ben, Jana, Willowturg, Missouri, Alexander, Wolf River, Buckingham, and Limberturg.

In the absence of any of the above varieties, which are called "vintage apples," any variety of winter apples is preferable to a summer variety. With the single exception of winesap apples, all the others are improved by suitable blending of varieties so that the desired qualities of flavor, acidity, sweetness, and astringency are brought to a balance.

The selected fruits are washed, if not already clean, rasped to a pulp, rather than crushed, and the pulp is pressed. The yield of juice varies with the apples and the type of pressing equipment used from 2-4 gallons per bushel. The juice is run into barrels or vats and either allowed to ferment naturally, or seeded with a pure culture of wine yeast at the rate of one pint of culture per fifty gallons of juice. The fermentation must proceed at an even lower temperature than that of wine, namely 50°-60° F. in order to avoid injury to the product. The fermentation proceeds in the usual violent manner, an abundant foam of yeast cells, pectins and albuminoids rising over the liquid as the reaction goes on. Then when most of the sugar of the must has been exhausted, in about a month, the foam subsides and the insoluble materials which it carried settle to the bottom of the barrel. At this time the batch is racked in a manner similar to wine (see p. 170) and allowed to carry on a quiet second fermentation at a lower temperature, *ca.* 45°-50° F. This second fermentation requires from three to six months and should not be entirely complete even then. Hence, if the cider is racked and bottled at this stage it will carry on a little further fermentation in the bottle, producing a beverage with some of the sparkle and life of champagne. A cider made with care, in the manner described, will be sound and stable, and unlikely to acetify (turn to vinegar). It may happen, however, that some clarification is needed. Formerly, skim milk in the proportion of one quart to about fifty gallons of cider was the preferred method of fining. Nowadays modern filtering equipment with such assistants as purified diatomaceous earth (kieselguhr) is used.

CHAPTER XII

LIQUEURS AND CORDIALS

General Statement.—Liqueurs and Cordials constitute a group of alcoholic beverages of a somewhat exotic nature. They are usually made from rectified alcohol, refined cane sugar and flavoring and aromatic substances extracted from fruits, herbs, seeds and roots. On account of their high content of sugar they are rarely consumed in any quantity and serve either as appetizers or as after dinner relishes.

Liqueurs as a class are very largely of foreign origin and manufacture and their terminology is somewhat confusing. In this country the names "liqueur" and "cordial" are practically interchangeable. Abroad, liqueurs generally are products made on the continent and especially in France, while cordials are products originating in the United Kingdom or elsewhere. Another possible distinction is that the liqueurs as a group are more perfume-like in character and exclude the cordials which are made with sharper flavors such as caraway, etc.

Classification.—The aim of all cordial and liqueur manufacture is a product in which the various separate constituents are so blended and united that only a summation is tasted by the drinker rather than a number of discordant single flavors. The varying degrees of success with which this object has been achieved and also the variations in concentration of the liqueur in alcohol, flavor and sugar have resulted in the recognition, especially in France, of a number of grades of liqueur, as follows:

1. Ordinaires	1. Average
a. Ordinaires	a. Single Strength
b. Liqueurs doubles	b. Double Strength
2. Demi-fines	2. Good
3. Fine	3. Very good
4. Superfine	4. Excellent

These grades are independent of the process of manufacture although it may be stated that the highest grades as to smoothness of flavor can in general only be made by the distillation process.

Manufacture.—There are three general methods by which cordials can be made:

    1. The distillation process.
    2. The infusion process.
    3. The essence process.

In brief statement, the distillation process consists in macerating the selected aromatic flavoring substances in alcohol for a fixed period. The liquid is then distilled and the aroma and flavor of the herbs, seeds, fruits, etc., will be found in the distillate. This is then sweetened and colored and may also be diluted and blended with alcohol and water, and other materials as required.

Certain aromas and flavors do not lend themselves to extraction by distillation and in these cases the infusion method is resorted to. In this process the aromatic substances are steeped in a solution of alcohol and sugar to which they impart their flavoring and aromatic principles. The solution may be colored and is then strained to separate the marc or solid residue.

The *"liqueurs par infusion"* do not have the fine bouquet, flavor and taste found in the *"liqueurs par distillation,"* with the exception of infusions of red fruits. These form a group of very fine liqueurs when they are made according to the best methods of the art. Typical of the finest are Cherry Brandy, Guignolet (brandy from black cherries) and Cassis (brandy from black currants).

In the essence process, essential oils, either natural or synthetic, are added to the alcohol, which is then sweetened and colored. This kind of liqueur is generally of inferior quality as compared with the others and should only be made under exceptional circumstances or when a cheap product is required.

Whatever general type of manufacture is selected, a special art is required to produce fine quality products. A series of operations are involved which must be conducted with skill, care, intelligence and knowledge because the characteristics of the finished product depend very largely on the technique of preparation, independent of the variation in quality of the raw materials. This last is a difficulty which always confronts the liqueur manufacturer. No standard formula can be relied on to produce a liqueur of unvarying quality for the reason that the herbs or seed, etc., which flavor it may not have grown under like conditions, or have ripened equally, etc.

The series of operations involved in the preparation of liqueurs may include most or all of the following:

> Infusion (Maceration)
> Distillation
> Blending
> Coloring
> Clarification
> Filtration
> Aging (True or Accelerated)

Infusion is the process by which the flavoring ingredients are extracted from their natural raw materials and brought into solution in the alcohol-water mixture desired. The details of the process depend very largely on the material which is being extracted. The strength of the alcohol used may vary according to the solubility of the flavor in diluted alcohol and also according to the solubility of such undesired materials as resins, bitter principles, etc., which may be present in the herbs. Similar considerations dictate in each special case whether the extraction shall be performed hot or cold or whether it shall be carried on for days or only a few hours.

Distillation for liqueur purposes is usually on a small scale employing a pot still, which, however, in the most modern practice may be equipped with a reflux condenser to permit partial extraction at a boil. Whenever extraction is carried on in the still it is desirable that the latter be equipped with a steam or hot water jacket to avoid the possibility of burning as by direct heat.

Blending includes the addition of sweetening, and coloring matters, other flavors, smoothening and softening agents, etc., to the distilled flavored alcohol. In France it is very often carried out in a hermetically sealed cylindrical copper vessel called a *conge*, see Fig. 45.

Note that it is fitted with a sight-glass; *i.e.*, there is an opening about three inches wide running down the side of the vessel and in this opening is a glass on which is etched a scale marked off in liters. By means of this sight-glass it is possible to determine the exact proportions of alcohol, syrup and water in the *conge* and also to observe what is happening to them.

Filling, mixing and emptying are all done by hand in the small establishments. Where the output is greater, the tanks containing the ingredients are connected to the blending machines by piping and the feed is under air pressure so that it is only necessary to open each valve for the liquid to run into the blender. While this takes place the operator reads the quantity on the sight-glass. Stirring is usually by means of a mechanical agitator although sometimes it is accomplished manually.

Perfect blending requires the mixing together of the various ingredients until they form an intimate and homogeneous whole. In carrying out this operation the following rules will prove of value:

(1) Always add the sugar in the form of a syrup. It is better to prepare the syrup by dissolving the sugar in hot water, rather than cold water, as this seems to favor intimate mixing and results in a liqueur of finer quality and smoothness.

(2) Always blend cold so as to avoid any evaporation of alcohol and aromatics and to prevent any spoilage which may ensue.

(3) Observe one of the following orders of addition: (a) put the aromatic spirit in first, (b) add the extra alcohol and stir for about ten minutes, (c) add the syrup and stir again, (d) add the required quantity of water and stir again in order to thoroughly incorporate the various ingredients. Some liqueur manufacturers favor a reverse order of procedure: (a) water, (b) sugar and glucose, (c) alcohol and alcoholic tinctures, (d) aromatic spirits, etc. Once the blending has been completed the coloring is added and stirred in.

The liqueur is allowed to rest for two or three days after mixing in order to give the ingredients time to blend thoroughly together. Thereafter, it is sampled to determine whether it has the desired combination of characteristics. If it is unsatisfactory, the operator must make such additions and modifications as his experience suggests.

In the United States especially and also in many places abroad it has been found that the use of a closed blending tank is not essential, particularly for the spicier liqueurs sometimes classed as cordials. An open, copper-lined tank is used and agitation supplied either mechanically or manually. The order of addition of the ingredients remains important, however. This probably derives its weight partly from the necessity of preventing the precipitation of alcohol-soluble flavors from a strong solution by too great dilution with water; and partly to allow the escape of air dislodged from solution before the flavors are attacked and oxidized. Whatever method of blending is followed the finer liqueurs require aging to unite their constituents firmly in a perfect blend and give them smoothness.

Aging of liqueurs depends only slightly on chemical reaction, but more on the effect of time to cause the desired union of flavors. Hence it may very well take place either in bottles or casks. The time required is seldom more than a few days to a few months. As with all changes in which time is involved, the elevation of temperature causes acceleration of the change. Hence, in the home of liqueurs, France, a special technique has been developed and given a name, *"Tranchage"* for the accelerated aging of liqueurs. The process is not universally applicable since it causes rancidity in some liqueurs, *e.g.*, anisette and crême de menthe. It also spoils chartreuse whose volatile oils will only commingle with time rather than heat. With care, however, it is an excellent and much used method.

Tranchage is accomplished by heating the liqueur gradually to a temperature of 70° to 90° C. in a hermetically sealed vessel. Heat is supplied either by a water or a steam jacket. When the set temperature has been reached the heating medium is withdrawn and the liqueur allowed to cool slowly.

The treatment is usually given in an apparatus called a *"conge à trancher,"* see Fig. 45. This vessel is fitted with a safety valve, a thermometer, a sight-glass and a steam coil, and is the type mostly used in medium sized plants. The larger manufacturers use a cylindrical tank fitted with legs, a large, quick-closing man-hole on top and a safety valve, and an inlet tap for the compressed air. Below is an exhaust valve for the air.

The usual sight-glass with scale graduated in litres is on the side. The bottom is fitted with valves for drawing off the liqueur, either by gravity into jugs or under pressure. There

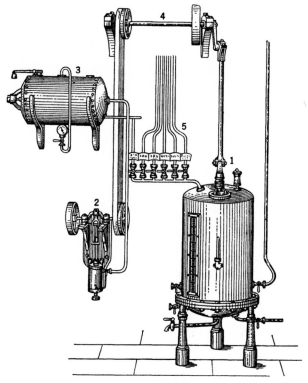

FIG. 45.—Liqueur blending equipment (conge à trancher).

is a separate tap for drawing off wash water. A steam coil is provided for heating and a mechanical agitator for mixing.

The method of operation is as follows: The feed taps for the ingredients are opened successively and the required quanti-ties of each liquid are run in. Then the contents of the vessel are agitated. When blending is completed the steam valve is opened and the mixture is heated until the temperature reaches the maximum allowable. Then the liqueur is either run off into casks under air pressure or is allowed to rest in the conge, whence it is run off, colored, filtered, and either barrelled or bottled.

Coloring of liqueurs is done to add what the jargon of the advertising profession calls "eye appeal" to the liqueur. Unless it is done with knowledge of the properties of the color used it is a risky business. On occasion the stronger colors may alter the taste and break up the harmonious blend of aromatics in the liqueur. Again the colors may be affected by light in storage and bleach or precipitate in the bottle. Infusions of red fruit tend to bleach to pinks and the violet ones tend to darken. The yellows tend to turn brown. Many of these changes can be averted by suitable skill or by proper selection of colors. Above all, where artficial aging is used, it is desirable to add color after the liqueur has cooled so that the effect of heat on the color will be avoided. With some vegetable colors the addition of about 0.01-0.02% of alum to the liqueur is claimed to give permanence and stability.

Clarification or fining of liqueurs is practiced not only to give them limpidity and brilliancy so that they are agreeable to the eye, but also to render them immune against changes caused by substances which they may hold in suspension.

Clarification methods precipitate these insoluble substances so that they can be removed by filtration. Both these operations are preferably carried out after the liqueur has completely cooled following tranchage, or better still after resting and settling for several days.

Various substances are used: albumen, white of egg, fish glue, gelatin, and skimmed milk, or the modern filter aids such as talc, asbestos, kieselguhr, etc.

Sample procedures: To fine one hectoliter (25 gals.) of liqueur are as follows:

Take three whites of egg and whip them up in a liter (quart) of water; pour it into the liqueur; stirring all the time; allow it to rest and settle for 24 to 48 hours; then decant.

Fining by means of white of egg works very well with cloudy

or milky liqueurs. It also works well with liqueurs made by the infusion process, but in this case only one white of egg must be used to avoid altering the coloring material, which is partially precipitated by the albumen.

Fish glue or isinglass is more often used. It works very well with strongly alcoholic liqueurs but its preparation is somewhat lengthy and calls for considerable care.

The best method of procedure is as follows:

Macerate the fish glue for 24 hours in ten times its weight of water, taking care to renew the water two or three times because the glue will otherwise putrefy and acquire a nauseating odor. When the glue becomes soft, wet, and white, it is put into a mortar and pounded for some time so as to disintegrate it and separate all fibres. In this condition, small quantities of fresh water are added to it gradually, with stirring until a milky suspension or solution results. This liquid is strained through silk or fine linen cloth. The coarse undisintegrated particles of glue left upon the silk or linen strainer are put back into the mortar and pounded again. Water is then added in the same manner as before and the whole procedure is gone through again until very little residue remains. The suspension is now stirred vigorously and a solution of tartaric acid is added, the stirring is continued until the glue goes into solution.

The final product is not a white, limpid, easily flowing liquid but a kind of thin, transparent jelly which should be free of every trace of animal fiber. The materials should be used in about the following proportions:

Fish glue	10 grams—⅓ oz.
Tartaric acid (dissolved in half a liter of water—1 pt.).......................	1 gram—15 gr.—
Water	1 to 1½ liters— 1 to 1½ quarts

for one hectoliter (25 gals.) of liqueur. The jelly solution is poured into the liqueur which is then well stirred up and finally allowed to rest for two or three days.

Another method replaces water as a solvent by either white wine or water to which some vinegar has been added. In this case it is not necessary to add tartaric acid. Ten per cent of alcohol added to the dissolved fish glue will preserve it from putrefying in the event that the glue solution is made up for some time before using.

To fine with gelatin soften 30 grams (1 oz.) in a liter (1 quart) of warm water; add the mixture to the liqueur; stir in vigorously and allow to rest for several days. This type of clarification agent is best adapted for white liqueurs of low alcohol content.

Milk is also a good clarification agent for white liqueurs of low alcohol content. It should be added in the proportions of one liter (1 quart) of milk to each hectoliter (25 gals.) of liqueur. It is advisable to boil the milk and the liqueur should be well stirred while the addition is made. Milk makes a particularly good fining agent for the curaçaos.

Always fine cold because hot fining results in the liqueur acquiring an albuminous taste which is very difficult to eliminate.

Filtration of liqueurs is done in order to give them the final "polish" that ensures brilliant clarity and absence of turbidity. The means by which it is accomplished range from a simple felt or flannel bag to more mechanical filters of larger capacity. In any type of filter the true filtration is done by some powdered material added to the liqueur which builds up a cake on the meshes of the filter apparatus and holds back the slimy suspended matter without clogging. Hence it is always necessary to return the first runnings of the filter one or more times until an efficient coating of filtering medium has been produced and the filtrate is absolutely clear. The felt bag filter is used in the same manner as the domestic jelly bag. That is, it is suspended with a hoop to keep its mouth open and filled with a bucket. It drains into another bucket. In spite of its primitiveness it is an effective filter for small lots of material. The next type of apparatus used for filtration is a copper cone fitted with a faucet at the bottom. To use this filter close the faucet and line the cone with filter paper or preferably pack the bottom with pure cellulose or filter paper reduced to pulp with a little water. The cone is filled with liqueur and the faucet opened. Figure 46 shows such a cone filter.

Figures 47 and 48 show a slightly more advanced type of filter and its method of application. It consists of a tinned

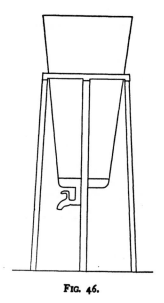

Fig. 46.

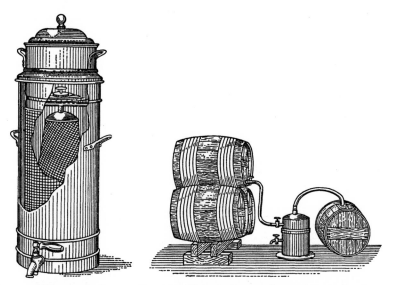

Fig. 47.—Wire mesh liqueur filter.

Fig. 48.—Method of connecting and operating liqueur filter shown in Fig. 47.

copper cylinder fitted with a faucet at the bottom and valves at the side and top. The cover seals it hermetically. The interior consists of two metallic screens, one with horizontal and vertical shoot and warp wires; the other, a cone, with diagonal shoot and warp. It is necessary to disperse some kind of filtering material in the liqueur, so that this material deposits on the screens and aids in the production of a clear filtrate. Paper pulp, asbestos wool, talc, diatomaceous earth, etc., are satisfactory for this purpose. The method of application is shown in Fig. 48. If necessary, it is possible to arrange the filter and casks in battery form, so that the receiving barrel of one filter serves as the feeding barrel of another filter.

Similar filters are used in which the capacity is increased by supporting the wire filter cloth over a thin hollow rectangular frame and placing a number of these in a container so that the liquid flows freely into the outer chamber, and after passing through the filtering surface is collected from the interior of the frame and drained into a storage vessel.

CURAÇAO

In order to summarize the actual application of the general process outline given above, the manufacture of Curaçao is detailed here because this liqueur has probably become the one best acclimated on American soil and is made here in larger volume than any other. The Curaçao fruit is one of the family of bitter oranges. They grow chiefly in the West Indies and the price for genuine Curaçao peels is quite high so that distillers generally substitute up to 50% of other bitter orange peels in the cheaper liqueurs.

Manufacture.—The outstanding characteristic of Curaçao liqueurs is a mild bitter taste derived from the maceration of the fresh Curaçao peel in 190 proof alcohol. Very little of this extract needs to be incorporated into the finished liqueur on account of its intense bitterness. The complete process of making Curaçao liqueur as stated by Wolff (Spirits (1934) II, No. 6, 73) is as follows:

FORMULA TO MAKE 100 GALLONS
CURAÇAO TRIPLE SEC. 40% ALCOHOL BY VOLUME
EXTRA FINEST QUALITY

a.
- 24 oz. Extra thin genuine fresh Curaçao peels
- 12 oz. Extra thin fresh Orange peels
- Grind and macerate for 2 days with
- 2.5 gal. Alcohol 190 proof (95%)
- Draw off 2 gallons of Extract. To the remaining macerate add:

b.
- 20 lb. Extra thin fresh Curaçao peels
- 15 lb. Extra thin fresh Orange peels
- 10 oz. Mace
- 2 oz. Cloves
- 38 gal. Alcohol 190 proof
- 40 gal. Distilled water

Digest for six hours at very gentle heat, place in a steam jacketed still, add another 5 gallons water, and distill slowly for 2 hours with partial reflux until all the alcohol is driven over. Rectify the raw distillate to 35% alcohol by volume and filter clear over Kieselguhr to remove terpenes. Clean still and then rectify the filtered distillate to 58 gallons, 60% alcohol by volume.

The extract and distillate are blended as follows:

BLENDING FORMULA

2 gal. Extract a.
58 gal. Rectified distillate 60% b.
1 gal. Genuine Jamaica Rum 74%
4 gal. Grape distillate 60%
2 gal. Port wine
5 gal. Glucose 42° B.
18 gal. Syrup made from
    250 lb. best grade sugar
    25 lb. milk sugar
    1 lb. Citric Acid C.P.
    13 gal. Distilled water
Caramel color as needed.

This formula will yield 100 gallons.

A cheaper product is made as follows:

a′
   1.5 lb. Extra thin Curaçao peel
   Macerate for two days with 1.5 gal. alcohol and draw off
     1 gallon extract.

To the remaining macerate add:

b′
20	lb.	Extra thin Curaçao peel
10	lb.	Dried expulped Curaçao peel
10	oz.	Mace
2.5	oz.	Cinnamon
2.5	oz.	Cloves
13	gal.	Alcohol 190 proof (95% vol.)
15	gal.	Distilled water

Direction for distillation same as preceding formula except that 20 gallons of rectified distillate at 60% are obtained.

BLENDING FORMULA

1	gal.	Extract a′.
20	gal.	Rectified distillate 60% b′.
0.5	gal.	Genuine Jamaica Rum 74%
2.5	gal.	Grape distillate 60%
24.5	gal.	Alcohol 190 proof (95%)
2	gal.	Port wine
23	gal.	Syrup made from

		300	lb.	sugar
		20	lb.	milk sugar
		1	lb.	Citric acid C.P.
		29.5	gal.	Distilled water

Caramel color as needed.

Newly made Curaçao liqueurs have a raw unpleasant taste of peel which the addition of milk sugar helps to overcome. Heating the liqueur in vacuo to 135° F. also accelerates aging the product.

LIQUEUR FORMULAE

There follows a selected list of formulae for the manufacture of liqueurs. In the case of many liqueurs a number of alternative formulae are cited according to the method of manufacture and the grade or quality of product. In using these formulae, the warning must be observed that no amount of direction can substitute safely for care, skill and experience.

In explanation of the following formulae it is also important to note that the words "spirit of" refer to an alcohol distillate from the flavoring material. The term essence refers to a solution of the essential oil in alcohol. Directions for coloring have been omitted, as the user will naturally follow his judgment in this matter. The section on coloring in this chapter and the list of colors in Chapter IV may be helpful in this connection.

ABSINTHE
First Quality

Wormwood...........................	28 lb.
Hyssop.............................	6 lb. 8 oz.
Lemon balm.........................	6 lb. 8 oz.
Anis (green).......................	40 lb.
Chinese aniseed....................	12 lb.
Fennel.............................	16 lb.
Coriander..........................	8 lb.
Alcohol (90%)......................	80 gal.
Water..............................	25 gal.

Macerate for 48 hours. Distill. Color with an infusion of wormwood and green herbs.

CREAM OF ABSINTHE
First Quality "Synthetic"

Essence of absinthe................	45 min.
" " English peppermint........	45 min.
" " anis......................	4 dr.
" " sweet fennel..............	1 dr.
" " distilled lemon..........	4 dr.
Alcohol (85%)......................	4 gal.
Sugar..............................	45 lb.
Water..............................	to make 10 gal.

ABSINTHE
Average Quality

Essence of absinthe................	45 min.
" " English peppermint........	45 min.
" " anis, green...............	4 dr.
" " lemon.....................	4 dr.
" " fennel....................	1 dr.
Alcohol (85%)......................	2.5 gal.
Sugar..............................	10 lb.
Water..............................	to make 10 gal.

ALKERMES DE FLORENCE
Elixir of Life of Florence

Essence of calamus....................	22 min.
" " Chinese cinnamon.............	15 min.
" " cloves.......................	40 min.
" " nutmeg......................	22 min.
" " roses.......................	30 min.
Extract of jasmin....................	4 dr.
" " anis.......................	4 dr.
Alcohol (85%).......................	4 gal.
Sugar...............................	45 lb.
Water...............................	to make 10 gal.

Color with cochineal

ANGELICA LIQUEUR
Excellent Quality

Angelica root........................	10 lb.
" seed........................	8 lb.
Coriander seed.......................	1 lb.
Fennel..............................	1 lb.
Alcohol (90%).......................	28 gal.

Macerate, distill and rectify to 36 gallons after addition of water. Add 400 lb. of sugar in syrup and make to 100 gallons with distilled water.

Very Good Grade

Spirit of angelica root...............	10 gal.
" " " seeds................	10 gal.
Alcohol (85%).......................	13.5 gal.
Sugar...............................	342 lb.
Water...............................	to make 100 gal.

Good Grade

Spirit of angelica roots..............	2 qt. 25 oz.
" " " seeds................	2 qt. 25 oz.
Alcohol (85%).......................	1 gal. 3 pt.
Sugar...............................	20 lb.
Water...............................	to make 10 gal.

Average Grade—Double Strength

Spirit of Angelica seeds..............	1 gal. 3 pt.
Alcohol (85%).......................	3 gal. 5 pt.
Sugar...............................	20 lb.
Water...............................	to make 10 gal.

Average Grade—Single Strength

Spirit of angelica seeds....................	3 qt.	12 oz.
Alcohol (85%)...........................	1 gal.	6 pt.
Sugar..................................	10 lb.	
Water.................................	to make 10 gal.	

ANISETTE

Highest Quality—Paris Type

Chinese aniseed.........................	16 lb.
Bitter almonds..........................	4 lb.
Green anis..............................	16 lb.
Coriander..............................	2 lb.
Fennel.................................	1 lb.
Angelica roots..........................	4 oz.
Lemon peel.............................	No. 80
Orange peel............................	No. 80
Alcohol (90%)..........................	34 gal.

Macerate in alcohol for 24 hours, add 19 gallons of water, distill. Add 19 gallons of water and rectify to draw off 36 gallons of good product. Dissolve 400 lb. of refined sugar in 24 gallons of water and cool. Add this to the distillate and add:

Infusion of orris........................	4 oz.
Orange flower water.....................	1 gal.
Chinese aniseed water...................	8 oz.
Clove water............................	1.5 oz.
Nutmeg water..........................	1.5 oz.

Add enough water to make up to 100 gallons.

Alternative

Anis de Tours..........................	8 lb.
Anis d'Albi.............................	8 lb.
Badiane (Chinese Aniseed)...............	4 lb.
Ceylon cinnamon.......................	1 lb.
Orris..................................	12 oz.
Cloves.................................	3 oz.
Angelica root...........................	3 oz.
Dictame de Crete (marjoram).............	12 oz.
Coriander seed.........................	12 oz.
Almonds (sweet).......................	4 lb.

Lemon peels.	No. 80
Orange peels.	No. 20
Nutmegs.	No. 40
Alcohol (85%).	40 gal.

Macerate for 24 hours, add water, distill, rectify, sweeten to 45°. Add 1 gallon of orange flower oil.

Anisette—*Bordeaux Type*

Badiane (Chinese aniseed) seeds.	16 lb.
Anise (green).	4 lb.
Fennel.	4 lb.
Coriander.	4 lb.
Sassafras.	4 lb.
Musk seed.	14 oz.
Lemon peel.	4 lb.
Alcohol (90%).	35 gal.

Macerate for 48 hours in alcohol. Add 30 to 40 gallons of water. Distill. Add 35 gallons of alcohol and 20 gallons of water and rectify. Dissolve 400-500 lb. of sugar in hot water. Mix the syrup with the alcohol. This should give a total volume of 100 gallons. Filter. Add 1 or 2 gallons of orange flower oil.

Very Good Grade

Spirit of anise (prepared as above).	2.5 gal.
Orange flower water.	1 pt.
Infusion of orris.	3 oz.
Alcohol (85%).	3 gal.
Sugar.	19 lb.
Water.	to make 10 gal.

Good Grade

Spirit of anise.	3 qt.
Orange flower water.	12 oz.
Alcohol (85%).	2.5 gal.
Sugar.	20 lb.
Water.	to make 10 gal.

These liqueurs should be colored red.

Anisette (*with Glucose*)

Spirit of anise	3 qt.
Orange flower water	3 qt.
Alcohol (90%)	1.75 gal.
Sugar	10 lb.
Glucose (36° B)	2.75 gal.
Water	to make 10 gal.

Average Grade—Double Strength

Spirit of anise	3.5 qt.
Alcohol (85%)	4.25 gal.
Sugar	20 lb.
Water	to make 10 gal.

With Glucose

Spirit of anise	3.5 qt.
Alcohol (90%)	1.5 gal.
Sugar	12 lb.
Glucose (36° B)	3.75 gal.
Water	to make 10 gal.

Single Strength

Spirit of anise	0.5 gal.
Alcohol (85%)	2 gal.
Sugar	10 lb.
Water	to make 10 gal.

With Glucose

Spirit of anise	0.5 gal.
Alcohol (90%)	2 gal.
Sugar	5 lb.
Glucose (36° B)	1.75 gal.
Water	to make 100 gal.

"SYNTHETIC" ANISETTE

Highest Grade

Essence of Chinese aniseed	1 oz.
" " anise	2 dr.
" " sweet fennel	1 dr.
" " coriander	5 min.
" " sassafras	45 min.

Extract of orris.......................... 0.75 oz.
"   " amber (not musk).............. 1 dr.
Alcohol (85%)........................... 4 gal.
Sugar................................... 45 lb.
Water................................... to make 10 gal.

Very Good Grade

Essence of Chinese aniseed................ 6 dr.
"   " anise......................... 2 dr.
"   " sweet fennel................... 45 min.
"   " coriander..................... 5 min.
"   " sassafras..................... 30 min.
Extract of orris......................... 4 dr.
"   " amber (not musk)............... 45 min.
Alcohol (85%)........................... 3.25 gal.
Sugar................................... 30 lb.
Water................................... to make 10 gal.

Ordinary Grade

Essence of aniseed....................... 3 dr.
"   " Chinese aniseed................ 3 dr.
"   " sweet fennel................... 40 min.
"   " coriander..................... 4 min.
Alcohol (85%)........................... 2.5 gal.
Sugar................................... 10 lb.
Water................................... to make 10 gal.

Benedictine

This product is made only by a company once the Benedictine Monks of Fecamp. Through hundreds of years they have kept its composition secret and have now a copyright on the name and bottle. The formula cited is an imitation which closely reproduces the original.

Cloves..................................... ¼ oz.
Nutmeg..................................... ¼ oz.
Cinnamon................................... ⅓ oz.
Mixture of peppermint, fresh angelica roots and
    alpine mugwort........................ 3½ oz.
Aromatic calamus.......................... 2 oz.
Cardamom, minor........................... 7 oz.
Flowers of arnica......................... 1 oz.

Cut up and crush the materials and macerate for two days

in 4 gallons of 85 per cent alcohol. Add 3 gallons of water and draw off 4 gallons. Distill. Add syrup made with 32 lb. of sugar and 2 gallons of water. Make the volume up to 10 gallons. Color yellow and filter.

BLACK CURRANT BRANDY (LIQUEUR)
(Crême de Cassis)
Excellent Quality

Infusion of black currants (first)............	4 gal. 1 qt.
Spirit of strawberries.....................	2 qt.
Alcohol (85%).........................	2 qt. 1 pt.
Sugar.................................	40 lb.
Water.................................	to make 10 gal.

In the formula just cited, as well as in the next following formulae reference is made to different infusions. To prepare these infusions the following directions are given: Steep or macerate the fruits or other flavoring ingredients in alcohol. When extraction appears complete, decant the solvent and use as the first infusion. A second and third infusion may be made from the same fruits but on account of the comparative weakness of these in extract they must be employed in the ratios of 1 : 2 and 1 : 3 to replace a first infusion. In balancing the formula correction must be made in the amount of neutral spirits added to compensate for the alcoholic content of the volume of infusion used.

Very Good Grade
(Ratafia de Cassis)

Infusion of black currants (first)..........	3 gal. 5 pt.
" " strawberries.................	3 qt. 12 oz.
Alcohol (85%).........................	1 gal
Sugar................................	30 lb.
Water...............................	to make 10 gal.

Good Grade

Infusion of black currants (first)..........	2 gal. 3 pt.
Vin de Roussillon......................	3 qt. 12 oz.
Infusion of wild cherries................	3 pt.
" " strawberries.................	3 pt.
Alcohol (85%).........................	2 gal. 3 qt.
Sugar................................	20 lb.
Water to make total volume up to 10 gallons.	

Average Grade—Double Strength

Infusion of black currants (first)..........	2 gal.
Alcohol (85%).........................	2 gal. 3 pt.
Sugar.................................	20 lb.
Water................................	to make 10 gal.

Average Grade—Single Strength

Infusion of black currants................	1 gal. 2 qt.
Alcohol (85%).........................	1 gal. 1 qt.
Sugar.................................	10 lb.
Water................................	to make 10 gal.

CREAM OF CELERY
(Crême de Celeri)
Very Good Grade

Spirit of celery........................	2 gal.
Alcohol (85%).........................	1 gal. 1 qt.
Sugar.................................	35 lb.
Water................................	to make 10 gal.

Good Grade

Spirit of celery........................	1 gal. 1 qt.
Alcohol (85%).........................	1 gal. 5 pt.
Sugar.................................	20 lb.
Water................................	to make 10 gal.

Good Grade "Synthetic"

Essence of celery.......................	2 dr.
Alcohol (85%).........................	3 gal. 1 pt.
Sugar.................................	35 lb.
Water................................	to make 10 gal.

CHARTREUSE
See remarks under Benedictine
Green Chartreuse

Melisse citronne, dry (lemon balm)........	8 lb.
Hyssop flowers........................	2 lb.
Dry peppermint........................	2 lb.
Alpine mugwort........................	2 lb.
Balsamite.............................	1 lb.
Thyme................................	7 oz.

Angelica leaves	1 lb. 5 oz.
" roots	1 lb. 5 oz.
Arnica flowers	3 oz.
Bourgeons de peuplier-baumier	3 oz.
Chinese aniseed	2 oz.
Mace	2 oz.
Alcohol (90%)	58½ gal.

Macerate for 24 hours, add water and rectify to draw off 60 gallons; add sugar syrup made by dissolving 200 lb. of sugar in hot water, mix and make total volume up to 100 gallons with water; age by heating; then color green with blue color and saffron or caramel, according to the desired shade.

Rest, fine and filter.

Yellow Chartreuse

Lemon balm	4 lb.
Hyssop flowers	1 lb.
Alpine mugwort	1 lb.
Angelica leaves	2 lb.
" roots	2 lb.
Arnica flowers	2 oz.
Chinese aniseed	2 oz.
Mace	2 oz.
Coriander	12 lb.
Aloes	4 oz.
Cardamom, minor	4 oz.
Cloves	3½ oz.
Alcohol (90%)	38½ gal.
Refined sugar	200 lb.
Water, to make up to 100 gallons.	

Proceed as above, coloring yellow with saffron.

White Chartreuse

Lemon balm	4 lb.
Hyssop flowers	1 lb.
Alpine mugwort	1 lb.
Angelica leaves	2 lb.
" roots	13 oz.
Chinese aniseed	13 oz.
Mace	4 oz.
Cloves	4 oz.

Nutmeg...............................	4 oz.
Cardamom, minor......................	4 oz.
Calamus, aromatic....................	4 oz.
Tonka beans..........................	1½ oz.
Alcohol (90%)........................	49 gal.
Refined sugar........................	300 lb.

Water to make up to 100 gallons.

Prepare as above.
The monks age their product for two or three years.

First Quality "Synthetic"

Essence of lemon......................	15 min.
" " hyssop......................	15 min.
" " angelica....................	1 dr.
" " English peppermint...........	2 dr.
" " chinese aniseed..............	15 min.
" " nutmeg.....................	15 min.
" " cloves.....................	15 min.
Alcohol (85%)........................	4 gal.
Sugar................................	45 lb.
Water...............................	to make 10 gal.

Color: yellow or green.

CHERRY CORDIAL
(Ratafia de cerises de Grenoble)
First Quality, Grenoble Type

Infusion of cherries....................	2½ gal.
" " wild cherries.................	1½ gal.
Spirit of apricot stones.................	5 pt.
" " strawberries..................	3 pt.
Sugar................................	40 lb.
Water...............................	to make 10 gal.

Angers' Type
Guignolet d'Angers

Infusion of cherries....................	2 gal.
" " wild cherries.................	2 gal.
Alcohol (85%)........................	1 gal.
Sugar................................	40 lb.
Water...............................	to make 10 gal.

Very Good Grade

Infusion of cherries.	3½ gal.
" " wild cherries.	3 qt. 12 oz.
Spirit of apricot stones.	3 qt.
Alcohol (85%).	1 qt. 22 oz.
Sugar.	30 lb.
Water.	to make 10 gal.

Good Grade

Infusion of cherries.	3 gal.
" " wild cherries.	2 qt.
Spirit of apricot stones.	2 qt.
Alcohol (85%).	1 qt. 22 oz.
Sugar.	20 lb.
Water.	to make 10 gal.

CREAM OF COCOA
(Crême de Cacao)
Highest Quality

Cocoa (Caracas).	20 lb.
Cocoa (maragnan).	20 lb.
Cloves.	7 oz.
Mace.	8 oz.
Vanilla.	4 oz.
Alcohol (85%).	40 gal.

Roast the cocoa, grind it. Macerate in alcohol the cocoa, cloves, mace, and vanilla and distill after 48 hours. Rectify. Add 1 gallon of tincture of vanilla and 400 lb. of refined sugar which has been dissolved in a sufficient quantity of water to bring the total quantity of liqueur to 100 gallons. The tincture of vanilla adds, in addition to flavor, a light yellow color which is much admired.

CREAM OF COFFEE
(Crême de Moka)
Very Good Grade

Spirit of moka (spirit of coffee).	2½ gal.
Alcohol (85%).	6 pt.
Sugar.	35 lb.
Water.	to make 10 gal.

Good Grade

Moka water (aqueous extract of coffee)....	2 gal.
Alcohol (85%).........................	2 gal. 6 pt.
Sugar.................................	20 lb.
Water.................................	to make 10 gal.

CURAÇAO

The best as well as most detailed directions for making Curaçao will be found on p. 201 *et seq.* in this section. However, a number of alternative formulae have been included here to guide in the preparation of different grades of the liqueur. In each of the formulae cited below the addition of caramel color is required.

Highest Quality Triple Sec.

Dutch curaçao peel.....................	64	lb.
Curaçao reeds.........................	32	lb.
Orange peels...........................	No.	6
Lemon peels...........................	No.	4
Alcohol (85%).........................	50	gal.
Distill, rectify. Add:		
Infusion of oranges...................	3	gal.
" " curaçao reeds..............	2	gal.
Sweeten with		
White sugar.......................	200	lb.
Raw white sugar...................	80	lb.
Color and filter.		

Very Good Grade

Spirit of Dutch curaçao.................	2½	gal.
" " oranges......................	3	qt.
Infusion of curaçao....................	3	oz.
Sugar.................................	35	lb.
Water.................................	to make 10 gal.	

Good Grade

Spirit of curaçao......................	1	gal 1 qt.
Infusion of curaçao....................	4	oz.
Alcohol (85%).........................	1½	gal.
Sugar.................................	20	lb.
Water.................................	to make 10 gal.	

Average Grade—Double Strength

Spirit of curaçao......................	1 gal.
Alcohol (85%)........................	4 gal.
Sugar................................	20 lb.
Water................................	to make 10 gal.

(With Glucose)

Spirit of curaçao......................	1 gal.
Alcohol (90%)........................	2½ gal.
Sugar................................	13 lb.
Glucose (36 B)........................	3 gal 3 qt.
Water................................	to make 10 gal.

Average Grade—Single Strength

Spirit of curaçao......................	3 qt. 12 oz.
Alcohol (85%)........................	1 gal. 3 qt.
Sugar................................	10 lb.
Water................................	to make 10 gal.

Proceed as above and color with a little caramel.

(With Glucose)

Spirit of curaçao......................	3 qt. 12 oz.
Alcohol (90%)........................	1 gal. 1 qt.
Sugar................................	4½ lb.
Glucose (36 B)........................	1 gal. 7 pt.
Water................................	to make 10 gal.

Highest Quality "Synthetic"

Essence of distilled curaçao..............	1 oz. 2 dr.
" " Portugal, distilled..............	4 dr.
Infusion of bitter curaçao...............	sufficient quantity
Alcohol (85%)........................	4 gal.
Sugar................................	45 lb.
Water................................	to make 10 gal.

Very Good Grade "Synthetic"

Essence of distilled curaçao..............	1 oz.
" " Portugal, distilled............	2 dr. 30 min.
" " cloves......................	30 min.
Infusion of bitter curaçao...............	sufficient quantity
Alcohol (85%)........................	3 gal. 1 qt.
Sugar................................	36 lb.
Water................................	to make 10 gal.

Average Grade "Synthetic"

Essence of curaçao	4 dr.
" " Portugal, distilled	1 dr. 30 min.
" " cloves	20 min.
Alcohol (85%)	2½ gal.
Sugar	10 lb.
Water	to make 10 gal.

DESSERT LIQUEUR
(Liqueur de Dessert)

Angelica seed	6 lb.
" root	4 lb.
Calamus (aromatic)	1 lb. 4 oz.
Ceylon cinnamon	1 lb. 4 oz.
Myrrh	12 oz.
Cloves	10 oz.
Aloes	7 oz.
Vanilla	8 oz.
Nutmegs	No. 20
Saffron	¾ oz.
Alcohol (86%)	40 gal.
Crystallized sugar	400 lb.

Water: sufficient to make up to 100 gal. total volume.

Macerate in alcohol 48 hours. Distill. Rectify, color with tincture of saffron. Filter.

GARUS' ELIXIR
(Elixir de Garus)

Spirit of aloes	14 oz.
" " myrrh	14 oz.
" " saffron	14 oz.
" " Chinese cinnamon	14 oz.
" " cloves	8 oz.
" " nutmeg	7 oz.
Orange flower water	14 oz.
Alcohol (85%)	2 gal. 3 pt.
Sugar	20 lb.
Water	to make 10 gal.

Color golden with saffron and a little caramel.

GOLDEN ELIXIR
(*Eau de vie de Dantzick*)

Ceylon cinnamon......................	8 lb.
Ripe figs.............................	8 lb.
Cumin...............................	1 lb. 12 oz.
Musk seed...........................	14 oz.
Mace................................	1 lb.
Cloves...............................	1 lb.
Lemon peel..........................	1 lb.
Alcohol (86%)........................	40 gal.
Refined white sugar...................	392 lb.

Macerate in alcohol 48 hours. Distill. Rectify. Dissolve sugar in sufficient water to make total volume up to 100 gallons. Settle. Filter. Add I leaf of gold per gallon. Agitate and bottle.

HENDAYE'S ELIXIR
(*Eau de vie de Hendaye*)
Very Good Grade

Spirit of anis.........................	1 qt.
" " coriander.....................	1 qt.
" " bitter almonds.................	1 qt.
" " angelica root..................	½ gal.
" " cardamom, major..............	7 oz.
" " " minor..............	7 oz.
" " lemon.......................	14 oz.
" " oranges......................	½ gal.
Infusion of orris.......................	3 oz.
Alcohol (85%)........................	1½ gal.
Sugar...............................	35 lb.
Water...............................	to make 10 gal.

Kirschwasser Liqueur

Kirsch, fine (true cherry brandy) (50%)....	15 gal.
Spirit of nuts (cherry stone or bitter almond)	10 gal.
Orange flower water...................	1 gal.
Alcohol (90%)........................	1½ gal.
Sugar...............................	400 lb.
Water...............................	to make 100 gal.

Huile de Kirschwasser—"Synthetic"

Essence of nuts........................	4 dr.
" " French neroli oil..............	40 min.
Alcohol (85%).........................	4 gal.
Sugar.................................	45 lb.
Water.................................	to make 10 gal.

CRÈME DE MENTHE
Very Good Grade

Spirit of peppermint....................	2½ gal.
Alcohol (85%).........................	3 qt.
Sugar.................................	35 lb.
Water.................................	to make 10 gal.

Color green with mixture of blue and yellow colors if desired.

Good Grade

Peppermint water......................	1 gal.	
Alcohol (85%).........................	2 gal.	3 qt.
Sugar.................................	20 lb.	
Water.................................	to make 10 gal.	

Alternative Formula

Peppermint water......................	1 gal.	1 qt.
Alcohol (85%).........................	5 gal.	
Sugar.................................	20 lb.	
Water.................................	to make 10 gal.	

Average Grade

Peppermint water......................	3 qt.	12 oz.
Alcohol (85%).........................	2½ gal.	
Sugar.................................	10 lb.	
Water.................................	to make 10 gal.	

Very Good Grade "Synthetic"

Essence of English peppermint..........	5 dr.	
Alcohol (85%).........................	3 gal.	1 qt.
Sugar.................................	35 lb.	
Water.................................	to make 10 gal.	

Good Grade "Synthetic"

Essence of English peppermint............	3 dr. 30 min.
Alcohol (85%)........................	2 gal. 3 qt.
Sugar................................	20 lb.
Water...............................	to make 10 gal.

NUT BREW
(Brou de Noix)
Good Grade

Infusion of nuts.......................	3 gal.
Spirit of nutmeg......................	5 oz.
Alcohol (85%)........................	1½ gal.
Sugar................................	30 lb.
Water...............................	to make 10 gal.

Color with caramel.

Average Grade—Double Strength

Infusion of nuts.......................	4 gal. 1 qt.
Spirit of nutmegs......................	7 oz.
Alcohol (85%)........................	2½ gal.
Sugar................................	20 lb.
Water...............................	to make 10 gal.

Color with caramel.

Average Grade—Single Strength

Infusion of nuts, aged...................	2 gal. 1 pt.
Spirit of nutmeg......................	3 oz.
Alcohol (85%)........................	1 gal. 3 pt.
Sugar................................	44 lb.
Water...............................	to make 10 gal.

Color with caramel.

NOYAUX
(Eau de Noyaux)
Finest Quality

Apricot stones........................	16 lb.
Cherry stones.........................	12 lb.
Dried peach leaves....................	4 lb.
Myrrh...............................	1 lb. 10 oz.
Alcohol (90%)........................	40 gal.
Water...............................	to make 100 gal.

Crush the stones, macerate, add water and distill.

CRÊME DE NOYAUX
Very Good Quality

Spirit of apricot stone....................	1 gal. 5 pt.
" " bitter almonds..................	1 pt.
Orange flower water.....................	3 qt.
Alcohol (85%).........................	3 qt.
Sugar.................................	35 lb.
Water................................	to make 10 gal.

Good Grade

Spirit of apricot stones..................	1 gal. 3 pt.
Alcohol (85%).........................	3 gal. 5 pt.
Sugar.................................	20 lb.
Water................................	to make 10 gal.

Average Grade

Spirit of apricot stones..................	7 pt.
Alcohol (85%).........................	1 gal. 5 pt.
Sugar.................................	10 lb.
Water................................	to make 10 gal.

Proceed as above.

ONE HUNDRED AND SEVEN YEARS
Cent Sept Ans

Spirit of lemon........................	1½ gal.
Rose water............................	6 gal.
Alcohol (90%).........................	21 gal.
Sugar.................................	100 lb.
Water................................	to make 100 gal.

Color strongly red with orchil.

ORANGE FLOWER CREAM
(Crême de Fleurs d'Oranger)
Very Good Grade

Spirits of orange flowers................	1 gal.
Orange flower water.....................	½ gal.
Alcohol (85%).........................	2 gal. 1 qt.
Sugar.................................	35 lb.
Water................................	to make 10 gal.

Good Grade

Orange flower water.....................	7 pt.
Alcohol (85%).........................	2 gal. 6 pt.
Sugar.................................	20 lb.
Water.................................	to make 10 gal.

Double Strength

Orange flower water.....................	1 gal.
Alcohol (85%).........................	5 gal.
Sugar.................................	20 lb.
Water.................................	to make 10 gal.

Highest Quality "Synthetic"

Essence of French neroli oil..............	2 dr.
Orange flower water.....................	1 pt. 9 oz.
Alcohol (85%).........................	4 gal.
Sugar.................................	45 lb.
Water.................................	to make 10 gal.

Average Grade "Synthetic"

Essence of neroli of Paris................	1 dr. 16 min.
Alcohol (85%).........................	2 gal. 3 qt.
Sugar.................................	20 lb.
Water.................................	to make 10 gal.

PERFECT LOVE
(Parfait Amour)
Very Good Grade

Spirit of lemon........................	1 qt. 7 oz.
" " oranges.......................	1 qt. 7 oz.
" " coriander.....................	3 pt.
" " anis..........................	1 qt.
Alcohol (85%).........................	2 gal.
Sugar.................................	35 lb.
Water.................................	to make 10 gal.

Color this or the following with orchil.

Good Grade

Spirit of lemon........................	1 qt. 7 oz.
" " coriander.....................	3 pt.
Alcohol (85%).........................	2 gal. 1 pt.
Sugar.................................	20 lb.
Water.................................	to make 10 gal.

Average Grade

Spirit of lemon	1 pt.	8 oz.
" " coriander	1 pt.	8 oz.
Alcohol (85%)	2 gal.	1 pt.
Sugar	10 lb.	
Water	to make 10 gal.	

Average Grade "Synthetic"

Essence of distilled lemons	4 dr.	30 min.
" " cedar	1 dr.	30 min.
" " coriander	6 min.	
Alcohol (85%)	2 gal.	5 pt.
Sugar	10 lb.	
Water	to make 10 gal.	

PINEAPPLE LIQUEUR
(Crême d'Ananas)
Finest Quality

Pineapple (fresh)	6 lb. 7 oz.
Alcohol (85%)	4 gal.

Crush and infuse pineapple in alcohol for 8 days. Filter through silk cloth. Add sugar dissolved in 2¼ gallons of water. Add 7 oz. infusion of vanilla. Color yellow with caramel.

PUNCH LIQUEUR

Brandy (58%)	5	oz.
Tafia (55%)	2	qt.
Spirit of lemon, concentrated	1⅓	oz.
Citric acid	⅔	oz.
Hyswen tea	1⅔	oz.
Burnt sugar (4° B)	15	lb.
Water	4	gal.

QUINCE BRANDY—Ratafia de Coings

Sweet juice of ripe quinces	5	pt.
Spirit of cloves	7	oz.
Alcohol (85%)	2½	gal.
Sugar	10	lb.
Water	to make 10 gal.	

Color yellow with caramel.

Rose Liqueur
(*Huile de Roses*)
Good Grade

Rose water............................	1 gal. 1 qt.
Alcohol (85%).........................	5 gal
Sugar.................................	20 lb.
Water.................................	to make 10 gal.

Average Grade "Synthetic"

Essence of roses.......................	11 oz.
Alcohol (85%).........................	2 gal. 3 qt.
Sugar.................................	20 lb.
Water.................................	to make 10 gal.

Strawberry Cordial
(*Ratafia de Frambroises*)
Finest Quality

Infusion of strawberries.................	3 gal.
" " cherries....................	1 gal.
Alcohol (85%).........................	1 gal.
Sugar.................................	40 lb.
Water.................................	to make 10 gal.

Average Grade

Infusion of strawberries.................	1 gal. 2 qt.
" " black currant...............	2 qt.
Alcohol (85%).........................	1 gal. 1 qt.
Sugar.................................	10 lb.
Water.................................	to make 10 gal.

Strawberry Brandy
Average Grade—Double Strength

Spirit of strawberries (80%).............	16 gal.
Alcohol (90%).........................	8 gal. 1 qt.
Sugar.................................	200 lb.
Water.................................	to make 100 gal.

Trappistine
(See note under "Benedictine")

Grand absinthe.......................	5½ oz.
Angelica..............................	5½ oz.
Peppermint...........................	11 oz.

Cardamom............................ 5½ oz.
Melisse (balm)....................... 4 oz.
Myrrh............................... 2½ oz.
Calamus............................. 2½ oz.
Cinnamon............................ ½ oz.
Cloves.............................. ½ oz.
Mace................................ ¼ oz.
Alcohol (85%)....................... 4 gal. 2 qt.
Sugar............................... 30 lb.
Water............................... to make 10 gal.

Follow the general methods for Benedictine.

VESPETRO
Very Good Grade

Spirit of amber seeds.................. 1 pt.
"   " dill........................... 1 qt.
"   " anis.......................... 2 qt.
"   " caraway...................... 2 qt.
"   " coriander..................... 2 qt.
"   " daucus....................... 1 qt.
"   " fennel........................ 1 qt. 1 pt.
Alcohol (85%)....................... 1 gal. 1 qt.
Sugar............................... 35 lb.
Water............................... to make 10 gal.

Good Grade

Spirit of amber seed.................. 7 oz.
"   " dill........................... 21 oz.
"   " anis .... 1 qt. 1 pt.
"   " caraway...................... 1 qt.
"   " coriander..................... 1 qt.
"   " daucus....................... 21 oz.
"   " fennel........................ 1 qt.
Alcohol (85%)....................... 1 qt. 1 pt.
Sugar............................... 20 lb.
Water............................... to make 10 gal.

Average Grade

Spirit of amber seed.................. 14 oz.
"   " dill........................... 14 oz.
"   " anis.......................... 28 oz.

Spirit of caraway...................... 14 oz.
"   " coriander...................... 28 oz.
"   " daucus........................ 14 oz.
"   " fennel........................ 14 oz.
Alcohol (85%)........................ 4 gal. 1 pt.
Sugar................................. 20 lb.
Water................................ to make 10 gal.

Best Grade "Synthetic"

Essence of anis........................ 4 dr.
"   " caraway.................... 3 dr.
"   " sweet fennel................. 1 oz.
"   " coriander................... 30 min.
"   " distilled lemon.............. 2 dr.
Alcohol (85%)........................ 4 gal.
Sugar................................. 45 lb.
Water................................ to make 10 gal.

Average Grade "Synthetic"

Essence of anis........................ 3 dr.
"   " black currants............... 2 dr.
"   " sweet fennel................. 40 min.
"   " coriander................... 50 min.
"   " distilled lemon.............. 1 dr.
Alcohol (85%)........................ 2 gal. 3 qt.
Sugar................................. 20 lb.
Water................................ to make 10 gal.

CREAM OF VANILLA
Very Good Grade

Infusion of vanilla..................... 3 qt.
Alcohol (85%)........................ 2 gal. 2 qt.
Sugar................................. 35 lb.
Water................................ to make 10 gal.
Color with cochineal or orchil.

Good Grade

Infusion of vanilla..................... 2 qt.
Alcohol (85%)........................ 2 gal. 1 qt.
Sugar................................. 20 lb.
Water................................ to make 10 gal.

Average Grade—Double Strength

Infusion of vanilla......................	1 qt.
Alcohol (85%).........................	4 gal. 3 qt.
Sugar................................	20 lb.
Water...............................	to make 10 gal.

Average Grade—Single Strength

Infusion of vanilla......................	1 pt.
Tincture of storax calamite..............	4 oz.
Alcohol (85%).........................	2 gal. 2 qt.
Sugar................................	10 lb.
Water...............................	to make 10 gal.

BITTERS

These are a special group of liqueurs used for their tonic properties and in small portions to flavor other beverages. In general, their manufacture is simple and quality depends on the proper selection of materials and care rather than intricate processing. A few formulae are cited below from the vast number available. These are selected to be sufficiently illustrative. The remainder, the matching of any preparation now on the market is rather a matter for the master of the art than choice from a receipt book.

Angostura Bitters

Angostura bark........................	3¼ lb.
Red sandalwood.......................	3 lb.
Liquorice (wood)......................	1 lb.
Chinese cinnamon......................	¾ lb.
Ginger root...........................	10 oz.
Galanga root..........................	10 oz.
Cardamom............................	10 oz.
Cloves...............................	12 oz.
Orange peel..........................	6 oz.
Mace................................	3 oz.

Cut the materials up finely and macerate them for 8 days in 6 gallons of alcohol (50%). Stir frequently. Add 40 gallons of alcohol (95%) and 8 gallons of sugar syrup. Add water to make up to 100 gallons total volume. Color reddish brown with caramel and tincture of cochineal.

Distilled Bitters

Orange peel	12	lb.
Dutch curaçao bark	12	lb.
Gentian (chopped)	6	oz.
Cinchona bark	2	lb.
Calamus	1	lb. 4 oz.
Cardamom	½	oz.
Lemon peel	2½	lb.
Columba	1¼	oz.
Tangerine peel	6	oz.
Alcohol (96%)	68	gal.
Water	100	gal.

Macerate for 48 hours. Distill to recover 75 gallons at 80%.

Add:

Distillate	75	gal.
Caramel	5	"
Sugar syrup	10	"
Water	110	"
	200	gal.

Fine and filter.

Unicum Bitters

Sugar	80	lb.
Honey	64	lb.
Absinthe	5	oz.
Calamus root	2	oz.
Cinnamon bark	1¼	oz.
Ginger	2	oz.
Orange peel	1¼	oz.
Lemon peel	1	oz.
Centaury	2½	oz.
Gentian	2	oz.
Cinchona bark (red)	1	oz.
Angelica root	1	oz.
Lemon balm	1¼	oz.
Spearmint	1	oz.
Rhubarb	1	oz.
Angostura bark	2½	oz.

Macerate 3-7 days in 100 gallons 42% alcohol. Draw off, fine, and filter.

VERMOUTH

(Sweet or Italian Type)

Absinthe..............................	1 lb.
Gentian...............................	1⅓ oz.
Angelica root..........................	8 oz.
Blessed thistle........................	1 lb.
Calamus root..........................	1 lb.
Starwort..............................	1 lb.
Centaury..............................	1 lb.
Forget-me-not.........................	1 lb.
Cinnamon.............................	12 oz.
Nutmeg...............................	2 oz.
Fresh cut oranges......................	No. 24
Sweet white wine......................	93½ gal.
Alcohol (85%).........................	5¼ gal.

Macerate 5 days. Draw off and fine. Let stand 8 days and fine again. The product is then ready to bottle. Isinglass is preferred for fining.

Vermouth—Dry or French Type

Coriander.............................	4 lb.
Bitter orange peel......................	2 lb.
Orris root (powder)....................	2 lb.
Cinchona bark (red)...................	1 lb. 4 oz.
Calamus..............................	1 lb. 4 oz.
Absinthe..............................	1 lb.
Blessed thistle........................	1 lb.
Star wort.............................	1 lb.
Centaury..............................	1 lb.
Germander............................	1 lb.
Cinnamon.............................	14 oz.
Cloves................................	7 oz.
Quassia...............................	3½ oz.
Dry white wine........................	100 gal.

Grind or crush the herbs, etc. Macerate 5-6 days. Draw off and fine. Let stand 15 days and add 2 gallons of bitter almond shell extract (see below) and three gallons grape brandy. The bitter almond shell extract is made by macerating 1 part bitter almond shells in 2 parts of 85% alcohol for 2 months.

Vermouth—Madeira Type

Absinthe	1 lb.
Angelica root	8 oz.
Blessed thistle	1 lb.
Lung moss	1 lb.
Veronica	1 lb.
Rosemary	1 lb.
Rhubarb	4 oz.
Cinchona bark (red)	1 lb. 12 oz.
Powdered orris root	2 lb.
Curaçao extract (see below)	½ gal.
Madeira wine	91 gal.
Grape sugar	3½ gal.
Old brandy	5½ gal.

Macerate 3 days. Draw off and fine. Age 8 days and fine again. The curaçao extract is prepared by macerating 1 part curaçao peels in 2 parts of 85% alcohol for 8-10 days.

CHAPTER XIII

THE ANALYSIS OF ALCOHOLIC BEVERAGES

INTERPRETATION

General Statement.—Materials in general and alcoholic beverages in particular may be subjected to chemical analysis for any of a great variety of reasons. These include among others (1) analysis by a manufacturer to determine uniformity of each batch of product with preceding batches; (2) analysis to determine the existence of adulteration in the product; (3) analysis to determine compliance with standards of quality such as the Federal Pure Food Standards; (4) analysis to determine the identity of the material (*i.e.*, compliance with definitions); (5) and analysis for the purpose of duplication of the material. Consideration must always be given to the purpose of the analysis before actual selection of the determinations to be made and the methods of making them. Obviously, a manufacturer can check his product from day to day by one or a few simple tests. On the other hand, analyses made to determine the identity or to duplicate a product must necessarily be as complete and as exact as the analytical art will permit. With this possibility in view, the methods of analysis cited in Chapter XIV have been reprinted, without exception, from the Official and Tentative Methods of Analysis of the Association of Official Agricultural Chemists, 3rd ed. 1930. Grateful acknowledgment is here made to the Association for permission to copy. These methods not only have official standing in the courts and with governmental administrative bodies, but they have been written after careful collaborative study so that they are both exact and complete in detail.

Despite the accuracy and reproductibility of the results obtainable by the methods cited, the analytical chemist must confess that his satisfaction of purposes 4 and 5 listed above, is difficult and often impossible of completion. This difficulty arises partly out of the inherent variability of many factors which enter into the composition of alcoholic beverages. Among these are noted especially the raw materials and the bio-chemical reactions of fermentation. Part of the difficulty is also due to the fact that the distinguishing characteristics of alcoholic beverages, flavor,

Fig. 49.—Modern distillery laboratory. (Courtesy American Wine and Liquor Journal.)

smoothness, aroma, etc., are intangible by the analyst and can only be known by sensing them.

However, notwithstanding the difficulties stated, the analyst has compiled data regarding the average composition of wines and distilled beverages. Comparison of the results obtained by the analysis of any given sample with the data so compiled will, therefore, indicate the approach of the sample to the norm for the kind of material it represents and may on occasion so emphasize the abnormality of the material as to establish its non-agreement with the definition.

Definitions.—There can be no discussion of the meaning of analysis of alcoholic beverages without prior agreement on the meaning of the names applied to the beverages, *i.e.*, definitions of the beverages. Since the definitions and standards used by the Food and Drug Administration of the U. S. Department of Agriculture and the labeling requirements of the Federal Alcohol Control Board have the force of law in this country, as well as the merit of presenting definitions in as accurate language as possible, they have been adopted by the authors and are presented here.

Definition of Whiskey.—The Department of Agriculture defines only *medicinal whiskey* and requires that it shall conform to the definition contained in the U. S. Pharmacopoeia. This definition reads as follows: "Whiskey is an alcoholic liquid obtained by the distillation of the fermented mash of wholly or partly malted cereal grains, and containing not less than 47 per cent and not more than 53 per cent by volume of C_2H_5OH at 15.56°C. It must have been stored in charred wood containers for a period of not less than four years.

The pharmacopoeia also sets up certain standards of identity and purity. These are:

Acidity (Calculated as acetic)—36-120 parts per 100,000
Esters (Calculated as ethyl acetate)—30-123 parts per 100,000
Solids (Extract)—not over 500 parts per 100,000
Color—To pass Marsh test for caramel.
Freedom from denaturants such as wood alcohol, diethylphthalate, formaldehyde, etc.

The Federal Alcohol Control Administration in a regulation (Series 4) dated June 13, 1934 defines a number of classes and types of whiskey which include:

Types:
Neutral whiskey, Straight whiskey, Straight rye whiskey, Straight bourbon whiskey, Blended whiskey, Blended rye whiskey, Blended bourbon whiskey, a blend of straight rye whiskeys, a blend of straight bourbon whiskeys, Spirit whiskey, Scotch whiskey, Irish whiskey, Blended Scotch whiskey, Blended Irish whiskey, Special types of whiskey.

The separate definitions of these are:

(a) *Neutral whiskey* is any alcoholic distillate from a fermented mash of grain, distilled at more than 160° proof, and less than 190° proof.

(b) *Straight whiskey* is any alcoholic distillate produced from a fermented mash of grain, distilled at not exceeding 160° proof and withdrawn from the cistern room of the distillery between 110° proof and 80° proof, and produced by the same distillery from the same type of materials, and as part of the same season's distillation, whether or not such proof is reduced prior to bottling.

(c) *Straight rye whiskey and Straight bourbon whiskey* are straight whiskey distilled from a fermented mash of grain in which rye or corn, respectively, are the principal materials.

(d) *Blended whiskey* is a mixture of straight whiskeys, or of straight whiskey or whiskeys and neutral whiskey, or of straight whiskey or whiskeys and neutral spirits distilled from grain, which contains at least 20% of 100° proof straight whiskey by volume.

(e) *Blended rye whiskey, and blended bourbon whiskey* are blended whiskeys in which the whiskey or whiskeys are all rye or all bourbon, respectively.

(f) *A blend of straight whiskeys, A blend of straight rye whiskeys, and A blend of straight bourbon whiskeys* are mixtures composed only of straight whiskeys, straight rye whiskeys, or straight bourbon whiskeys, respectively.

(g) *Spirit whiskey* is a mixture of straight whiskey or whiskeys and neutral whiskey, or of straight whiskey or whiskeys and neutral spirits distilled from grain, which contains at least 5% and less than 20% of 100° proof straight whiskey or straight whiskeys by volume.

(h) *Scotch whiskey* is a distinctive product of Scotland (1) composed of not less than — per cent by volume of straight whiskey or whiskeys distilled therein from a fermented mash of barley malt, and of neutral whiskey distilled therein, and (2) manufactured and blended in compliance with the laws and regulations of the United Kingdom, and (3) containing no whiskey less than three years old.

(i) *Irish whiskey* is a distinctive product of Ireland (1) composed of spirits distilled at approximately 171° proof from a fermented mash of malted barley and unmalted barley and other grains, with or without the addition of other spirits similarly distilled in other seasons by the same distillery, and with or without the addition of not more than — per cent by volume of neutral whiskey distilled at a higher proof, and (2) manufactured (including blending if practiced) in accordance with the laws and regulations of that division of Ireland in which manufactured, and (3) containing no whiskey less than three years old.

(j) *Blended Scotch whiskey or blended Irish whiskey* is Scotch whiskey or Irish whiskey that is in fact a mixture of whiskeys.

(k) *Special types of whiskey* (1) Any person producing any distilled spirits which, as a result of treatment by a chemical or mechanical process, possess the taste, aroma, characteristics and chemical composition of any whiskey for which standards of identity are herein prescribed, may petition the Administration for permission to designate such distilled spirits as whiskey of some new type, and the Administration may take such action on such petition as it deems fair and reasonable. (2) Any whiskey of any class or type prescribed in paragraphs (b) to (g) above, inclusive, produced in a foreign country, shall be designated by the name of the country in which produced, together with the applicable designation prescribed in such paragraphs.

OTHER DISTILLED LIQUORS

The F. A. C. A. classes and definitions are:

Gins

(a) *"Distilled gin" and "compound gin," without appropriate qualifying words,* are distilled gin and compound gin, respectively, in which the predominant flavor is derived from juniper berries.

(b) *Distilled gin* is the product obtained by distilling juniper berries or other similar flavoring materials with neutral spirits.

(c) *Compound gin* is the product obtained by mixing distilled gin or gin essence or similar gin flavoring material with neutral spirits.

(d) *"London Dry," "Hollands," "Plymouth," "Geneva," "Old Tom," "Buchu,"* and *"Sloe"* gin are the types of gin known to the trade under such generic designations, and "distilled gin" or "compound gin," whichever is appropriate, shall accompany such designations.

BRANDIES

(a) *Brandy* is the alcoholic distillate obtained solely from the fermented juice of fruit, distilled under such conditions that the characteristic bouquet or volatile flavoring and aromatic principles are retained in the distillate.

(b) *"Brandy" without appropriate qualifying words, and "Grape Brandy"* are the distillates obtained from grape wine or wines under the conditions set forth in (a).

(c) *Apple Brandy, Peach Brandy, or other fruit brandies* are distillates obtained from the fermented juice of the respective fruits under the conditions set forth in (a).

(d) *Cognac and Cognac Brandy* is grape brandy distilled in the Cognac region of France, which is entitled to be designated as "cognac" by the laws and regulations of the French government.

RUM

(a) *Rum* is any alcoholic distillate obtained solely from the fermented juice of sugar cane, sugar cane molasses, or other sugar cane by-products, in such a manner that the distillate possesses the taste, aroma, characteristic and chemical composition generally attributed to rum, and known to the trade as such.

WINES

The Department of Agriculture definitions of wine have been readopted since the repeal of prohibition from those promulgated on June 12, 1914. They read as follows:

1. *Wine* is the product made by the normal alcoholic fermentation of the juice of sound ripe grapes, and the usual cellar

treatment, and contains not less than 7 per cent or more than 16 per cent of alcohol, by volume, and, in 100 cubic centimeters (20°C.) not more than 0.1 gram of sodium chloride nor more than 0.2 gram of potassium sulphate; and for red wine not more than 0.14 gram, and for white wine not more than 0.12 gram of volatile acids produced by fermentation and calculated as acetic acid. Red wine is wine containing the red coloring matter of the skins of grapes. White wine is wine made from white grapes or the expressed fresh juice of other grapes.

2. *Dry wine* is wine which the fermentation of the sugars is practically complete, and which contains, in 100 cubic centimeters (20°C.), less than 1 gram of sugars and for dry red wine not less than 0.16 gram of grape ash and not less than 1.6 grams of sugar-free grape solids, and for dry white wine not less than 0.13 gram of grape ash and not less than 1.4 grams of sugar-free grape solids.

3. *Fortified dry wine* is dry wine to which brandy has been added but which conforms in all other particulars to the standard of dry wine.

4. *Sweet wine* is wine which the alcoholic fermentation has been arrested and which contains, in 100 cubic centimeters (20°C.), not less than 1 gram of sugars, and for sweet red wine not less than 0.16 gram of grape ash, and for sweet white wine not less than 0.13 gram of grape ash.

5. *Fortified sweet wine* is sweet wine to which wine spirits have been added. By act of Congress, "sweet wine" used for making fortified sweet wine and "wine spirits" used for such fortification are defined as follows (sec. 43, act of Oct. 1, 1890, 26 Stat. 621; as amended by Sec. 68, act of Aug. 27, 1894, 28 Stat. 568; as amended by sec. 1, act of June 7, 1906, 34 Stat. 215; as amended by sec. 2, act of Oct. 22, 1914, 38 Stat. 747; as amended by sec. 402 (c), act of Sept. 8, 1916, 39 Stat. 785; and as further amended by sec. 617 act of Feb. 25, 1919, 40 Stat. 1111):

That the wine spirits mentioned in section 42 is the product resulting from the distillation of fermented grape juice, to which water may have been added prior to, during or after fermentation, for the sole purpose of facilitating the fermentation and economical distillation thereof, and shall be held to include the product from grapes or their residues commonly known as grape brandy, and shall include commercial grape brandy which may have been colored with burnt sugar or caramel; and the pure sweet wine which may be fortified with wine spirits under the provisions of this act is fermented or partially fermented grape juice only, with the usual cellar treatment, and shall contain no other substance whatever introduced before, at the time of, or after fermentation, except as herein expressly provided: *Provided,* That the addition of pure boiled or condensed grape must or pure crystallized cane or beet sugar, or pure dextrose sugar containing, respectively, not less than 95 per centum or actual sugar, calculated on a dry basis, or water, or any or all of them, to the pure grape juice before fermentation, or to the fermented product of such grape juice, or to both, prior to the fortification herein provided for, either for the purpose of perfecting sweet wines according to commercial standards or for mechanical purposes, shall not be excluded by the definition of pure sweet wine aforesaid: *Provided, however,* That the cane or beet sugar, or pure dextrose sugar added for sweetening purposes shall not be in excess of 11 per centum of the weight of the wine to be fortified: *And provided furthur,* That the addition of water herein authorized shall be under such regulations as the Commissioner of Internal Revenue, with the approval of the Secretary of the Treasury, may from time to time prescribe: *Provided, however,* That records kept in accordance with such regulations as to the percentage of saccharine, acid, alcoholic, and added water content of the wine offered for fortification shall be open to inspection by any official of the Department of Agriculture thereto duly authorized by the Secretary of Agriculture; but in no case shall such wines to which water has been added be eligible for fortification under the provisions of this act, where the same, after fermentation and before fortification, have an alcoholic strength of less than 5 per centum of their volume.

6. *Sparkling wine* is wine in which the afterpart of the fermentation is completed in the bottle, the sediment being disgorged, and its place supplied by wine or sugar liquor and/or

dextrose liquor, and which contains, in 100 cubic centimeters (20°C.), not less than 0.12 gram of grape ash.

7. *Modified wine, ameliorated wine, corrected wine,* is the product made by the alcoholic fermentation, with the usual cellar treatment, of a mixture of the juice of sound, ripe grapes with sugar and/or dextrose, or a sirup containing not less than 65 per cent of these sugars, and in quantity not more than enough to raise the alcoholic strength after fermentation to 11 per cent by volume.

Food Inspection Decision 156

As a result of investigations carried on by this Department and of the evidence submitted at a public hearing given on November 5, 1913, the Department of Agriculture has concluded that gross deceptions have been practiced under Food Inspection Decision 120. The department has also concluded that the definition of wine in Food Inspection Decision 109 should be modified so as to permit correction of the natural defects in grape musts and wines due to climatic or seasonal conditions.

Food Inspection Decisions 109 and 120 are, therefore, hereby abrogated and, as guide for the officials of this Department in enforcing the Food and Drugs Act, wine is defined to be the product of the normal alcoholic fermentation of the juice of fresh, sound, ripe grapes, with the usual cellar treatment.

To correct the natural defects above mentioned the following additions to musts or wines are permitted:

In the case of excessive acidity, neutralizing agents which do not render wine injurious to health, such as neutral potassium tartrate or calcium carbonate;

In the case of deficient acidity, tartaric acid;

In the case of deficiency in saccharine matter, condensed grape must or a pure dry sugar.

The foregoing definition does not apply to sweet wines made in accordance with the Sweet Wine Fortification Act of June 7, 1906 (34 Stat. 215).

A product made from pomace, by the addition of water, with or without sugar or any other material whatsoever, is not entitled to be called wine. It is not permissible to designate such a product as "pomace wine," nor otherwise than as "imitation wine."

CORDIALS AND LIQUEURS

Cordials are defined by the Department of Agriculture as follows:

Food Inspection Decision No. 125—July 7, 1910
The Labeling of Cordials

"The term 'cordial' is usually applied to a product, the alcohol content of which is some type of a distilled spirit, commonly neutral spirits of brandy. To this is added sugar and some type of flavor. The flavor is sometimes derived directly by the addition of essential oils, again by use of synthetic flavors, and also by the treatment of some vegetable product with the alcoholic spirit to extract the flavoring ingredients. It is likewise the general custom to color cordials. When a cordial is colored in such a way as to simulate the color of the fruit, flavor, plant, etc., the name of which it bears, the legend 'Artificially Colored' in appropriate size type shall appear immediately beneath the name of the cordial, as is required by Regulation 17. Where the color used is not one which simulates the color of a natural product, the name of which is borne by the liqueur, then the legend as to the presence of artificial color need not be used. For example, creme de menthe which is artificially colored green should be labeled 'Artificially Colored.' On the contrary, chartreuse, whether green or yellow, need bear no such legend for color.

"When the flavoring material is not derived in whole directly from a flower, fruit, plant, etc., the name of any such flower, fruit, plant, etc., should not be given to any cordial or liqueur unless the name is preceded by the word 'Imitation.' "

The F. A. C. A. definition of cordials and liqueurs is:

(a) *Cordials and Liqueurs* are the products obtained by distilling fruits, flowers, plants, leaves, roots, or other flavoring materials, except gin flavoring materials, with brandy or neutral spirits, and to which sugar has been added; or the products obtained by mixing fruit juices, essential oils and flavoring materials, other than gin flavoring materials, with brandy or with neutral spirits and to which sugar has been added.

Imitations.—The attitude of the Department of Agriculture has always been that any product which simulated those for which it has definitions and standards, but does not fully comply therewith, is an imitation. The F. A. C. A. has actually set forth definitions of imitations.

(a) *Imitation whiskey* is any distilled spirits containing rye or bourbon essential oils or essences, or any distilled spirits colored or flavored in imitation of whiskey; and the designation "Whiskey" shall not be used unless immediately preceded by "Imitation."

(b) *Imitation cordials and liqueurs.*—When the flavoring material of a cordial or liqueur is not derived in whole directly from a fruit, flower, plant, leaf, root or other flavoring material, the name of any such fruit, flower, plant, leaf, root or other flavoring material shall not be given to the cordial or liqueur unless the name is immediately preceded by the word "Imitation."

(c) *Imitations,* other than (a) and (b) above, are distilled spirits colored or flavored in imitation of any class or type of distilled spirits defined in these Regulations, and the name of such class or type of distillated spirits shall not be used unless immediately preceded by the word "Imitation."

Finally, the F. A. C. A. has set forth certain regulations of interest, regarding geographical names and additions of coloring, etc.

Section I. (a) *Geographical and distinctive names.* The name for distilled spirits which have a geographical name, or which are distinctive products of a particular place or country shall not be given to the product of any other place or country, unless such name is immediately followed by the word "type," and unless such product in fact conforms to such distilled spirits except as to age.

(b) This section shall not apply to designations which by reason of usage and common knowledge have lost their geographical significance to such an extent that they have become generic, provided the approval of the Administration is obtained prior to using such designation.

Section 2. *Coloring and Flavoring Materials.* The addition of harmless coloring or flavoring materials, such as burnt sugar and blending materials, in a total amount not in excess of 2½% of the distilled spirits by volume, shall not alter the classification of any distilled spirits as defined in these Regulations, provided such coloring or flavoring materials are not used to imitate any class or type of distilled spirits for which standards of identity are established herein. This section shall not affect cordials or liqueurs.

It will have been noted by the reader that the definitions just cited are in general based on the principle of stating that a product bearing a given name must be made from the materials commonly used in the manufacture of that product, by the commonly understood processes of manufacture. *It is only as a corollary that one may deduce that the composition of the product must conform to the general average of the type within reasonable limits.* In order that the reader may judge for himself the fact of this compliance, a selection of the published analytical data regarding liquors is presented here. The tabulated analyses of whiskey were made at the end of the first decade of this century. The "pure food" movement both here and abroad was then at its crest, and the question, "What is whiskey" was considered at length by legal bodies. The analyses presented were generally used as evidence in public hearings.

ANALYSES OF WHISKEY

British.—Shidrowitz (Royal Commission of Whiskey and other Potable Spirits. Minutes of Evidence, Vol. 1, pages 409, 410, 1909) reports analyses of Scotch and Irish pot and patent still whiskeys and of American Bourbons and Ryes as follows:

TABLE X.—ANALYSES OF PATENT STILL WHISKIES (SCHIDROWITZ)

I.—*Scotch Samples Sold and Consumed Unblended. Sold under Name of Distillery*

	Alcohol per cent by volume	Extract per cent	Total acid	Non-volatile acid	Ethers	Higher alcohols (A.-M.)	Aldehydes	Furfural
Sample A.—Labelled "Three year old grain whiskey. Cameron Bridge Distillery, Fife."—The name and address of the retailer (in Glasgow) also appears on the label. Sold in bottle. Purchased September, 1905;	45.48	0.11	16	4	52	58	13	0.18
Sample B.—Labelled "Distilled in 1884. A perfectly pure spirit for medicinal purposes. 'Cambus' grain whiskey."—The name of the retailer (in Glasgow) also appears on the label. Purchased September, 1905;	44.79	0.10	32	6	44	62	26	0.31
Sample C.—Taken from Prof. Dewar's Private Cask at the Royal Institution, November, 1905. (This is *Cambus* whiskey)	57.06	0.12	50	40	45	47	9	Trace

Note.—I have personally seen at the retailers in Glasgow (as early as 1904) samples labelled as A and B.

II.—*Scotch Patent Still Whiskies Made from Malt Only (Schidrowitz)*

	Alcohol per cent by volume	Extract per cent	Total acid	Non-volatile acid	Ethers	Higher alcohols (A.-M.)	Aldehydes	Furfural
Distillery No. 49 (J. Soc. Chem. Ind., June, 1905). —New Whiskey	94.77	......	3	......	42	70	3	Nil
Distillery No. 30.—New Whiskey	62.1	......	8	......	44	88	9	Nil

Note.—Except where otherwise stated, results are in grams per 100 litres of absolute alcohol.

TABLE X (Continued)—Set of Samples from Scotch Patent Still Distillery (Schidrowitz)
(The produce of this distillery had not previously been examined)

	Alcohol per cent by volume	Extract per cent	Total acid	Non-volatile acid	Ethers	Higher alcohols (A.-M.)	Aldehydes	Furfural
(a) New Whiskey	60.34	......	4	......	25	65	2	0.07
(b) 4 months Sherry Wood	60.76	0.059	24	10	26	31	4	Trace
(c) 2 years Plain Wood	59.32	0.026	14	3	25	100	5	
(d) 2 years Sherry Wood	59.68	0.244	58	34	40	44	7	0.15

IV.—Irish Patent Still Whiskies (Schidrowitz)

	Alcohol per cent by volume	Extract per cent	Total acid	Non-volatile acid	Ethers	Higher alcohols (A.-M.)	Aldehydes	Furfural
Distillery A.—New	70.76	......	17	......	25	38	......	Trace
" "	69.87	......	21	......	41	50	......	Trace
" 2 years	70.72	0.01	10	4	24	48	3	
Distillery B.—New	71.94	......	3	......	19	52	......	
" 1½ years	69.34	0.007	9	1	25	46	1	Trace
Distillery C.—New	71.6	......	......	33	33	50	11	Nil

Note.—Except where otherwise stated, results are in grams per 100 litres of absolute alcohol.

TABLE X (*Continued*).—ANALYSES OF IRISH POT STILL WHISKIES (SCHIDROWITZ)

	Alcohol per cent by volume	Extract per cent	Total acid	Non-volatile acid	Ethers	Higher alcohols (A.-M.)	Aldehydes	Furfural
Distillery A.—New	71.72	......	7	......	34	145	12	5.5
" 13 years plain wood	57.08	0.04	29	8	38	185	68	3.3
Distillery B.—New	74.11	......	6	......	28	233	8	4.1
" 13 years plain	60.47	0.05	32	8	47	264	21	4.4
Distillery C.—14 years sherry	63.42	0.29	87	44	87	226	32	4.5
Distillery D.—Mature	46.64	0.13	67	26	59	147	16	2.6
AMERICAN WHISKIES (SCHIDROWITZ)								
Bourbon Whiskey A.—5 years	49.40	0.14	126	40	99	197	11	2.2
" B.—Mature	47.78	0.56	122	30	78	129	14	3.0
Rye Whiskey A.—7 years	56.31	0.18	140	49	134	277	20	3.9
" B.—Mature	49.32	0.16	160	49	141	268	18	3.4
" C.—Mature	46.08	0.16	135	31	125	187	21	3.9
" D.—Mature	46.56	0.22	82	21	71	150	13	3.6
" E.—10 years	44.37	0.60	70	18	79	98	11	2.7
Doubtful	45.99	0.21	63	16	69	110	14	2.2

Note re American Whiskies.—All the above samples were retail, bottle goods, excepting Rye A, which was received from Dr. Wiley. Both the Bourbon samples and Rye samples A, B, and C, are pot still whiskies, the others probably blends.

Note.—Except where otherwise stated, results are in grams per 100 litres of absolute alcohol.

TABLE X (Continued)

The average composition of New and Old Bourbon, and New and Old Rye whiskies (American), based on a large number of analyses of samples examined in the laboratory of the *Department of Agriculture*, U.S.A., are given below. These analyses were kindly transmitted to me by Dr. Wiley, chief chemist of the Department.

		Extract	Ash	Total acid	Non-volatile acid	Ethers	Alde-hydes	Furfural	Higher alcohols (A.-M.)
New Rye Whiskies (19 samples)	Average	12.0	5.2	11.4	2.2	50.6	17.8	2.4	266.4
	Maximum	27.0	13.6	28.2	2.4	132.8	86.0	5.2	378.4
	Minimum	2.0	0.0	2.6	0.0	8.8	0.0	0.0	178.0
Old Rye Whiskies (76 samples)	Average	363.2	17.8	147.6	21.4	137.0	27.6	4.6	352.0
	Maximum	456.0	38.2	312.4	33.4	287.4	47.2	14.0	585.8
	Minimum	226.5	2.0	72.8	13.4	60.0	4.4	0.8	146.6
New Bourbon Whiskies (18 samples)	Average	16.0	5.2	16.4	0.0	45.0	7.8	Except in 5 cases	229.4
	Maximum	110.0	35.4	39.2	0.0	112.8	36.2	Traces only—6.8;	343.4
	Minimum	2.8	0.0	3.6	0.0	1.8	0.0	6.4; 1.0; 12.8; 4.0	70.6
Old Bourbon Whiskies (75 samples)	Average	308.6	21.8	137.6	16.8	103.6	19.4	3.8	287.2
	Maximum	456.6	161.6	203.8	24.0	209.0	46.4	16.2	517.0
	Minimum	94.6	0.0	69.4	4.6	39.0	7.2	1.4	50.4

Note.—Except where otherwise stated, the figures represent grams per 100 litres of absolute alcohol, i.e., parts per 100,000 parts of alcohol.

TABLE X (*Continued*).—EXAMINATION OF 10 AUTHENTICATED SAMPLES (FROM 9 DIFFERENT DISTILLERIES) OF SCOTCH POT STILL WHISKIES BY DR. TEED'S METHOD.

Nature of Sample	Alcohol per cent by volume	Grams per 100 Litres Absolute Alcohol					Total coefficient
		Volatile acid	Ethers	Higher alcohols (Teed)	Aldehydes	Furfural	
1. Highland Malt (New)............	63.5	16	78	200	11	2.4	307.4
2. Highland Malt (Mature).........	60.8	37	73	208	9	2.4	329.4
3. Highland Malt (New)............	63.5	11	89	201	7	2.1	310.1
4. Lowland Malt (New).............	63.3	16	64	153	10	3.6	246.6
5. Lowland Malt (Mature).........	59.1	40	102	187	28	4.5	361.5
6. Highland Malt (New)............	63.7	15	52	325	6	1.6	399.6
7. Lowland Malt (New).............	63.1	15	52	363	14	2.6	426.6
8. Campbeltown (Mature) *........	59.8	32	63	405	40	8.0	548.0
8A. Campbeltown (Mature) *........	60.9	72	70	383	30	8.0	563.0
9. Campbeltown (New).............	62.1	23	61	325	36	6.1	451.1

The above constitute *all* the authenticated Scotch pot still whiskies which I have examined by this method. The figures for the "coefficient" vary from 246.6 to 563.0, and 5 samples are below, 5 above, the proposed 380 standard.

* Samples 8 and 8A were from the same distillery, and of the same age (6 years). Sample 8 was stored in " plain," sample 8A in " sherry " wood.

The maximum and minimum limits obtained by Schidrowitz on 58 samples of Scotch whiskey regardless of age are shown in Table XI. (Royal Comm. on Whiskey, etc., Minutes of Evidence, Vol. 1, p. 416, 1909.)

TABLE XI.—MAXIMUM AND MINIMUM RESULTS OF ANALYSES ON FIFTY-EIGHT SAMPLES OF SCOTCH WHISKEY. ANALYSES BY SCHIDROWITZ.

Whiskey	Total acid	Non-volatile acid	Ether	Higher alcohols, color-imetric method	Higher alcohols, Allen-Marquardt method	Alde-hydes	Fur-fural
Highland Malts...	10–83	0–35	33–185	328–864	112–235	4–66	1.6–6.3
Lowland Malts....	6–60	0–16	27–87	189–897	82–228	8–54	0–5.2
Campbeltowns....	12–100	0–28	53–140	357–930	160–259	11–85	2.4–8.0
Islays..........	15–36	0–33	40–86	620–740	155–200	17–40	3.8–5.2
Grains..........	3–69	0–26	20–55	39–400	33–80	tr.–17	0–0.9

Note: results are calculated to a basis of parts per 100,000 of absolute alcohol.

Another series of analyses made by Tatlock on 75 samples of Scotch, Irish and American whiskies are shown in Table 12 (Tatlock, Royal Comm. on Whiskey, etc., Minutes of Evidence, Vol. 1, p. 431, 1909.)

TABLE XII.—MAXIMUM, MINIMUM, AND AVERAGE ANALYSES OF 75 SAMPLES OF WHISKEY, 31 MALT-POT-STILL (HIGHLAND), 9 MALT-POT-STILL (LOWLAND), 4 MALT-POT-STILL (ISLAY), 2 MALT-POT-STILL (CAMPBELTOWN), 2 MALT-POT-STILL (IRISH), 4 AMERICAN, 7 GRAIN PATENT STILL, AND 16 BLENDS.

	Highlands, 31 samples	Lowlands, 9 samples	Islays, 4 samples	Campbeltowns, 2 samples	Irish, 2 samples	American, 4 samples	Grains, 7 samples	Blends, 16 samples
Ethers:								
Maximum	154.16	85.80	78.05	85.62	44.22	119.52	50.90	92.05
Minimum	45.01	34.52	48.31	61.42	38.42	111.06	22.04	33.56
True average	83.06	66.19	61.51	73.52	41.32	115.99	33.26	58.96
Higher alcohols:								
Maximum	190.40	240.66	161.70	206.70	169.10	177.34	68.47	111.25
Minimum	73.51	103.49	130.20	139.58	154.93	102.55	32.47	48.71
True average	144.61	140.40	144.55	173.14	162.01	156.99	45.32	80.76
Volatile acids:								
Maximum	89.06	39.86	32.41	56.35	49.23	132.11	32.83	100.13
Minimum	21.76	4.70	17.11	31.16	4.55	65.24	2.86	13.12
True average	41.78	17.35	25.86	43.75	26.89	99.43	16.67	40.61
Aldehydes:								
Maximum	35.06	33.59	31.74	33.54	21.12	32.85	12.95	26.60
Minimum	3.21	4.70	19.82	32.45	4.34	17.01	1.59	1.26
True average	19.48	20.42	28.40	32.99	12.73	26.53	7.66	9.29
Furfural:								
Maximum	7.34	5.85	6.29	3.77	3.02	3.86	.50	3.17
Minimum	1.58	2.65	3.25	3.23	2.73	1.95	.10	.43
True average	4.19	4.39	5.02	3.50	2.87	3.00	.25	1.63
Total secondary products:								
Maximum	382.35	314.81	294.99	385.98	269.62	452.92	130.38	267.64
Minimum	218.94	218.52	225.94	267.84	222.04	328.76	68.04	106.22
True average	294.41	282.03	265.37	326.91	245.83	396.89	102.08	191.25
Higher alcohols, Dr. Teed's process:								
Maximum	400.00*	350.00†	285.00	260.00	320.00‡	320.00	268.00§	290.00¶
Minimum	140.00	190.00	200.00	230.00	320.00	160.00	50.00	105.00
True average	225.00	269.00	244.00	245.00	320.00	230.00	140.00	190.57
Total secondary products, Dr. Teed's process:								
Maximum	566.67	480.55	410.21	439.28	387.11	595.58	342.75	411.12
Minimum	250.44	328.18	324.06	358.26	387.11	386.21	124.79	231.20
True average	411.08	396.68	364.57	398.77	387.11	474.95	200.09	318.90

* 20 samples. † 8 samples. ‡ 1 sample only. § 16 samples. ¶ 8 samples.

ANALYSES OF WHISKEY

American.—There are available two extensive sets of analyses of American Whiskies completed in the years 1909–1912. The first set made by Crampton and Tolman (J. A. C. S. V. XXX (1908), 98 et. seq.) was made in an investigation of the effect of aging on whiskey. They drew samples each year for 4-8 years from the same barrels of whiskey stored in warehouses and subjected these to extensive comparison. A summary of the results follows.

It is of great importance in the interpretation of whiskey analyses to note the conclusions obtained by Crampton and Tolman from their investigation.

"1. There are important relationships among the acids, esters, color, and solids in a properly aged whiskey, which will differentiate it from artificial mixtures and from young spirit.

2. All of the constituents are undergoing changes as the aging process proceeds, and it is evident that the matured whiskey is the result of these combined changes.

3. The amount of higher alcohols increases in the matured whiskey only in proportion to the concentration.

4. Acids and esters reach an equilibrium, which is maintained after about three or four years.

5. The characteristic aroma of American whiskey is derived almost entirely from the charred package in which it is aged.

6. The rye whiskies show a higher content of solids, acids, esters, etc., than do the Bourbon whiskies, but this is explained by the fact that heated warehouses are almost universally used for the maturing of rye whiskies, and unheated warehouses for the maturing of Bourbon whiskies.

7. The improvement in flavor of whiskies in charred packages after the fourth year is due largely to concentration.

8. The oily appearance of a matured whiskey is due to material extracted from the charred package, as this appearance is almost lacking in whiskies aged in uncharred wood."

TABLE XIII.—AVERAGE, MAXIMA AND MINIMA DATA ON RYE WHISKIES.
Calculated to 100 Proof.

Age	Data	Original proof	Color	Solids	Acids	Esters	Aldehydes	Furfural	Fusel oil
New	Average	101.2	0.0	13.3	4.4	16.3	5.4	1.0	90.4
	Maximum	102.0	0.0	30.0	7.2	21.8	15.0	1.9	161.8
	Minimum	100.0	0.0	5.0	2.0	4.3	0.7	Trace	61.8
1 yr.	Average	102.5	8.8	119.7	46.6	37.0	7.0	1.8	111.5
	Maximum	104.0	13.8	171.0	60.5	64.8	15.5	3.3	194.0
	Minimum	101.0	7.2	93.0	33.1	6.8	2.8	0.4	80.4
2 yrs.	Average	104.9	11.6	144.7	51.9	54.0	10.5	2.2	112.4
	Maximum	109.0	16.7	199.0	75.6	75.1	18.7	5.7	214.0
	Minimum	100.0	8.8	121.0	44.3	41.5	5.4	0.7	83.4
3 yrs.	Average	107.7	13.2	171.4	62.7	61.5	12.5	1.5	112.7
	Maximum	112.0	18.3	224.0	81.8	79.8	20.8	6.1	202.0
	Minimum	104.0	11.4	145.0	52.3	47.6	6.5	0.7	79.0
4 yrs.	Average	111.2	14.0	185.0	65.9	69.3	13.9	2.8	125.1
	Maximum	118.0	18.9	238.0	83.8	89.1	22.1	6.7	203.5
	Minimum	105.0	11.6	156.0	58.6	57.7	6.4	0.7	83.8
5 yrs.	Average	113.8	15.9	206.5	67.6	75.0	15.0	3.2	128.1
	Maximum	125.0	19.2	251.0	92.6	105.5	22.4	7.7	254.2
	Minimum	108.0	13.2	170.0	59.4	60.5	6.6	1.4	86.8
6 yrs.	Average	118.0	17.0	223.1	72.4	80.4	14.6	3.3	145.5
	Maximum	132.0	21.2	284.0	95.8	109.0	22.3	8.3	245.3
	Minimum	110.0	13.7	193.0	67.1	64.0	7.3	0.7	99.2
7 yrs.	Average	121.4	18.0	242.2	76.7	84.2	15.5	3.2	145.2
	Maximum	141.0	22.7	306.0	100.0	114.9	25.2	8.5	264.5
	Minimum	111.0	14.6	195.0	60.9	64.6	7.5	0.8	99.2
8 yrs.	Average	123.8	18.6	256.0	82.9	89.1	16.0	3.4	154.2
	Maximum	132.0	24.2	339.0	112.0	126.6	26.5	9.2	280.3
	Minimum	112.0	13.7	214.0	73.7	68.4	7.9	0.8	109.0

Calculated to 100 Proof.

Age	Data	Proof	Color	Solids	Acids	Esters	Aldehydes	Furfural	Fusel oil
New	Average	101.1	0.0	26.5	10.0	18.4	3.2	0.7	100.9
	Maximum	104.0	0.0	161.0	29.1	53.2	7.9	2.0	171.3
	Minimum	100.0	0.0	4.0	12.0	13.0	1.0	Trace	*71.3*
									42.0
1 yr.	Average	101.8	7.1	99.6	41.1	28.6	5.8	1.6	110.1
	Maximum	103.0	10.9	193.0	55.3	55.9	8.6	7.9	173.4
	Minimum	100.0	5.4	61.0	24.7	77.2	2.7	Trace	*58.0*
			4.6	*54.0*	*7.2*	*10.4*	*2.7*	Trace	*42.8*
2 yrs.	Average	102.2	4.6	54.0	7.2	10.4	2.7	1.6	42.8
	Maximum	104.0	8.6	126.8	45.6	40.0	8.4	1.6	108.9
	Minimum	100.0	11.8	214.0	61.7	59.8	12.0	9.1	197.1
			6.9	81.0	25.5	24.4	5.9	0.4	*86.2*
yrs.	Average	103.0	5.7	78.0	23.3	11.2	5.9	0.4	42.8
	Maximum	106.0	10.0	149.3	54.3	48.1	10.5	1.7	112.4
	Minimum	100.0	13.8	245.0	64.8	73.0	22.1	9.5	221.8
			8.9	*95.0*	38.4	27.2	5.9	0.6	*88.0*
4 yrs.	Average	104.3	7.0	90.0	32.1	12.1	5.9	0.6	43.5
	Maximum	108.0	10.8	151.9	58.4	53.5	11.0	1.9	123.9
	Minimum	100.0	14.8	249.0	73.0	80.6	22.2	9.6	237.1
			8.6	101.0	40.4	28.2	6.9	0.8	*95.0*
5 yrs.	Average	106.1	7.4	92.0	40.4	13.8	6.9	0.8	43.5
	Maximum	113.0	12.3	173.3	56.3	55.9	11.4	1.9	125.3
	Minimum	101.0	16.7	280.0	78.9	87.2	23.1	9.6	243.4
			11.8	*125.0*	48.2	27.7	7.1	0.8	*98.0*
6 yrs.	Average	107.9	8.4	114.0	42.7	17.3	7.1	0.8	45.1
	Maximum	116.0	13.1	185.1	67.1	64.0	11.9	1.8	135.3
	Minimum	102.0	17.5	287.0	81.0	83.9	23.3	9.5	240.0
			12.0	132.0	53.6	30.4	7.7	0.9	*98.1*
7 yrs.	Average	109.6	9.8	127.0	45.0	17.9	7.7	0.9	44.6
	Maximum	120.0	13.9	200.9	71.9	63.3	12.4	1.9	137.2
	Minimum	103.0	19.4	309.0	86.4	90.0	26.7	8.3	243.4
			11.8	140.0	60.6	37.1	7.7	0.9	*98.2*
8 yrs.	Average	111.1	10.1	134.0	49.0	21.3	7.7	0.9	46.6
	Maximum	124.0	14.2	210.3	76.4	65.6	12.9	2.1	143.5
	Minimum	102.0	20.9	326.0	91.4	93.6	28.8	10.0	241.8
			12.3	*152.0*	64.1	37.7	8.7	1.0	*110.0*
			10.5	141.0	53.7	22.1	8.7	1.0	47.6

Note: Since some of the minimum figures shown are derived from an abnormal sample, the next to minimum results are included in the table.

"9. The 'body' of a whiskey, so-called, is due largely to the solids extracted from the wood."

Adams J. Ind. & Eng. Ch. 3, 647 (1911) reports the results of an "Investigation to detect substitution of spirits for aged whiskey." His analytical results were similar to those of Crampton and Tolman (*loc. cit.*), but his conclusion merits repetition. This is particularly the case because the question was passed on by a Federal Court and his conclusions approved by the Court. He states "In conclusion it should be stated that in work of this kind, the acids, esters and color form the points which should be used to determine the authenticity of the contents of a package of whiskey. The content of solids, higher alcohols, aldehydes and furfural will assist in arriving at a conclusion, but should not be relied on solely as can be done in the case of the acids, esters, and color."

Analyses of Typical Brandies.—Girard and Cuniasse ("Manuel Pratique de L'Analyse des Alcools," 1899) state that the sum of the secondary constituents, referred to as the "coefficient of impurity," is seldom less than 300 in genuine brandy made from wine. They give various analyses of brandies from which the following have been chosen:

TABLE XV.—ANALYSIS OF EAUX-DE-VIE DE VIN OF KNOWN ORIGIN.*

	Cozes, 1874	Gemozac, 1893	Gemozac, 1896	Champagne, 15–20 yrs.
Acidity....................	201.0	100.4	29.0	111.8
Aldehydes.................	46.0	32.3	11.5	30.0
Furfural...................	0.4	0.8	1.2	1.6
Esters.....................	95.1	139.1	101.3	137.2
Higher alcohols............	254.0	221.7	260.0	244.0
Total secondary constituents *.	596.5	494.3	403.0	524.6
Density at 15..............	0.9571	0.9075	0.9022	.....
Alcohol, by volume..........	37.0	64.5	66.0	59.0
Extract, per 100 cc..........	....	....	....	0.2

* Expressed as parts per 100,000 of absolute alcohol.

TABLE XVI.—ANALYSIS OF EAUX-DE-VIE OF KNOWN ORIGIN. *

(*Analyzed in 1896*)

	Saintonge, 1880	Saintonge, 1896	Armagnac, 40 yrs.
Acidity........................	105.7	17.5	146.7
Aldehydes......................	27.9	23.9	31.4
Furfural.......................	2.3	2.6	0.7
Esters.........................	167.0	61.9	125.5
Higher alcohols................	159.8	259.8	203.5
Total secondary constituents *......	462.7	365.7	507.8
Density at 15..................	0.9157	0.8947	.....
Alcohol by volume.............	59.0	68.2	49.8
Extract, per 100 cc...........	....	Nil	0.18

* Expressed as parts per 100,000 of absolute alcohol.

These analyses tend to demonstrate the following changes brought about by aging:

(1) Increase in acidity
(2) Increase in aldehydes
(3) Increase in esters
(4) Decrease in furfural
(5) Decrease in higher alcohols
(6) Decrease in total alcohol

Other analyses of unidentified brandies listed as commercial cognacs and thought to be wine brandy cut with rectified alcohol showed total secondary constituents ranging from 202 to 283. These were compared with analyses of industrial alcohol (alcools d'industrie) showing total secondary constituents ranging from 9 to 40.9.

Ordonneau (Compt. rend. 102, 217) subjected 100 litres of 25-year-old brandy to fractional distillation and reports the following:

TABLE XVII.—ANALYSIS OF 25-YEAR OLD BRANDY.

	Grams per Hectolitre
Aldehyde................................	3.0
Normal propyl alcohol....................	40.0
Normal butyl alcohol.....................	218.6
Amyl alcohol............................	83.8
Hexyl alcohol...........................	0.6
Heptyl alcohol..........................	1.5
Acetic ester............................	35.0
Propionic, butyric and caproic esters.....	3.0
Oenanthic ester (about).................	4.0
Acetal and amines......................	Tr.

W. Collingwood Williams (J. Soc. Chem. Ind., 1907, *26*, 499) gives results of analyses of 28 samples of Jamaica rum as follows:

TABLE XVIII.—ANALYSES OF JAMAICA RUM.*

	Min.	Max.	Average
Ordinary type. 21 samples			
Alcohol, vol. per cent........................	68.6	82.1	79.1
Total solids, gms/100 cc.....................	0.1	1.16	0.43
Total acid as acetic..........................	30	155	78.5
Volatile acid.................................	21	146	61
Esters.......................................	88	1058	366.5
Higher alcohols..............................	46	150	98.5
Furfural.....................................	1.0	11.5	4.5
Aldehydes...................................	5.0	30.0	15.3
Flavored rum. 7 samples			
Alcohol, vol. per cent........................	66.1	80.6	77.3
Total solids, gms/100 cc.....................	Nil	0.61	0.31
Total acid as acetic..........................	45	145	102.5
Volatile acid.................................	39	137	95.5
Esters.......................................	391	1204	768.5
Higher alcohols..............................	80	144	107
Furfural.....................................	2.7	12.0	5.2
Aldehydes...................................	13.0	37.5	20.7

* Results expressed as grams per 100 litres of alcohol (except the alcohol and solids).

J. B. Harrison (Official Gazette, Demerara, Oct. 19, 1904, Extract, 2,093) Government Analyst, believes that a characteristic commercial Demerara rum would yield 70 to 80 parts of esters per 100,000 of alcohol. He gives figures for various Demerara rums as follows.

TABLE XIX.—ANALYSES OF DEMERARA RUM.

Origin	AVERAGE VALUES	
	Esters	Vol. acid
Distilleries in Demerara County..........................	54.7	26.5
" " Essequilo "	79.5	33.0
" " Berbice "	78.3	37.0
Continuous and Coffey stills............................	44.9	18.4
Vat stills...	69.9	33.1

Bonis (Ann. Falsif. 1909, *12*, 521) gives the following results of analyses of Martinique rum:

TABLE XX.—ANALYSES OF MARTINIQUE RUM.

	Vol. acid	Esters	Aldehydes	Higher alcohols	Furfural
High quality	201	443	92	68	8.8
	201	91	59	385	5.3
	174	93	32	425	11.0
	165	62	34	339	0.9
Average quality	173	83	20	244	0.5
	145	118	23	167	6.3
Poor quality	197	95	16	97	3.8
	158	90	15	143	0.1
	53	51	10	280	0.7

On fixed acids, the first sample showed 2.2 per cent and the balance ranged from 0.37 to 0.95 per cent.

Girard and Cuniasse ("Analyse des Alcools et des spiritueux") claim the average proportion of esters and other secondary products comprised in the non-alcohol coefficient found in ordinary rums of commerce known to be genuine, are shown in the following table, expressed as parts per 100,000 parts of absolute alcohol:

TABLE XXI.—AVERAGE PROPORTION OF ESTERS AND OTHER SECONDARY PRODUCTS AND
THE NON-ALCOHOL COEFFICIENT OF COMMERCIAL RUM.

	Min.	Max.	Average
Acidity (as acetic acid)............................	158.4	400.0	250.8
Aldehydes (as acetaldehyde)......................	0.3	54.5	24.4
Furfural..	1.2	50.0	7.1
Esters (as acetic ester)............................	105.7	443.1	229.5
Higher alcohols (isobutyl standard)...............	52.0	308.6	140.4
Coefficient of secondary products.................	432.7	919.0	652.5

Analyses of Gin.—Since gin consists of a highly rectified alcohol or spirit with flavor, little can be learned from its analysis. However, Vasey ("Analysis of Potable Spirits," p. 85) cites the following results:

TABLE XXII.

Volatile acids....... 0.0 grams per 100 liters of absolute alcohol
Esters............. 37.3 " " " " " " "
Aldehydes.......... 1.8 " " " " " " "
Furfural........... 0.0 " " " " " " "
Fusel oil.......... 44.6 " " " " " " "

Analyses of Wine.—There are reprinted here two extensive tabulations of the analyses of European and American wines respectively. The European Wine analyses (Table XXIII) were compiled by König and are copied from Leach ("Food Inspection and Analysis," 4th ed., 1920, p. 717). The American Wine Analyses (Table XXIV) were compiled by Bigelow and are copied from Leach (*loc. cit.*, p. 718).

Bioletti ("Principles of Wine Making," California Ag. Exp. Sta., Bul. No. 213) summarizes the composition of wines as follows:

The alcohol and acid in natural wines vary in an inverse ratio, in such a way that the volume percentage of alcohol added to the grams per liter of acid as sulfuric make a sum lying be-

TABLE XXIII.—AVERAGE COMPOSITION OF EUROPEAN WINES.

Grams per 100 cc.

	Number analyzed	Specific gravity	Alcohol	Extract	Total acids as tartaric	Volatile acids as acetic	Tartaric acid	Sugar	Glycerin	Ash	Potassium bitartrate	Free tartaric acid	Nitrogen	Phosphoric acid (P$_2$O$_5$)	Sulphoric acid (SO$_2$)	Carbon dioxide (CO$_2$)
RED DRY WINES																
Bordeaux (Claret)	44	0.9958	8.16	2.42	0.58	0.10	...	0.23	0.73	0.25	...	...	...	0.029	...	...
North Italy	76	0.9958	9.25	2.57	0.76	0.11	...	0.16	0.70	0.25	0.257	0.050	0.015	0.030	0.034	...
Central Italy (Chianti, etc.)	35	0.9946	10.29	2.91	0.72	0.12	...	0.12	0.84	0.26	0.159	0.043	0.021	0.030	...	...
Hungary	47	0.9952	9.15	2.62	0.68	...	...	...	0.85	0.22	...	...	0.033	0.036	0.026	...
Tyrol	88	0.9934	8.91	2.16	0.58	0.07	...	0.15	0.69	0.23	...	...	0.035	0.028	0.023	...
WHITE DRY WINES																
Moselle and Saar	187	0.9963	7.36	2.31	0.77	0.05	0.34	0.20	0.66	0.16	0.171	0.071	0.061	0.033	0.017	...
Rhine and Main	68	0.9977	8.12	2.91	0.77	0.05	0.18	0.23	0.85	0.20	0.130	0.017	...	0.045	0.014	...
Lower Rhine	17	0.9959	7.99	2.39	0.75	0.04	...	0.15	0.72	0.15	...	...	...	0.028	0.017	...
Rhine-Hesse	116	0.9900	7.42	2.15	0.58	0.04	0.19	0.08	0.63	0.22	...	...	...	0.025	0.008	...
Pfalz	129	0.9946	8.54	2.26	0.64	0.05	0.19	0.13	0.71	0.21	...	...	...	0.032	0.022	...
Franconia	319	0.9972	7.01	2.17	0.69	0.05	0.21	0.07	0.64	0.19	...	0.015	...	0.032	0.022	...
Alsace	242	0.9961	6.44	1.92	0.64	...	0.23	0.09	0.53	0.22	0.134	0.026	...	0.026	...	...
France (Sauterne)	5	0.9963	9.48	3.03	0.66	0.09	...	0.84	0.97	0.25	...	...	...	0.032	0.036	...
Hungary	86	0.9950	8.97	2.45	0.55	...	...	0.25	0.94	0.20	...	...	0.016	0.048	0.028	...
SWEET WINES																
Spain (Sherry)	25	0.9932	16.09	4.06	0.41	...	...	2.40	0.51	0.46	0.071	...	...	0.028	0.186	...
Portugal (Port)	15	1.0088	16.18	8.25	0.42	0.09	...	6.04	0.34	0.22	...	...	...	0.035	0.023	...
Madeira	13	0.9996	14.47	5.23	0.49	...	...	2.95	0.67	0.25	...	...	...	0.052	0.067	...
Tokay, Real	51	1.0354	11.19	12.72	0.60	0.10	...	9.01	1.11	0.27	0.079	...	...	0.070	0.015	...
Tokay, Commercial	57	1.0767	9.93	23.76	0.65	0.16	...	19.80	0.69	0.35	...	...	...	0.058	0.044	...
SPARKLING WINES (CHAMPAGNE, ETC.)																
France and Germany (Dry)	27	0.9925	10.42	2.36	0.61	...	0.25	0.53	0.71	0.14	...	...	...	...	...	0.857
France and Germany (Sweet)	32	1.0347	9.50	12.88	0.63	...	...	10.92	0.70	0.15	...	...	...	0.022	0.026	0.628

TABLE XXIV.—CALIFORNIA WINES.

	Number of samples	Specific gravity	Per cent alcohol by volume	Grams per 100 cc.		Glycerin-alcohol ratio	Total acids	Grams per 100 cc.	Polariz-ation	Grams per 100 cc.				
				Alcohol	Glycerin			Extract		Reducing sugar	Potassium sulphate	Proteids	Tannin and coloring matter	Ash
RED WINES														
Bordeaux, or claret type....	81													
Maximum....		1.0020	15.09	11.97	0.852	8.7 : 100	0.888	3.81	−2.1	0.628	0.1570	0.5544	0.358	0.429
Minimum....		0.9900	8.92	7.09	0.330	3.4 : 100	0.368	2.09	−0.5	0.040	0.0470	0.1865	0.064	0.209
Rhine type....	6													
Maximum....		0.9970	13.82	10.96	......	......	0.718	3.34	......	......	......	......	0.349	
Minimum....		0.9940	11.00	8.73	......	......	0.358	2.69	......	......	......	......	0.211	
Red Burgundy type....	25													
Maximum....		0.9962	15.48	12.29	0.656	7 : 100	0.762	3.46	−1.7	0.418	0.2515	0.4482	0.328	0.416
Minimum....		0.9911	8.00	6.35	0.461	4.7 : 100	0.408	2.10	−0.5	0.030	0.0455	0.1864	0.033	0.188
Southern French type....	92													
Maximum....		1.0050	19.28*	15.30	......	......	0.834	6.88*	......	......	......	......	0.344	0.430
Minimum....		0.9900	8.07	6.40	......	......	0.201	1.91	......	......	......	......	0.050	0.202
WHITE WINES														
Rhine-wine type....	87													
Maximum....		1.0020	14.57	11.57	0.971	10.2 : 100	0.788	4.38	−3.5	0.626	0.1771	0.5164	......	0.447
Minimum....		0.9883	5.00	3.98	0.474	4.6 : 100	0.327	1.51	−0.2	0.060	0.0633	0.0989	......	0.140
Sauterne type....	55													
Maximum....		1.0157	15.20	12.07	0.904	7.7 : 100	0.766	6.78	+0.6	3.559	0.1648	0.3809	0.087	0.368
Minimum....		0.9892	8.23	6.53	0.178	1.8 : 100	0.377	1.69	−18.6	0.069	0.0453	0.0867	0.015	0.050
Southern French type....	63													
Maximum....		0.9888	22.18	17.00	0.918	14 : 100	0.656	4.56	−3.4	0.936	......	0.3162	0.045	0.290
Minimum....		0.9882	8.07	6.40	0.318	3.4 : 100	0.219	1.09	−0.1	0.069	......	0.0988	0.034	0.148
Port type....	50													
Maximum....		1.0429	22.19	17.61	0.707	4.5 : 100	0.700	17.22	−27.1	13.559	0.1861	0.9379	1.066	0.394
Minimum....		0.9875	10.38	8.24	0.163	1.1 : 100	0.181	2.43	−14.4	0.228	0.0594	0.1893	0.059	0.222
Sherry and Madeira type....	66													
Maximum....		1.0560	21.85	17.34	0.936	6.8 : 100	0.798	19.66	−23.1	17.210	0.1196	0.4751	0.350	0.436
Minimum....		0.9866	8.22	6.52	0.324	2 : 100	0.253	1.31	−0.5	0.119	0.0504	0.0859	0.021	0.156

* Fortified.

tween 13 and 17. This is what is known as the acid: alcohol ratio, and is used for the detection of watering. Water can be added only to very sweet grapes without exceeding the limits of this ratio and then only in very limited quantities.

The alcohol and extract vary directly and in such proportions that the number representing the extract in grams per hundred cc. multiplied by the factor 4.5 gives a figure equal to or greater than the alcohol in grams per 100 cc. With white wines, in which the extract is normally lower, the factor, 6.5 is used in the same way. This is known as the alcohol: extract ratio and is used for the detection of the addition of sugar to the must.

Analyses of Cordials and Liqueurs.—As can be expected from the wide range of proportions cited in Chapter XII for making cordials their composition as shown by analysis also varies greatly. However, Leach (*loc. cit.*, p. 787) cites the following results compiled by König as typical.

TABLE XXV.—ANALYSES OF LIQUEURS.

	Specific gravity	Alcohol by volume	Alcohol by weight	Extract	Cane sugar	Other extractives	Ash
Absinthe..........	0.9116	58.93	....	0.18	.....	0.32	
Benedictine........	1.0709	52.	38.5	36.00	32.57	3.43	0.043
Ginger............	1.0481	47.5	36.0	27.79	25.92	1.87	0.141
Crême de Menthe...	1.0447	48.0	36.5	28.28	27.63	0.65	0.068
Anisette de Bordeaux	1.0847	42.0	30.7	34.82	37.44	0.38	0.040
Curaçao...........	1.0300	55.0	42.5	28.60	28.50	0.10	0.040
Kummel............	1.0830	33.9	24.8	32.02	31.18	0.84	0.058
Angostura.........	0.9540	49.7	....	5.85	4.16	1.69	
Chartreuse........	1.0799	43.18	....	36.11	34.35	1.76	

More generally these beverages may be classed according to their sugar and alcohol content as follows:

TABLE XXVI.

	Alcohol	Sugar
Average grade......................	20-25%	10%
Good grade........................	25-30%	20%
Very good grade....................	30-40%	35%
Best grade........................	40-50%	45%

CHAPTER XIV

ANALYSIS OF ALCOHOLIC BEVERAGES

METHODS

(Reprinted by special permission from Official and Tentative Methods of Analysis of the Association of Official Agricultural Chemists, 3rd ed., 1930.)

WINES

1. PHYSICAL EXAMINATION—TENTATIVE

Note and record the following: (1) Whether the container is "bottle full"; (2) the appearance of the wine, whether it is bright or turbid and whether there is any sediment; (3) condition when opened, whether still, gaseous, or carbonated; (4) color and depth of color; (5) odor, whether vinous, acetous, pleasant, or foreign; and (6) taste, whether vinous, acetous, sweet, dry, or foreign.

2. PREPARATION OF SAMPLE—OFFICIAL

If gas is contained in the wine, remove it by pouring the sample back and forth in beakers.

Filter the wine, regardless of appearance, before analyzing and determine immediately the specific gravity and such ingredients as alcohol, acids, and sugars, which are liable to change through exposure.

3. SPECIFIC GRAVITY—OFFICIAL

Determine the specific gravity at 20/4° (in vacuo) by means of a pycnometer as follows: Carefully clean the pycnometer by filling with a saturated solution of CrO_3 in H_2SO_4, allowing to stand for several hours, emptying, and rinsing thoroughly with H_2O. Fill the pycnometer with recently boiled distilled H_2O

previously cooled to 16–18°, place in a water bath cooled to the same temperature and allow the bath to warm slowly to 20°. Adjust the level of the H_2O to the proper point on the pycnometer, put the perforated cap or stopper in place, remove from the bath; wipe dry with a clean cloth, and after allowing to stand for 15–20 minutes, weigh. Empty, rinse several times with alcohol and then with ether, remove the ether fumes, allow the instrument to become perfectly dry, and weigh. Ascertain the weight of contained H_2O at 20° in air (W of the formula below) by subtracting the weight of the empty pycnometer from its weight when full. Cool the sample to 16–18°, adjust the level of the liquid to the proper point on the pycnometer, put the perforated cap or stopper in place, wipe dry, and weigh as before. Ascertain the weight of the contained sample at 20° in air (S of the formula below) by subtracting the weight of the empty pycnometer from its weight when filled with the sample. Calculate the specific gravity in vacuo by the following formula:

$$G = \frac{S + 0.00105W}{1.00282W} \text{, in which}$$

G = corrected specific gravity of sample at 20/4° in vacuo;
W = weight of contained H_2O at 20° in air; and
S = weight of contained sample at 20° in air.

4. ALCOHOL—OFFICIAL

(a) *By volume.*—Measure 100 cc. of the liquid at 20° into a 300–500 cc. distillation flask and add 50 cc. of H_2O. Attach the flask to a vertical condenser by means of a bent tube, distil almost 100 cc., and make to a volume of 100 cc. at 20°. (Foaming, which sometimes occurs especially with young wines, may be prevented by the addition of a small quantity of tannin.) To determine the alcohol in wines that have undergone acetous fermentation and contain an abormal quantity of acetic acid, exactly neutralize the portion taken with NaOH solution before distillation. (This is unnecessary, however, for wines of normal taste and odor.) Determine the specific gravity of the distillate at 20°/4° as directed under 3 and obtain the corresponding percentage of alcohol by volume from Tables A3-A5.

(b) *Grams per 100 cc.*—From the specific gravity of the distillate, obtained under (a), ascertain the corresponding alcohol content in g per 100 cc. from Tables A3-A5.

(c) *By weight.*—Divide the number of g in the 100 cc. of distillate, as obtained in (b), by the weight of the sample as calculated from its specific gravity.

(d) *By immersion refractometer.*—Verify the percentages of alcohol, as determined under (a) and (c), by ascertaining the immersion refractometer reading of the distillate and obtaining the corresponding percentages of alcohol from Table A6.

GLYCEROL IN DRY WINES

5. Method I (By Direct Weighing)—Official

Evaporate 100 cc. of the wine in a porcelain dish on a water bath to a volume of about 10 cc. Treat the residue with about 5 g. of fine sand and 4–5 cc. of milk of lime (containing 15 g. of CaO per 100 cc.) for each g. of extract present and evaporate almost to dryness. Treat the moist residue with 50 cc. of alcohol, 90% by volume; remove the substance adhering to the sides of the dish with a spatula; and rub the whole mass to a paste. Heat the mixture on a water bath, with constant stirring, to incipient boiling and decant the liquid through a filter into a small flask. Wash the residue repeatedly by decantation with 10 cc. portions of hot 90% alcohol until the filtrate amounts to about 150 cc. Evaporate the filtrate to a sirupy consistency in a porcelain dish on a hot, but not boiling, water bath; transfer the residue to a small glass-stoppered, graduated cylinder with 20 cc. of absolute alcohol; and add 3 portions of 10 cc. each of anhydrous ether, shaking thoroughly after each addition. Let stand until clear, pour off through a filter, and wash the cylinder and filter with a mixture of 2 parts of absolute alcohol to 3 parts of anhydrous ether, also pouring the wash liquor through the filter. Evaporate the filtrate to a sirupy consistency, dry for an hour at the temperature of boiling H_2O, weigh, ignite, and weigh again. The loss on ignition gives the weight of glycerol.

6. Method II (By Oxidation with Dichromate)—Official

Evaporate 100 cc. of the wine in a porcelain dish on a water bath, the temperature of which is maintained at 85–90°, to a volume of 10 cc. Treat the residue with about 5 g. of fine sand and 5 cc. of milk of lime (containing 15 g. of CaO per 100 cc.). Proceed from this point as directed under 7 and 8, beginning with the clause "evaporate almost to dryness with frequent stirring," except to dilute the solution of glycerol after treatment with $(Ag_2)CO_3$ and Pb-acetate to a volume of 100 cc. instead of 50 cc. Observe the precautions given concerning the temperature at which all evaporations are to be made.

7. REAGENTS

(a) *Strong potassium dichromate solution.*—Dissolve 74.55 g. of dry, recrystallized $K_2Cr_2O_7$ in H_2O; add 150 cc. of H_2SO_4; cool; and dilute with H_2O to 1 liter at 20° C. 1 cc. of this solution = 0.01 g. of glycerol. Owing to the high coefficient of expansion of this strong solution it is necessary to make all volumetric measurements of the solution at the same temperature as that at which it was diluted to volume.

(b) *Dilute potassium dichromate solution.*—Measure 25 cc. of the strong $K_2Cr_2O_7$ solution at 20° into a 500 cc. volumetric flask and dilute to the mark with H_2O at room temperature. 20 cc. of this solution = 1 cc. of (a).

(c) *Ferrous ammonium sulfate solution.*—Dissolve 30 g. of crystallized ferrous ammonium sulfate in H_2O, add 50 cc. of H_2SO_4, cool, and dilute with H_2O to 1 liter at room temperature. 1 cc. of this solution = approximately 1 cc. of (b). As its value changes slightly from day to day, it must be standardized against (b) whenever used.

(d) *Potassium ferricyanide indicator.*—Dissolve 1 g. of crystallized $K_3Fe(CN)_6$ in 50 cc. of H_2O. This solution must be freshly prepared.

(e) *Milk of lime.*—Introduce 150 g. of CaO, selected from clean hard lumps, prepared preferably from marble, into a large porcelain or iron dish; slake with H_2O, cool, and add sufficient H_2O to make 1 liter.

(f) *Silver carbonate.*—Dissolve 0.1 g. of Ag_2SO_4 in about 50 cc. of H_2O, add an excess of Na_2CO_3 solution, allow the precipitate to settle, and wash with H_2O several times by decantation until the washings are practically neutral. This reagent must be freshly prepared immediately before use.

8.   DETERMINATION

Make evaporations on a water bath maintained at a temperature of 85–90°. The area of the dish exposed to the bath should not be greater in circumference than that covered by the liquid inside.

Evaporate 100 cc. of the vinegar to 5 cc., add 20 cc. of H_2O, and again evaporate to 5 cc. to expel acetic acid. Treat the residue with about 5 g. of 40-mesh sand and 15 cc. of the milk of lime and evaporate almost to dryness, with frequent stirring, avoiding the formation of a dry crust or evaporation to complete dryness. Treat the moist residue with 5 cc. of H_2O; rub into a homogeneous paste; add slowly 45 cc. of absolute alcohol, washing down the sides of the dish to remove adhering paste; and stir thoroughly. Heat the mixture on a water bath, with constant stirring, to incipient boiling; transfer to a suitable vessel; and centrifugalize. Decant the clear liquid into a porcelain dish and wash the residue with several small portions of hot alcohol, 90% by volume, by aid of the centrifuge. (If a centrifuge is not available, decant the liquid through a folded filter into a porcelain dish. Wash the residue repeatedly with small portions of hot 90% alcohol, twice by decantation, and then by transferring all the material to the filter. Continue the washing until the filtrate amounts to 150 cc.) Evaporate to a sirupy consistency; add 10 cc. of absolute alcohol to dissolve this residue; and transfer to a 50 cc. glass-stoppered cylinder, washing the dish with successive small portions of absolute alcohol until the volume of the solution is 20 cc. Add 3 portions of 10 cc. each of anhydrous ether, shaking thoroughly after each addition. Let stand until clear, pour off through a filter, and wash the cylinder and filter with a mixture of 2 volumes of absolute alcohol and 3 of anhydrous ether. If a heavy precipitate has formed in the cylinder, centrifugalize at low speed, decant the clear liquid, and wash 3 times with 20 cc. portions of the alcohol-ether mixture, shaking the mixture thoroughly each time and separating the precipitate by means of the centrifuge. Wash the paper with the alcohol-ether mixture and evaporate the filtrate and washings on the water bath to about 5 cc., add 20 cc. of H_2O, and again evaporate to 5 cc.; again add 20 cc. of H_2O and evaporate to 5 cc.; finally add 10 cc. of H_2O and evaporate to 5 cc.

These evaporations are necessary to remove all the ether and alcohol, and when conducted at 85–90° they result in no loss of glycerol if the concentration of the latter is less than 50%.

Transfer the residue with hot H_2O to a 50 cc. volumetric flask, cool, add the Ag_2CO_3 prepared from 0.1 g. of Ag_2SO_4, shake, and allow to stand 10 minutes. Then add 0.5 cc. of basic Pb-acetate solution, shake occasionally, and allow to stand 10 minutes. Make up to the mark, shake well, filter, rejecting the first portion of the filtrate, and pipet 25 cc. of the clear filtrate into a 250 cc. volumetric flask.

Add 1 cc. of H_2SO_4 to precipitate the excess of Pb and then 30 cc. of Reagent (a). Add carefully 24 cc. of H_2SO_4, rotating the flask gently to mix the contents and avoid violent ebullition, and then place in a *boiling* water bath for exactly 20 minutes. Remove the flask from the bath, dilute, cool, and make up to the mark at room temperature. The quantity of strong dichromate solution used must be sufficient to leave an excess of about 12.5 cc. at the end of the oxidation, the quantity given above (30 cc.) being sufficient for ordinary vinegar containing about 0.35 g. or less of glycerol per 100 cc.

Standardize the ferrous ammonium sulfate solution against Reagent (b) by introducing from the respective burets approximately 20 cc. of each of these solutions into a beaker containing 100 cc. of H_2O. Complete the titration, using Reagent (d) as an outside indicator. From this titration calculate the volume (F) of the ferrous ammonium sulfate solution equivalent to 20 cc. of the dilute and therefore, to 1 cc. of Reagent (a).

In place of Reagent (b) solution substitute a buret containing the oxidized glycerol with an excess of Reagent (a) and ascertain how many cc. are equivalent to (F) cc. of the ferrous ammonium sulfate solution and, therefore, to 1 cc. of Reagent (a). Then 250, divided by this last equivalent, = the number of cc. of Reagent (a) present in excess in the 250 cc. flask after oxidation of the glycerol.

The number of cc. of Reagent (a) added, minus the excess found after oxidation, multiplied by 0.02, gives the grams of glycerol per 100 cc. of vinegar.

9. GLYCEROL IN SWEET WINES—OFFICIAL

With wines in which the extract exceeds 5 g. per 100 cc., heat 100 cc. to boiling in a flask and treat with successive small portions of milk of lime until the wine becomes first darker and then lighter in color. Cool, add 200 cc. of 95% alcohol, allow the precipitate to subside, filter, and wash with 95% alcohol. Treat the combined filtrate and washings as directed under 5 or 6.

10. GLYCEROL-ALCOHOL RATIO—OFFICIAL

Express this ratio as X: 100, in which X is obtained by multiplying the percentage weight of glycerol by 100 and dividing the result by the percentage of alcohol by weight.

EXTRACT

11. I. From the Specific Gravity of the Dealcoholized Wine—Official

Calculate the specific gravity of the dealcoholized wine by the following formula:

$S = G + 1 - A$, in which
S = specific gravity of the dealcoholized wine;
G = specific gravity of the wine, 3; and
A = specific gravity of the distillate obtained in the determination of alcohol, 4 (a).

From Table A2, ascertain the percentage by weight of extract in the dealcoholized wine corresponding to the value of S. Multiply the figure thus obtained by the value of S to obtain the g. of extract per 100 cc. of wine.

12. II. By Evaporation—Official

(a) *In dry wines, having an extract content of less than 3 grams per 100 cc.*—Evaporate 50 cc. of the sample on a water bath to a sirupy consistency in a 75 cc. flat-bottomed Pt dish, approximately 85 mm. in diameter. Heat the residue for 2–5 hours in a drying oven at the temperature of boiling H_2O, cool in a desiccator, and weigh as soon as the dish and contents reach room temperature.

(b) *In sweet wines.*—If the extract content is between 3 and 6 g. per 100 cc., treat 25 cc. of the sample as directed under (a). If the extract exceeds 6 g. per 100 cc., however, the result, obtained as directed under 11, is accepted, and no gravimetric determination is attempted because of the inaccurate results obtained by drying levulose at a high temperature.

13. NON-SUGAR SOLIDS—OFFICIAL

Determine the non-sugar solids (sugar-free extract) by subtracting the quantity of reducing sugars before inversion, 14, from the extract, 11 or 12. If sucrose is present in the wine, determine the nonsugar solids by subtracting the sum of reducing sugars before inversion and the sucrose from the extract.

14. REDUCING SUGARS—OFFICIAL

(a) *Dry wines.*—Place 200 cc. of the wine in a porcelain dish; exactly neutralize with normal NaOH, calculating the quantity required from the determination of acidity, 44; and evaporate to about ¼ the original volume. Transfer to a 200 cc. flask, add sufficient neutral Pb-acetate solution to clarify, dilute to the mark with H_2O, shake, and pass through a folded filter. Remove the Pb with dry K-oxalate and determine, reducing sugars as directed under 15–18.

(b) *Sweet wines.*—With sweet wines, approximate the sugar content by subtracting 2 from the result in the determination of the extract and employ such a quantity of the sample that the aliquot taken for the Cu reduction shall not exceed 240 mg. of invert sugar. Proceed as directed under (a) except to take this smaller quantity of the sample for the determination.

15. *Munson and Walker General Method—Official*

REAGENTS

(a) *Asbestos.*—Digest the asbestos, which should be the amphibole variety, with HCl (1 + 3) for 2–3 days. Wash free from acid, digest for a similar period with 10% NaOH solution, and then treat for a few hours with hot alkaline tartrate solution (old alkaline tartrate solutions that have stood for some time may be used for this purpose) of the strength used in sugar determinations. Wash the asbestos free from alkali; digest for several hours with HNO_3 (1 + 3); and, after washing free from acid, shake with H_2O into a fine pulp. In preparing the Gooch crucible, make a film of asbestos ¼ inch thick and wash thoroughly with H_2O to remove fine particles of asbestos. If the precipitated Cu_2O is to be weighed as such, wash the crucible with 10 cc. of alcohol, then with 10 cc. of ether; dry for 30 minutes at 100°; cool in a desiccator; and weigh.

Soxhlet's modification of Fehling's solution.—Prepared by mixing immediately before use, equal volumes of (a) and (b).

(a) *Copper sulfate solution.*—Dissolve 34.639 g. of $CuSO_4.5H_2O$ in H_2O, dilute to 500 cc., and filter through prepared asbestos.

(b) *Alkaline tartrate solution.*—Dissolve 173 g. of Rochelle salts and 50 g. of NaOH in H_2O, dilute to 500 cc., allow to stand for 2 days, and filter through prepared asbestos.

16. PRECIPITATION OF CUPROUS OXIDE

Transfer 25 cc. of each of the $CuSO_4$ and alkaline tartrate solutions to a 400 cc. beaker of alkali-resistant glass and add 50 cc. of the reducing sugar solution, or if a smaller volume of sugar solution is used, add H_2O to make the final volume 100 cc.

Heat the beaker on an asbestos gauze over a Bunsen burner, regulate the flame so that boiling begins in 4 minutes, and continue the boiling for exactly 2 minutes. (It is important that these directions be strictly observed. To regulate the burner for this purpose it is advisable to make preliminary tests, using 50 cc. of the reagent and 50 cc. of H_2O before proceeding with the actual determination.) Keep the beaker covered with a watch-glass during the heating. Filter the hot solution at once through an asbestos mat in a porcelain Gooch crucible, using suction. Wash the precipitate of Cu_2O thoroughly with H_2O at a temperature of about 60° and either weigh directly as Cu_2O as directed under 17, or determine the quantity of reduced Cu as described under 18. Conduct a blank determination, using 50 cc. of the reagent and 50 cc. of H_2O, and if the weight of Cu_2O obtained exceeds 0.5 mg., correct the result of the reducing sugar determination accordingly. The alkaline tartrate solution deteriorates on standing, and the quantity of Cu_2O obtained in the blank increases.

17. DETERMINATION OF REDUCED COPPER

Direct Weighing of Cuprous Oxide

(This method should be used only for determinations in solutions of reducing sugars of comparatively high purity. In products containing large quantities of mineral or organic impurities, including sucrose, determine the Cu of the Cu_2O by one of the methods described under 18, since the Cu_2O is very likely to be contaminated with foreign matter.)

Prepare a Gooch crucible as directed under 15. Collect the precipitated Cu_2O on the mat as directed under 16 and wash thoroughly with hot H_2O, then with 10 cc. of alcohol, and finally with 10 cc. of ether. Dry the precipitate for 30 minutes in a water oven at the temp. of boiling H_2O, cool, and weigh. Calculate the weight of metallic Cu, using the factor 0.8882. Obtain from Table A7 the weight of invert sugar equivalent to the weight of Cu.

The number of mg. of Cu reduced by a given quantity of reducing sugar varies, depending upon whether or not sucrose is present. In the tables the absence of sucrose is assumed except in the entries under invert sugar, where, in addition to the column for invert sugar alone, there are given one column for mixtures of invert sugar and sucrose containing 0.4 g. of total sugar in 50 cc. of solution and one column for invert sugar and sucrose when the 50 cc. of solution contains 2 g. of total sugar. Two entries are also given under lactose and sucrose mixtures, showing proportions of 1 part lactose to 4 and 12 parts of sucrose, respectively.

18. REAGENTS

Volumetric Thiosulfate Method

Standard thiosulfate solution.—Prepare a solution of $Na_2S_2O_3$ containing 19 g. of pure crystals in 1 liter. Weigh accurately about 0.2 g. of pure Cu and place in a flask of 250 cc. capacity. Dissolve by warming with 5 cc. of a mixture of equal volumes of strong HNO_3 and H_2O. Dilute to 50 cc., boil to expel the red fumes, add a slight excess of strong Br water, and boil until the Br is completely driven off. Cool, and add a strong NaOH solution with agitation until a faint turbidity of $Cu(OH)_2$ appears (about 7 cc. of a 25% NaOH solution is required). Discharge the turbidity with a few drops of 80% acetic acid and add 2 drops in excess. (The solution should now occupy a volume of 50–70 cc.) Add 10 cc. of 30% KI solution. Titrate at once with the thiosulfate solution until the brown tinge becomes weak and add sufficient starch indicator to produce a marked blue coloration. Continue the titration cautiously until the color changes toward the end to a faint lilac. (If at this point the thiosulfate is added dropwise and a little time is allowed for complete reaction after each addition, no difficulty is experienced in determining the end point within a single drop.) 1 cc. of the thiosulfate solution = about 0.005 g. of Cu.

18a. DETERMINATION

After washing the precipitated Cu_2O, cover the Gooch with a watch-glass and dissolve the oxide by means of 5 cc. of warm HNO_3 $(1 + 1)$ poured under the watch-glass with a pipet. Collect the filtrate in a 250 cc. flask and wash the watch-glass and the Gooch free from Cu, using about 50 cc. of H_2O. Boil to expel red fumes; add a slight excess of Br water; boil off the Br completely; and proceed as directed under 18, beginning with "Cool and add a strong NaOH solution. . . ."

SUCROSE

19. *I. By Reducing Sugars Before and After Inversion—Official*

Determine the reducing sugars (clarification having been effected with neutral Pb-acetate, never with basic Pb-acetate) as directed under 15 and calculate to invert sugar from A7. Invert the solution as directed under 20 (b) or (c), or 22 (b) or (c); exactly neutralize the acid; and again determine the reducing sugars, but calculate them to invert sugar from the table referred to above, using the invert sugar column alone. Deduct the percentage of invert sugar obtained before inversion from that obtained after inversion and multiply the difference by 0.95 to obtain the percentage of sucrose. The solutions should be diluted in both determinations so that not more than 240 mg. of invert sugar is present in the quantity taken for reduction. It is important that all lead be removed from the solution with anhydrous powdered K-oxalate or Na_2CO_3 before reduction.

20. *II. By Polarization—Official*

Polarize before and after inversion in a 200 mm. tube, as directed under 20 or 22, a portion of the filtrate obtained under 14. In calculating the percentage of sucrose do not fail to take into consideration the relation of the weight of the sample contained in 100 cc. to the normal weight for the instrument.

(a) *Direct reading.*—Pipet one 50 cc. portion of the Pb-free filtrate into a 100 cc. flask, dilute with H_2O to the mark, mix well, and polarize in a 200 mm. tube. The result, multiplied by 2, is the direct reading (*P* of formula given below) or polarization before inversion. (If a 400 mm. tube is used, the reading equals *P*.)

(b) *Invert reading.*—First determine the quantity of acetic acid necessary to render 50 cc. of the Pb-free filtrate distinctly acid to methyl red indicator; then to another 50 cc. of the lead-free solution in a 100 cc. volumetric flask, add the requisite quantity of acid and 5 cc. of the invertase preparation, fill the flask with H_2O nearly to 100 cc., and let stand overnight (preferably at a temperature not less than 20°). Cool, and dilute to 100 cc. at 20°. Mix well and polarize at 20° in a 200 mm. tube. If the analyst is in doubt as to the completion of the hydrolysis, allow a portion of the solution to remain for several hours and again polarize. If there is no change from the previous reading, the inversion is complete, and the reading and temperature of the solution should be carefully noted. If it is necessary to work at a temperature other than 20°, which is permissible within narrow limits, the volumes must be completed and both direct and invert readings must be made at the same temperature. Correct the invert reading for the optical activity of the invertase solution and multiply by 2. Calculate the percentage of sucrose by the following formula:

$$S = \frac{100\,(P - I)}{142.1 + 0.073\,(m - 13) - t/2}, \text{ in which}$$

S = percentage of sucrose;
P = direct reading, normal soln.;
I = invert reading, normal soln.;
t = temp. at which readings are made; and
m = g. of total solids in 100 cc. of the invert soln. read in the polariscope.

Determine the total solids as directed under 11.

(c) *Rapid inversion at 55–60°.*—If more rapid inversion is desired, proceed as follows: Prepare the sample as directed under (a) and to 50 cc. of the Pb-free filtrate in a 100 cc. volumetric flask add glacial acetic acid in sufficient quantity to render the solution distinctly acid to methyl red. The quantity of acetic acid required should be determined before pipetting the 50 cc. portion. Then add 10 cc. of invertase solution, mix thoroughly, place the flask in a water bath at 55–60°, and allow to stand at that temperature for 15 minutes with occasional shaking. Cool, add Na_2CO_3 solution until distinctly alkaline to litmus paper, dilute to 100 cc. at 20°, mix well, and determine the polarization at 20° in a 200 mm. tube. Allow the solution to remain in the tube for 10 minutes and again determine the polarization. If there is no change from the previous reading, the mutarotation is complete. Carefully note the reading and the temperature of the solution. Correct the polarization for the optical activity of the invertase solution and multiply by 2. Calculate the percentage of sucrose by the formula given under (b).

(If the solution has been rendered so alkaline as to cause destruction of sugar, the polarization, if negative, will in general decrease, since the decomposition of fructose ordinarily is more rapid than that of the other sugars present. If the solution has not been made sufficiently alkaline to complete mutarotation quickly, the polarization, if negative, will in general increase. As the analyst gains experience he may omit the polarization after 10 minutes if he has satisfied himself that he is adding Na_2CO_3 in sufficient amount to complete mutarotation at once without causing any destruction of sugar during the period intervening before polarization.)

21. *II. By Polarization Before and After Inversion With Hydrochloric Acid—Official*

(In the presence of much levulose, as in honeys, fruit products, sorghum sirup, cane sirup, and molasses, the optical method for sucrose, requiring hydrolysis by acid, gives erroneous results.)

22. (a) *Direct reading.*—Proceed as directed under 20 (a).

(b) *Invert reading.*—Pipet a 50 cc. portion of the Pb-free filtrate into a 100 cc. flask and add 25 cc. of H_2O. Then add, little by little, while rotating the flask, 10 cc. of HCl (sp. gr. 1.1029 at 20/4° (or 24.85° Brix at 20°)). Heat a water bath to 70° and regulate the burner so that the temperature of the bath remains approximately at that point. Place the flask in the water bath, insert a thermometer, and heat with constant agitation until the thermometer in the flask indicates 67°. This preliminary heating period should require from 2½-2¾ minutes.

From the moment the thermometer in the flask indicates 67°, leave the flask in the bath for exactly 5 minutes longer, during which time the temperature should gradually rise to about 69.5°. Plunge the flask as once into H_2O at 20°. When the contents have cooled to about 35°, remove the thermometer from the flask, rinse it, and fill almost to the mark. Leave the flask in the bath at 20° for at least 30 minutes longer and finally make up exactly to volume. Mix well and polarize the solution in a 200 mm. tube provided with a lateral branch and a water jacket, maintaining a temperature of 20°. This reading must also be multiplied by 2 to obtain the invert reading. If it is necessary to work at a temperature other than 20°, which is permissible within narrow limits, the volumes must be completed and both direct and invert polarizations must be made at exactly the same temp.

Calculate sucrose by the following formula:

$$S = \frac{100\ (P - I)}{143 + 0.0676\ (m - 13) - t/2}, \text{ in which}$$

S = percentage of sucrose;
P = direct reading, normal soln.;
I = invert reading, normal soln.;
t = temp. at which readings are made; and
m = g. of total solids in 100 cc. of the invert soln. read in the polariscope.

Determine the total solids as directed under 11.

(c) *Inversion at room temperature.*—The inversion may also be accomplished as follows: (1) To 50 cc. of the clarified soln., freed from Pb, add 10 cc. of HCl (sp. gr. 1.1029 at 20/4° or 24.85° Brix at 20°) and set aside for 24 hours at a temp. not below 20°; or, (2) if the temp. is above 25°, set aside for 10 hours. Make up to 100 cc. at 20° and polarize as directed under (b). Under these conditions the formula must be changed to the following:

$$S = \frac{100\ (P - I)}{143.2 + 0.0676\ (m - 13) - t/2}.$$

23. COMMERCIAL GLUCOSE—OFFICIAL

Polarize a portion of the clarified filtrate after inversion in a 200 mm. jacketed tube at 87°, as directed under 24. In calculating the percentage of glucose do not fail to take into consideration the relation of the weight of the sample contained in 100 cc. to the normal weight for the instrument.

24. Commercial glucose cannot be determined accurately owing to the varying quantities of dextrin, maltose, and dextrose present in the product. However, in sirups in which the quantity of invert sugar is so small as not to affect appreciably the result, commercial glucose may be estimated approximately by the following formula:

$$G = \frac{(a - S)\ 100}{211}, \text{ in which}$$

G = percentage of commercial glucose solids;

a = direct polarization, normal soln.; and

S = percentage of cane sugar.

Express the results in terms of commercial glucose solids polarizing $+ 211°V$. (This result may be recalculated in terms of commercial glucose of any Baumé reading desired.)

25.   ASH—OFFICIAL

Proceed as directed below using the residue from 50 cc. of the wine.

Weigh a quantity of the substance representing about 2 g. of dry material and burn at a low heat, not exceeding dull redness, until free from C. If a C-free ash cannot be obtained in this manner, exhaust the charred mass with hot H_2O, collect the insoluble residue on an ashless filter, and burn the filter and contents to a white or nearly white ash. Add the filtrate, evaporate to dryness, and heat at dull redness until the ash is white or grayish white. Cool in a desiccator and weigh.

26.   ASH EXTRACT RATIO—OFFICIAL

Express results as $1 : X$, in which X is the quotient obtained by dividing the g. of extract per 100 cc. by the g. of ash per 100 cc.

27. ALKALINITY OF THE WATER-SOLUBLE ASH—OFFICIAL

Extract the ash obtained as directed under 25 with successive small portions of hot H_2O until the filtrate amounts to about 60 cc.

Cool the filtrate and titrate with 0.1 N HCl, using methyl orange indicator. Express the alkalinity in terms of the number of cc. of 0.1 N acid per 100 cc. of the wine.

28. ALKALINITY OF THE WATER-INSOLUBLE ASH—OFFICIAL

Ignite the filter and residue from 27 in the Pt. dish in which the wine was ashed and proceed as directed below. Express the alkalinity in terms of the number of cc. of 0.1 N acid required to neutralize the water-insoluble ash from 100 cc. of the wine.

Add an excess of 0.1 N HCl (usually 10-15 cc.) to the ignited insoluble ash in the Pt. dish, heat to boiling on an asbestos plate, cool, and titrate the excess of HCl with 0.1 N NaOH, using methyl orange indicator.

29. PHOSPHORIC ACID—OFFICIAL

Dissolve the ash obtained as directed under 25 in 50 cc. of boiling HNO_3 $(1+9)$, filter, wash the filter, and determine P_2O_5 in the combined filtrate and washings as directed below. If the ash ignites without difficulty, no free phosphoric acid need be suspected. Should there be any free acid, the ash remains black even after repeated leaching. In such cases, add Ca-acetate or a mixture containing 3 parts of Na_2CO_3 and 1 part of $NaNO_3$ to avoid loss of P_2O_5 before attempting to ash. Add NH_4OH in slight excess; and barely dissolve the precipitate formed with a few drops of HNO_3, stirring vigorously. If HCl or H_2SO_4 has been used as a solvent, add about 15 g. of crystalline NH_4NO_3 or a soln. containing that quantity. To the hot soln. add 70 cc. of molybdate soln. for every decigram of P_2O_5 present. Digest at about 65° for 1 hour, and determine whether or not the P_2O_5 has been completely precipitated by the addition of more molybdate soln. to the clear supernatant liquid. Filter, and wash with cold H_2O or preferably with the NH_4NO_3 soln. Dissolve the precipitate on the filter with NH_4OH ($1 + 1$) and hot H_2O and wash into a beaker to a volume of not more than 100 cc. Neutralize with HCl, using litmus paper or bromthymol blue as an indicator; cool; and from a buret add slowly (about 1 drop per second), stirring vigorously, 15 cc. of magnesia mixture for each decigram of P_2O_5 present. After 15 min. add 12 cc. of NH_4OH. Let stand until the supernatant liquid is clear (usually 2 hours), filter, wash the precipitate with the dilute NH_4OH until the washings are practically free from chlorides, dry, burn first at a low heat and ignite to constant weight, preferably in an electric furnace, at 950-1,000°; cool in a desiccator, and weigh as $Mg_2P_2O_7$. Calculate and report the result as percentage of P_2O_5.

30. SULFURIC ACID—OFFICIAL

Precipitate directly the H_2SO_4 in 50 cc. of the wine by means of 10% $BaCl_2$ soln. after acidifying with a small excess of HCl, and determine the resulting $BaSO_4$ as directed below. Allow the precipitate to stand for at least 6 hours before filtering. Report as SO_3, using the factor 0.3430.

Heat to boiling and add slowly in small quantities a 10% $BaCl_2$ soln. until no further precipitate is formed. Continue the boiling for about 5 min. and allow to stand for 5 hours or longer in a warm place. Decant the liquid through an ashless filter or an ignited and weighed Gooch crucible, treat the precipitate with 15-20 cc. of boiling H_2O, transfer to the filter, and wash with boiling H_2O until the filtrate is free from chlorides. Dry the precipitate and filter, ignite, and weigh as $BaSO_4$.

31. CHLORIDES—OFFICIAL

To 100 cc. of dry wine or 50 cc. of sweet wine, add sufficient Na_2CO_3 to make distinctly alkaline. Evaporate to dryness, ignite at a heat not above low redness, cool, extract the residue with hot H_2O, acidify the water extract with HNO_3 ($1 + 4$), and determine chlorides as directed under 32 or 34.

32. *I. Gravimetric Method*

To the soln. prepared as directed in 31 add a 10% AgNO₃ soln., avoiding more than a slight excess. Heat to boiling, protect from the light, and allow to stand until the precipitate is granular. Filter on a weighed Gooch crucible, previously heated to 140-150°, and wash with hot H_2O, testing the filtrate to prove excess of AgNO₃. Dry the AgCl at 140-150°, cool, and weigh. Report as percentage of Cl.

33. *II. Volumetric Method*

REAGENTS

(a) *Silver nitrate.*—Adjust to exact 0.1 *N* strength by standardizing against a 0.1 *N* NaCl soln. containing 5.846 g. of pure NaCl per liter.

(b) *Ammonium or potassium thiocyanate.*—0.1 *N*. Adjust by titrating against the 0.1 *N* AgNO₃.

(c) *Ferric indicator.*—A saturated soln. of ferric ammonium alum.

· (d) *Nitric acid.*—Free from lower oxides of N by diluting the usual pure acid with about ¼ volume of H_2O, and boiling until perfectly colorless.

34. DETERMINATION

To the soln. prepared as directed under 31, add a known volume of the 0.1 *N* AgNO₃ in slight excess. Stir well, filter, and wash the AgCl precipitate thoroughly. To the combined filtrate and washings add 5 cc. of the ferric indicator and a few cc. of the HNO₃ and titrate the excess of Ag with the 0.1 *N* thiocyanate until a permanent light brown color appears. From the number of cc. of 0.1 *N* AgNO₃ used, calculate the quantity of Cl. 1 cc. of 0.1 *N* AgNO₃ = 0.00355 g. of Cl.

35. TOTAL ACIDS—OFFICIAL

Measure 20 cc. of the wine into a 250 cc. beaker, heat rapidly to incipient boiling, and immediately titrate with 0.1 *N* NaOH soln. Determine the end point with neutral 0.05% azolitmin soln. as an outside indicator. Place the indicator in the cavities of a spot plate and spot the wine into the azolitmin soln. The end point is reached when the color of the indicator remains unchanged by the addition to the wine of a few drops of 0.1 N alkali.

In the case of wines that are artificially colored and therefore cannot be titrated satisfactorily in the above manner, it will be found helpful to use phenolphthalein powder (one part of phenolphthalein mixed with 100 parts of dry, powdered K_2SO_4) as an indicator. Place this indicator in the cavities of a spot plate and spot the wine into the powder. The end of the titration is indicated when the powder acquires a pink tint.

Express the result in terms of tartaric acid. 1 cc. of 0.1 N NaOH soln. = 0.0075 g. of tartaric acid.

VOLATILE ACIDS

36. Method I—Official

Heat rapidly to incipient boiling 50 cc. of the wine in a 500 cc. distillation flask and pass steam through until 15 cc. of the distillate requires only 2 drops of 0.1 N NaOH soln. for neutralization. Boil the H_2O used to generate the steam several minutes before connecting the steam generator with the distillation flask in order to expel CO_2. Titrate rapidly with 0.1 N NaOH soln., using phenolphthalein indicator. The color should remain about 10 seconds. Express the result as acetic acid. 1 cc. of 0.1 N NaOH soln. = 0.0060 g. of acetic acid.

37. Method II—Official

Introduce 10 cc. of the wine, previously freed from CO_2, into the inner tube of a modified Sellier distillation apparatus (Fig. 50); add a small piece of paraffin to prevent foaming; and adjust the tube and its contents in place within the larger flask, which contains 100 cc. of recently boiled H_2O. Connect with a condenser as illustrated in the figure and distil by heating the outer flask. When 50 cc. of the distillate has been collected, empty the receiver into a beaker and titrate with 0.1 N NaOH soln., using phenolphthalein indicator. Continue the distillation and titrate each succeeding 10 cc. of distillate until not more than 1 drop of standard alkali is required to reach the neutral point. Usually 80 cc. of distillate will contain all the volatile acids.

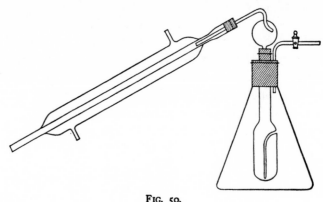

FIG. 50.

38. FIXED ACIDS—OFFICIAL

To obtain the quantity of fixed acids, expressed as tartaric acid, multiply the quantity of volatile acids by 1.25 and subtract this product from the total acids.

39. TOTAL TARTARIC ACID—OFFICIAL

Neutralize 100 cc. of the wine with N NaOH soln., calculating from the acidity, 44, the number of cc. of N alkali necessary for the neutralization. If the volume of the soln. is increased more than 10% by the addition of the alkali, evaporate to approximately 100 cc. Add to the neutralized soln. 0.075 g. of tartaric acid for each cc. of N alkali added and after the tartaric acid has dissolved add 2 cc. of glacial acetic acid and 15 g. of KCl. After the KCl has dissolved, add 15 cc. of 95% alcohol; stir vigorously until the K-bitartrate begins to precipitate; and let stand in an ice-box at 15-18° for at least 15 hours. Decant the liquid from the separated K-bitartrate on a Gooch crucible prepared with a very thin film of asbestos, or on filter paper in a

Büchner funnel. Wash the precipitate and filter 3 times with a few cc. of a mixture of 15 g. of KCl, 20 cc. of 95% alcohol, and 100 cc. of H_2O, using not more than 20 cc. of the wash soln. in all. Transfer the asbestos or paper and precipitate to the beaker in which the precipitation was made; wash the Gooch crucible or Büchner funnel with hot H_2O, using about 50 cc. in all; heat to boiling; and titrate the hot soln. with 0.1 N NaOH soln., using phenolphthalein indicator. Increase the number of cc. of 0.1 N alkali required by 1.5 cc. to allow for the solubility of the precipitate. 1 cc. of 0.1 N alkali is equivalent, under these conditions, to 0.015 g. of tartaric acid. To obtain the g. of total tartaric acid per 100 cc. of the wine, subtract the quantity of tartaric acid added from this result.

40.   FREE TARTARIC ACID AND CREAM OF TARTAR—OFFICIAL

Calculate the free tartaric acid and cream of tartar in the following manner:

Let A = total tartaric acid in 100 cc. of wine, divided by 0.015;
   B = total alkalinity of the ash (sum of C and D);
   C = alkalinity of water-soluble ash; and
   D = alkalinity of water-insoluble ash.
Then
   (1) If A is greater than B,
         Cream of tartar = 0.0188 $\times$ C, and
         Free tartaric acid = 0.015 $\times$ ($A - B$);
   (2) If A equals B or is smaller than B but greater than C
         Cream of tartar = 0.0188 $\times$ C, and
         Free tartaric acid = 0; and
   (3) If A is smaller than C,
         Cream of tartar = 0.0188 $\times$ A, and
         Free tartaric acid = 0.

41.   TANNIN AND COLORING MATTER—OFFICIAL

REAGENTS

(a) *Oxalic acid.*—0.1 N.   1 cc. = 0.00416 g. of tannin.

(b) *Standard potassium permanganate soln.* — Dissolve 1.333 g. of $KMnO_4$ in 1 liter of H_2O and standardize the soln. against (a).

(c) *Indigo soln.*—Dissolve 6 g. of Na-sulfindigotate in 500 cc. of H_2O by heating, cool, add 50 cc. of H_2SO_4, make up to 1 liter, and filter.

(d) *Purified boneblack.*—Boil 100 g. of finely powdered boneblack with successive portions of HCl ($+3$), filter, and wash with boiling H_2O until free from chlorides. Keep covered with H_2O.

42. DETERMINATION

Dealcoholize 100 cc. of the wine by evaporation and dilute with H_2O to the original volume. Transfer 10 cc. to a 2-liter porcelain dish and add about 1 liter of H_2O and exactly 20 cc. of the indigo soln. Add the standard $KMnO_4$ soln., 1 cc. at a time, until the blue color changes to green; then add a few drops at a time until the color becomes golden yellow. Designate the number of cc. of $KMnO_4$ soln. used as "a."

Treat 10 cc. of the dealcoholized wine, prepared as above, for 15 min. with boneblack; filter; and wash thoroughly with H_2O. Add 1 liter of H_2O and 20 cc. of the indigo soln. and titrate with $KMnO_4$, as above. Designate the number of cc. of $KMnO_4$ used as "b."

Then $a - b = c$, the number of cc. of the $KMnO_4$ soln. required for the oxidation of the tannin and coloring matter in 10 cc. of the wine.

43. CRUDE PROTEIN—OFFICIAL

Determine N in 50 cc. of the wine as directed below, and multiply the result by 6.25.

Place the sample in a digestion flask. Add approximately 0.7 g. of HgO, or its equivalent in metallic Hg, and 20-30 cc. of H_2SO_4 (0.1-0.3 g. of crystallized $CuSO_4$ may also be used in addition to the Hg, or in many cases, in place of it). Place the flask in an inclined position and heat below the boiling point of the acid until frothing has ceased. (A small piece of paraffin may be added to prevent extreme foaming.) Increase the heat until the acid boils briskly and digest for a time after the mixture is colorless or nearly so, or until oxidation is complete. (The digestion usually requires at least 2 hours.)

After cooling, dilute with about 200 cc. of H_2O, and add a few pieces of granulated Zn or pumice stone to prevent bumping, and 25 cc. of K_2S or $Na_2S_2O_3$ soln. with shaking. (If $Na_2S_2O_3$ is to be used, it should first be mixed with the NaOH so that they may be added together. When no Hg or HgO is used the addition of K_2S or Na_2S or $Na_2S_2O_3$ soln. is unnecessary.) Next add sufficient NaOH soln. to make the reaction strongly alkaline (50 cc. is usually sufficient), pouring it down the side of the flask so that it does not mix at once with the acid soln. Connect the flask to the condenser by means of a Kjeldahl connecting bulb, taking care that the tip of the condenser extends below the surface of the standard acid in the receiver; mix the contents by shaking; and distil until all NH_3 has passed over into a measured quantity of the standard acid. (The first 150 cc. of the distillate will generally contain all the NH_3.) Titrate with standard alkali soln., using the methyl red or cochineal indicator.

44. PENTOSANS—OFFICIAL

Proceed as directed under 45, 46, except to use 100 cc. of the wine and 43 cc. of HCl in beginning the distillation. Owing to the interference of sugars this determination can be made in dry wines only.

45. REAGENTS

(a) *Hydrochloric acid.*—Contains 12% by weight HCl. To 1 volume of HCl add 2 volumes of H_2O. Determine the percentage of acid by titration against standard alkali and adjust to proper strength by dilution or addition of more strong acid, as may be necessary.

(b) *Phloroglucin.*—Dissolve a small quantity of phloroglucin in a few drops of acetic anhydride, heat almost to boiling, and add a few drops of H_2SO_4. A violet color indicates the presence of diresorcin. A phloroglucin which gives more than a faint coloration may be purified by the following method:

Heat in a beaker about 300 cc. of the dilute HCl (a) and 11 g. of commercial phloroglucin, added in small quantities at a time, stirring constantly until it is nearly dissolved. Pour the hot soln. into a sufficient quantity of the same HCl (cold) to make the volume 1500 cc. Allow to stand at least over night, preferably several days, to permit the diresorcin to crystallize. Filter immediately before using. A yellow tint does not interfere with its usefulness. In using, add the volume containing the required quantity of phloroglucin to the distillate.

46. DETERMINATION

Place the sample in a 300 cc. distillation flask, together with 100 cc. of the dilute HCl and several pieces of recently ignited pumice stone. Place the flask on a wire gauze; connect with a condenser; and heat, rather gently at first, but then regulating so as to distil over 30 cc. in about 10 min. Pass the distillate through a small filter paper. Replace the 30 cc. distilled by a like quantity of the dilute acid, added by means of a separatory funnel in such a manner as to wash down the particles adhering to the sides of the flask, and continue the process until the distillate amounts to 360 cc. To the total distillate add gradually a quantity of phloroglucin dissolved in the dilute HCl and thoroughly stir the resulting mixture. The quantity of phloroglucin used should be about double that of the furfural expected. The soln. turns yellow, then green, and very soon there appears an amorphous greenish precipitate that grows darker rapidly, till it becomes almost black. Make the soln. up to 400 cc. with the dilute HCl and allow to stand over night.

Collect the amorphous black precipitate in a weighed Gooch crucible having an asbestos mat, wash carefully with 150 cc. of H_2O so that the H_2O is not entirely removed from the crucible until the very last, then dry for 4 hours at the temp. of boiling H_2O, cool, and weigh in a weighing bottle. The increase in weight is taken to be furfural phloroglucide. To calculate the furfural, pentose, or pentosan from the phloroglucide, use the following formulas given by Kröber:

(1) For a weight of phloroglucide, designated by "a" in the following formulas, *under* 0.03 g,

$$\text{Furfural} = (a + 0.0052) \times 0.5170.$$
$$\text{Pentoses} = (a + 0.0052) \times 1.0170.$$
$$\text{Pentosans} = (a + 0.0052) \times 0.8949.$$

In the above and also in the following formulas, the factor 0.0052 represents the weight of the phloroglucide that remains dissolved in the 400 cc. of acid soln.

(2) For a weight of phloroglucide "a" *between* 0.03 and 0.300 g., use the following formulas:

$$\text{Furfural} = (a + 0.0052) \times 0.5185.$$
$$\text{Pentoses} = (a + 0.0052) \times 1.0075.$$
$$\text{Pentosans} = (a + 0.0052) \times 0.8866.$$

(3) For a weight of phloroglucide "a" *over* 0.300 g.,

$$\text{Furfural} = (a + 0.0052) \times 0.5180.$$
$$\text{Pentoses} = (a + 0.0052) \times 1.0026.$$
$$\text{Pentosans} = (a + 0.0052) \times 0.8824.$$

47. GUM AND DEXTRIN—TENTATIVE

Evaporate 100 cc. of the wine to about 10 cc. and add 10 cc. of 95% alcohol. If gum or dextrin is present (indicated by the formation of a voluminous precipitate), continue the addition of alcohol, slowly and with stirring, until 100 cc. has been added. Let stand over night; filter; and wash with alcohol, 80% by volume. Dissolve the precipitate on the paper with hot H_2O, hydrolyze the filtrate and washings with HCl, and proceed as directed below.

48. (This method is intended only for such materials as raw starch, potatoes, etc., and includes as starch the pentosans and other carbohydrate bodies that undergo hydrolysis and are converted into reducing sugars on boiling with HCl.)

49. Heat the liquid for 2.5 hours with 200 cc. of H_2O and 20 cc. of HCl (sp. gr. 1.125) in a flask provided with a reflux condenser. Cool, and nearly neutralize with NaOH. Complete the volume to 250 cc., filter and determine the dextrose in an aliquot of the filtrate as directed under 16 and 17 or 18. The weight of the dextrose obtained multiplied by 0.90 gives the weight of starch.

50. NITRATES—TENTATIVE

(a) *White wine.*—Treat a few drops of the wine in a porcelain dish with 2-3 cc. of H_2SO_4 that contains about 0.1 g. of diphenylamine per 100 cc. The deep blue color formed in the presence of nitrates appears so quickly that it is not obscured, even in sweet wine, by the blackening produced by the action of H_2SO_4 on the sugar.

(b) *Red wine.*—Clarify with basic lead acetate, filter, remove the excess of Pb from the filtrate with Na_2SO_4, filter again, and treat a few drops of this filtrate as directed under (a).

51. COLORING MATTERS AND PRESERVATIVES

These subjects are treated extensively in Chapters XXI and XXXII of the Official and Tentative Methods of Analysis of the Association of Official Agricultural Chemists, 3rd ed. 1930, to which the student is referred. The following note is applicable to the question of preservatives in wine.

The detection of added boric acid is somewhat difficult because a small quantity of it is normally present in certain wines. Therefore a quantitative determination should be made. The determination of SO_2 must also be quantitative. A small quantity of salicylic acid is also normal in wine, and for that reason not more than 50 cc. of the sample should be used in testing for that preservative.

DISTILLED LIQUORS

52. SPECIFIC GRAVITY—OFFICIAL

Determine the specific gravity at 20/4° by means of a pycnometer, as directed under 3, or by means of a small, accurately graduated hydrometer.

53. ALCOHOL RY WEIGHT—OFFICIAL

Weigh 20-25 g. of the sample into a distillation flask, dilute with 100 cc. of H_2O, and distil nearly 100 cc. Weigh the distillate or make to volume at 20°. In either case determine the specific gravity as directed under 3. Obtain the corresponding percentage of alcohol by weight from Tables A3-A5; multiply this figure by the weight of the distillate; and divide by the weight of the sample taken to obtain the percentage of alcohol by weight.

The alcohol content of the distillate may be checked by determining the immersion refractometer reading and obtaining the percentage of alcohol from Table A6.

ALCOHOL BY VOLUME

54. *Method I—Official*

From the specific gravity of the distillate obtained under 53 ascertain the corresponding percentage of alcohol by volume from Tables A3-A5. Multiply this figure by the volume of distillate and divide by the volume of the sample (calculated from the specific gravity) to obtain the percentage of alcohol by volume in the original sample.

55. *Method II—Official*

Measure 25 cc. of the sample at 20° into a distillation flask, dilute with 100 cc. of H_2O, distil nearly 100 cc., make to volume at 20°, and determine the specific gravity as directed under 53. Obtain, from Tables A3-A5, the corresponding percentage of alcohol by volume in the distillate and multiply by 4 to obtain the percentage of alcohol by volume in the original substance.

The alcohol content of the distillate may be checked by determining the immersion refractometer reading and obtaining the percentage of alcohol from Table A6.

56. EXTRACT—OFFICIAL

Weigh, or measure at 20°, 100 cc. of the sample, evaporate nearly to dryness on a steam bath, transfer to a water oven, and dry at the temp. of boiling H_2O for 2.5 hours.

57. ASH—OFFICIAL

Proceed as directed under 25 using the residue from the determination of the extract, 56.

58. ACIDITY—OFFICIAL

Titrate 100 cc. of the sample (or 50 cc. diluted to 100 cc. if the sample is dark) with 0.1 N alkali, using phenolphthalein indicator. Express the result as acetic acid. 1 cc. of 0.1 N alkali = 0.0060 g. of acetic acid.

59. ESTERS—OFFICIAL *

Measure 100–200 cc. of the sample into a distillation flask; add 12.5–25 cc. of H_2O; and distil slowly 100–200 cc., depending upon the amount of sample taken, using a mercury valve to prevent loss of alcohol. Exactly neutralize the free acid in 50 cc. of the distillate with 0.1 N alkali and add a measured excess of 25–50 cc. of 0.1 N alkali. Then either boil for an hour under a reflux condenser, cool, and titrate with 0.1 N acid, or allow the soln. to stand over night in a stoppered flask with the excess of alkali, heat with a tube condenser for 30 min. at a temp. below the boiling point, cool, and titrate. Calculate the number of cc. of 0.1 N alkali used in the saponification of the esters as ethyl acetate. 1 cc. of 0.1 N alkali = 0.0088 g. of ethyl acetate. Run a blank, using water in place of the distillate, and make any necessary correction.

60. ALDEHYDES—OFFICIAL

REAGENTS

(a) *Aldehyde-free alcohol.*—Redistil 95% alcohol over NaOH or KOH; add 2–3 g. per liter of meta-phenylenediamine hydrochloride; digest at ordinary temp. for several days (or under a reflux condenser on a steam bath for several hours); and distil slowly, rejecting the first 100 cc. and the last 200 cc. of the distillate.

(b) *Sulfite-fuchsin soln.*—Dissolve 0.50 g. of pure fuchsin in 500 cc. of H_2O, add 5 g. of SO_2 dissolved in H_2O, make up to 1 liter, and allow to stand until colorless. As this soln. decomposes rapidly, prepare it in small quantities and keep at a low temp.

* The use of "100-200 cc." instead of "200 cc." and of "12.5-25 cc." instead of "25 cc." has been approved as official, first action.

(c) *Standard acetaldehyde soln.*—Prepare according to the directions of Vasey, as follows: Grind aldehyde ammonia in a mortar with anhydrous ether and decant the ether. Repeat this operation several times and dry the purified salt in a current of air and then in vacuo over H_2SO_4. Dissolve 1.386 g. of this purified aldehyde ammonia in 50 cc. of 95% alcohol, add 22.7 cc. of N alcoholic H_2SO_4, make up to 100 cc. and add 0.8 cc. of alcohol for the volume of the $(NH_4)_2SO_4$ precipitate. Allow the mixture to stand over night, and filter. This soln. contains 1 g. of acetaldehyde in 100 cc. and will retain its strength.

The standard found most convenient for use is 2 cc. of this strong aldehyde soln. diluted to 100 cc. with alcohol, 50% by volume. 1 cc. of this soln. = 0.0002 g. of acetaldehyde. Make up the soln. every day or so, as it loses strength.

61. DETERMINATION

Determine the aldehyde in the distillate prepared as directed under 65,57. Dilute 5–10 cc. of the distillate to 50 cc. with aldehyde-free alcohol, 50% by volume; add 25 cc. of the sulfite-fuchsin soln.; and allow to stand for 15 min. at 15°. The solns. and reagents should be at 15° when they are mixed. Prepare standards of known strength and blanks in the same way. The comparison standards found most convenient for use contain 0.1, 0.2, 0.3, 0.4, 0.5, and 0.6 mg. of acetaldehyde.

62. FURFURAL—OFFICIAL

REAGENT

Standard furfural soln.—Dissolve 1 g. of redistilled furfural in 100 cc. of 95% alcohol. Prepare standards by diluting 1 cc. of this soln. to 100 cc. with alcohol, 50% by volume. 1 cc. of this soln. contains 0.1 mg. of furfural. (The strong furfural soln. will retain its strength, but the dilute soln. will not.)

63. DETERMINATION

Dilute 10–20 cc. of the distillate, as prepared under 65, to 50 cc. with furfural-free alcohol, 50% by volume. Add 2 cc. of colorless aniline and 0.5 cc. of HCl (sp. gr. 1.125) and keep for 15 min. in a water bath at about 15°. Prepare standards of known strength and blanks in the same way. The comparison standards found most convenient for use contain 0.05, 0.1, 0.15, 0.2, 0.25, and 0.3 mg. of furfural.

64. FUSEL OIL—OFFICIAL*

REAGENTS

(a) *Purified carbon tetrachloride.*—Mix in a separatory funnel crude CCl_4 with $\frac{1}{10}$ its volume of H_2SO_4, shake thoroughly at frequent intervals, and allow to stand overnight. Wash free of acid and impurities with tap H_2O, remove the H_2O, add an excess of NaOH soln., and distil the CCl_4.

The refuse CCl_4 after titration is purified for further work by collecting in a large bottle, adding NaOH soln. ($1 + 1$), shaking, washing with tap H_2O until the washings are neutral to phenolphthalein, and distilling.

(b) *Oxidizing soln.*—Dissolve 100 g. of $K_2Cr_2O_7$ in 900 cc. of H_2O and add 100 cc. of H_2SO_4.

65. DETERMINATION

(1) To 50 cc. of the sample add 50 cc. of H_2O, then add 20 cc. of 0.5 N NaOH, and saponify the mixture by boiling for an hour under a reflux condenser; or, (2) mix 50 cc. of the liquid and 50 cc. of H_2O with 20 cc. of 0.5 N NaOH, allow to stand overnight at room temp., and distil directly. Connect the flask with a distillation apparatus, distil 90 cc., add 25 cc. of H_2O, and continue the distillation until an additional 25 cc. is collected.

Whenever aldehydes are present in excess of 15 parts per 100,000, add to the distillate 0.5 g. of metaphenylenediamine hydrochloride, boil·under a reflux condenser for an hour, distil 100 cc., add 25 cc. of H_2O, and continue the distillation until an additional 25 cc. is collected.

---

* In 65, lines 1 and 3, the use of "50 cc." instead of "100 cc." has been approved as official, first action.

Approximately saturate the distillate with finely ground NaCl and add saturated NaCl soln. until the specific gravity is 1.10. Extract this salt soln. 4 times with the purified CCl_4, using 40, 30, 20, and 10 cc., respectively, and wash the CCl_4 3 times with 50 cc. portions of saturated NaCl soln., and twice with saturated Na_2SO_4 soln. Transfer the CCl_4 to a flask containing 50 cc. of the oxidizing soln. and boil for 8 hours under a reflux condenser.

Add 100 cc. of H_2O and distil until only about 50 cc. remains. Add 50 cc. of H_2O and again distil until 35–50 cc. is left. Use extreme care to prevent the oxidizing mixture from burning and baking on the side of the distilling flask. The distillate should be water white; if it is colored discard it and repeat the determination. Titrate the distillate with 0.1 N NaOH, using phenolphthalein indicator. 1 cc. of 0.1 N NaOH = 0.0088 g. of amyl alcohol.

If preferred, use rubber stoppers in the saponification and first distillation, but use corks covered with tinfoil in the oxidation and second distillation. Renew the corks and tinfoil frequently.

Conduct a blank determination upon 100 cc. of CCl_4, beginning the blank at that point of the procedure immediately after the extraction and just before the washings with NaCl and Na_2SO_4 solns.

66. SUGARS—OFFICIAL

Proceed as directed under 14–24.

METHYL ALCOHOL

67. *Trillat Method—Official*

To 50 cc. of the sample add 50 cc. of H_2O and 8 g. of lime and fractionate by the aid of Glinsky bulb tubes. Dilute the first 15 cc. of the distillate to 150 cc., mix with 15 g. of $K_2Cr_2O_7$ and 70 cc. of H_2SO_4 (1 + 5), and allow to stand for 1 hour, shaking occasionally.

Distil, reject the first 25 cc., and collect 100 cc. Mix 50 cc. of the distillate with 1 cc. of redistilled dimethylaniline, transfer to a stout tightly stoppered flask, and keep on a bath at 70-80°

for 3 hours, shaking occasionally. Make distinctly alkaline with NaOH solution and distil off the excess of dimethylaniline, stopping the distillation when 25 cc. has passed over.

Acidify the residue in the flask with acetic acid, shake, and test a few cc. by adding 4 or 5 drops of a 1% suspension of PbO₂. If methyl alcohol is present, there occurs a blue coloration, which is increased by boiling. Ethyl alcohol thus treated yields a blue coloration which changes immediately to green, later to yellow, and becomes colorless when boiled.

68. *Riche and Bardy Method—Official*

The following method depends on the formation of methylaniline violet:

Place 10 cc. of the sample, previously redistilled over K_2CO_3 if necessary, in a small flask with 15 g. of I and 2 g. of red P. Keep in ice H_2O for 10–15 minutes or until action has ceased. Distil on a water bath into about 30 cc. of H_2O, the methyl and ethyl iodides formed. Wash with dilute alkali to eliminate free I. Separate the heavy, oily liquid that settles and transfer to a flask containing 5 cc. of aniline. If the action is too violent, place the flask in cold H_2O; if too slow, stimulate by gently warming the flask. After an hour boil the product with H_2O, cool, and add about 20 cc. of 15% NaOH solution; when the bases rise to the top as an oily layer, fill the flask up to the neck with H_2O and draw them off with a pipet. Oxidize 1 cc. of the oily liquid by adding 10 g. of a mixture of 100 parts of clean sand, 2 of NaCl, and 3 of $Cu(NO_3)_2$; mix thoroughly; transfer to a glass tube; and heat to 90° for 8–10 hours. Exhaust the product with warm alcohol, filter, and dilute to 100 cc. with alcohol. If the sample is free from methyl alcohol, the liquid has a red tint, but in the presence of 1% of methyl alcohol it has a distinct violet shade; with 2.5% the shade is very distinct and still more so with 5%. To detect more minute quantities of methyl alcohol, dilute 5 cc. of the colored liquid to 100 cc. with H_2O and dilute 5 cc. of this again to 400 cc. Heat the liquid thus obtained in a porcelain dish and immerse in it a fragment of white merino (free from S) for 30 minutes. If the alcohol is pure, the wool will remain white, but if methyl alcohol is present the fiber will become violet, the depth of tint giving a fairly approximate indication of the proportion of methyl alcohol.

69. *Immersion Refractometer Method—Official*

Determine by the immersion refractometer at 20° the refraction of the distillate obtained in the determination of alcohol. If, on reference to the table under A6, the refraction shows the percentage of alcohol agreeing with that obtained from the specific gravity, it may be assumed that no methyl alcohol is present. If, however, there is an appreciable quantity of methyl alcohol, the low refractometer reading will at once indicate the fact. If the absence from the solution of refractive substances other than H_2O and the alcohols is assured, this difference in refraction is conclusive of the presence of methyl alcohol.

The addition of methyl alcohol to ethyl alcohol decreases the refraction in direct proportion to the quantity present; hence the quantitative calculation is made readily by interpolation in the table under 72 of the figures for pure ethyl and methyl alcohol of the same alcoholic strength as the sample being used.

Example.—The distillate has a specific gravity of 0.97080, corresponding to 18.38% alcohol by weight, and has a refraction of 35.8 at 20° by the immersion refractometer; by interpolation in the refractometer table the readings of ethyl and methyl alcohol corresponding to 18.38% alcohol are 47.3 and 25.4, respectively, the difference being 21.9; $47.3 - 35.8 = 11.5$; $(11.5 \div 21.9)100 = 52.5$, showing that 52.5% of the total alcohol present is methyl alcohol.

70. COLORING MATTERS—TENTATIVE

See under 51.

71. WATER-INSOLUBLE COLOR—TENTATIVE

Evaporate 50 cc. of the sample just to dryness on a steam bath. Take up with approximately 15 cc. of cold H_2O, filter, and

72. TABLE AI.

SCALE READINGS ON ZEISS IMMERSION REFRACTOMETER AT 20°, CORRESPONDING TO EACH PER CENT BY WEIGHT OF METHYL AND ETHYL ALCOHOLS

Per cent alcohol by weight	Scale Readings		Per cent alcohol by weight	Scale Readings		Per cent alcohol by weight	Scale Readings		Per cent alcohol by weight	Scale Readings	
	Methyl alcohol	Ethyl alcohol		Methyl alcohol	Ethyl alcohol		Methyl alcohol	Ethyl alcohol		Methyl alcohol	Ethyl alcohol
0	14.5	14.5	25	29.7	60.1	50	39.8	90.3	75	29.7	101.0
1	14.8	16.0	26	30.3	61.9	51	39.7	91.1	76	29.0	101.0
2	15.4	17.6	27	30.9	63.7	52	39.6	91.8	77	28.3	100.9
3	16.0	19.1	28	31.6	65.5	53	39.6	92.4	78	27.6	100.9
4	16.6	20.7	29	32.2	67.2	54	39.5	93.0	79	26.8	100.8
5	17.2	22.3	30	32.8	69.0	55	39.4	93.6	80	26.0	100.7
6	17.8	24.1	31	33.5	70.4	56	39.2	94.1	81	25.1	100.6
7	18.4	25.9	32	34.1	71.7	57	39.0	94.7	82	24.3	100.5
8	19.0	27.8	33	34.7	73.1	58	38.6	95.2	83	23.6	100.4
9	19.6	29.6	34	35.2	74.4	59	38.3	95.7	84	22.8	100.3
10	20.2	31.4	35	35.8	75.8	60	37.9	96.2	85	21.8	100.1
11	20.8	33.2	36	36.3	76.9	61	37.5	96.7	86	20.8	99.8
12	21.4	35.0	37	36.8	78.0	62	37.0	97.1	87	19.7	99.5
13	22.0	36.9	38	37.3	79.1	63	36.5	97.5	88	18.6	99.2
14	22.6	38.7	39	37.7	80.2	64	36.0	98.0	89	17.3	98.9
15	23.2	40.5	40	38.1	81.3	65	35.5	98.3	90	16.1	98.6
16	23.9	42.5	41	38.4	82.3	66	35.0	98.7	91	14.9	98.3
17	24.5	44.5	42	38.8	83.3	67	34.5	99.1	92	13.7	97.8
18	25.2	46.5	43	39.2	84.2	68	34.0	99.4	93	12.4	97.2
19	25.8	48.5	44	39.3	85.2	69	33.5	99.7	94	11.0	96.4
20	26.5	50.5	45	39.4	86.2	70	33.0	100.0	95	9.6	95.7
21	27.1	52.4	46	39.5	87.0	71	32.3	100.2	96	8.2	94.9
22	27.8	54.3	47	39.6	87.8	72	31.7	100.4	97	6.7	94.0
23	28.4	56.3	48	39.7	88.7	73	31.1	100.6	98	5.1	93.0
24	29.1	58.2	49	39.8	89.5	74	30.4	100.8	99	3.5	92.0
									100	2.0	91.0

wash until the filtrate amounts to nearly 25 cc. To this filtrate add 25 cc. of absolute alcohol, or 26.3 cc. of 95% alcohol, and make up to 50 cc. with H₂O. Mix thoroughly and compare in a colorimeter with the original material. Calculate from these readings the percentage of color insoluble in H_2O.

73. COLORS INSOLUBLE IN AMYL ALCOHOL—TENTATIVE

Evaporate 50 cc. of the sample just to dryness on a steam bath. Dissolve the residue in H_2O and 95% alcohol and make to a volume of 50 cc., using a total volume of 26.3 cc. of 95% alcohol. Place 25 cc. of this solution in a separatory funnel and add 20 cc. of freshly shaken Marsh reagent (100 cc. of pure amyl alcohol, 3 cc. of sirupy H_3PO_4, and 3 cc. of H_2O), shaking lightly so as not to form an emulsion. Allow the layers to separate and repeat this shaking and standing twice. After the layers have separated completely draw off the lower or aqueous layer, which contains the caramel, into a 25 cc. cylinder and make up to volume with alcohol, 50% by volume. Compare this solution in a colorimeter with the untreated 25 cc. Calculate from this reading the percentage of color insoluble in amyl alcohol.

74. CARAMEL—TENTATIVE

Add 10 cc. of paraldehyde to 5 cc. of the sample in a test tube and shake. Add absolute alcohol, a few drops at a time, shaking after each addition until the mixture becomes clear. Allow to stand. Turbidity after 10 minutes is an indication of caramel.

MARSH TEST FOR ARTIFICIAL COLORS—TENTATIVE

(Caramel and Some Coal Tar Dyes)

75. REAGENT

Marsh Reagent.—Prepare as directed under 73.

76. DETERMINATION

Place 10 cc. of the sample in a 20 cc. test tube, add sufficient Marsh reagent to nearly fill the tube, and shake several times. Allow the layers to separate; if the lower layer is colored, it is an indication that the sample has been colored with caramel or a coal tar dye.

In the absence of any color, test 10 cc. of the sample in the same manner, using sufficient fusel oil, amyl alcohol or pentasol to nearly fill the tube, and shake several times. A deeply colored lower layer is an indication of a coal tar dye; its identity should be confirmed by using the methods under 51.

DETECTION OF METHANOL IN ALCOHOLIC BEVERAGES

77. REAGENT

(a) *Potassium permanganate solution.*—Dissolve 3 g. of $KMnO_4$ in 15 cc. of 85% H_3PO_4 and make up to 100 cc. with H_2O.

(b) *Oxalic acid solution.*—Dissolve 5 g. of oxalic acid in 100 cc. of H_2SO_4 (1 + 1).

(c) *Schiff's reagent.*—Dissolve 0.2 g. of Kahlbaum rosaniline hydrochloride in 120 cc. of hot H_2O, cool, add 2 g. of anhydrous Na_2SO_3 dissolved in 20 cc. of H_2O, and 2 cc. of HCl, make up to 200 cc., and store in well-filled glass-stoppered amber bottles.

78. DETERMINATION

Dilute the alcoholic beverage to 5% total alcohol by volume. Transfer 5 cc. of this solution to a 6-inch test tube; add 2 cc. of the $KMnO_4$ solution; and let stand 10 minutes. Remove the excess of $KMnO_4$ by the addition of 2 cc. of the oxalic acid solution. As soon as the $KMnO_4$ is decolorized add 5 cc. of Schiff's reagent. Mix thoroughly and let stand 10 minutes. If HCHO is present, the characteristic reddish purple color is produced.

Run blanks on pure ethyl alcohol and on ethyl alcohol containing about 1% of methanol.

CORDIALS AND LIQUEURS

No special methods for the examination of liqueurs and cordials have been included in the Official and Tentative Methods of Analysis of the Association of Official Agricultural Chemists. The reader will find that the methods described above are generally applicable to this purpose. The following are suggested as providing most of the determinations which will be required:

	Paragraph		*Paragraph*
Physical Examination.	1	Volatile Acids........	36, or 37
Specific Gravity......	3	Fixed Acids.........	38
Alcohol.............	4	Total Tartaric Acid...	39
Glycerol.............	9, 5, or 6	Tannin and Coloring	
Total Solids.........	11	Matter...........	41
Sugars..............	14–24	Gum and Dextrin.....	47
Non-sugar Solids.....	13	Coloring Matters and	
Ash.................	25	Preservatives.......	51
Ash-extract Ratio.....	26	Esters..............	59
Alkalinity of Water-		Aldehydes..........	60
soluble Ash........	27	Furfural............	61, 62
Alkalinity of Water-		Fusel Oil...........	63, 64
insoluble Ash.......	28	Methyl Alcohol......	67–69
Total Acids.........	35	Detection of Methanol.	77–78

TABLE A2.—DENSITIES OF SOLUTIONS OF CANE SUGAR AT 20° C.

Per cent sugar	Tenths of Per Cent					Per cent sugar
	0	1	2	3	4	
0	0.998234	0.998622	0.999010	0.999398	0.999786	0
1	1.002120	1.002509	1.002897	1.003286	1.003675	1
2	1.006015	1.006405	1.006796	1.007188	1.007580	2
3	1.009934	1.010327	1.010721	1.011115	1.011510	3
4	1.013881	1.014277	1.014673	1.015070	1.015467	4
5	1.017854	1.018253	1.018652	1.019052	1.019451	5
6	1.021855	1.022257	1.022659	1.023061	1.023463	6
7	1.025885	1.026289	1.026694	1.027099	1.027504	7
8	1.029942	1.030349	1.030757	1.031165	1.031573	8
9	1.034029	1.034439	1.034850	1.035260	1.035671	9
10	1.038143	1.038556	1.038970	1.039383	1.039797	10
11	1.042288	1.042704	1.043121	1.043537	1.043954	11
12	1.046462	1.046881	1.047300	1.047720	1.048140	12
13	1.050665	1.051087	1.051510	1.051933	1.052356	13
14	1.054900	1.055325	1.055751	1.056176	1.056602	14
15	1.059165	1.059593	1.060022	1.060451	1.060880	15
16	1.063460	1.063892	1.064324	1.064756	1.065188	16
17	1.067789	1.068223	1.068658	1.069093	1.069529	17
18	1.072147	1.072585	1.073023	1.073461	1.073900	18
19	1.076537	1.076978	1.077419	1.077860	1.078302	19
20	1.080959	1.081403	1.081848	1.082292	1.082737	20
21	1.085414	1.085861	1.086309	1.086757	1.087205	21
22	1.089900	1.090351	1.090802	1.091253	1.091704	22
23	1.094420	1.094874	1.095328	1.095782	1.096236	23
24	1.098971	1.099428	1.099886	1.100344	1.100802	24
25	1.103557	1.104017	1.104478	1.104938	1.105400	25
26	1.108175	1.108639	1.109103	1.109568	1.110033	26
27	1.112828	1.113295	1.113763	1.114229	1.114697	27
28	1.117512	1.117982	1.118453	1.118923	1.119395	28
29	1.122231	1.122705	1.123179	1.123653	1.124128	29
30	1.126984	1.127461	1.127939	1.128417	1.128896	30
31	1.131773	1.132254	1.132735	1.133216	1.133698	31
32	1.136596	1.137080	1.137565	1.138049	1.138534	32
33	1.141453	1.141941	1.142429	1.142916	1.143405	33
34	1.146345	1.146836	1.147328	1.147820	1.148313	34
35	1.151275	1.151770	1.152265	1.152760	1.153256	35
36	1.156238	1.156736	1.157235	1.157733	1.158233	36
37	1.161236	1.161738	1.162240	1.162742	1.163245	37
38	1.166269	1.166775	1.167281	1.167786	1.168293	38
39	1.171340	1.171849	1.172359	1.172869	1.173379	39
40	1.176447	1.176960	1.177473	1.177987	1.178501	40
41	1.181592	1.182108	1.182625	1.183142	1.183660	41
42	1.186773	1.187293	1.187814	1.188335	1.188856	42
43	1.191993	1.192517	1.193041	1.193565	1.194090	43
44	1.197247	1.197775	1.198303	1.198832	1.199360	44
45	1.202540	1.203071	1.203603	1.204136	1.204668	45
46	1.207870	1.208405	1.208940	1.209477	1.210013	46
47	1.213238	1.213777	1.214317	1.214856	1.215395	47
48	1.218643	1.219185	1.219729	1.220272	1.220815	48
49	1.224086	1.224632	1.225180	1.225727	1.226274	49

TABLE A2.—DENSITIES OF SOLUTIONS OF CANE SUGAR AT 20° C.—*Continued*

Per cent sugar	Tenths of Per Cent					Per cent sugar
	5	6	7	8	9	
0	1.000174	1.000563	1.000952	1.001342	1.001731	0
1	1.004064	1.004453	1.004844	1.005234	1.005624	1
2	1.007972	1.008363	1.008755	1.009148	1.009541	2
3	1.011904	1.012298	1.012694	1.013089	1.013485	3
4	1.015864	1.016261	1.016659	1.017058	1.017456	4
5	1.019851	1.020251	1.020651	1.021053	1.021454	5
6	1.023867	1.024270	1.024673	1.025077	1.025481	6
7	1.027910	1.028316	1.028722	1.029128	1.029535	7
8	1.031982	1.032391	1.032800	1.033209	1.033619	8
9	1.036082	1.036494	1.036906	1.037318	1.037730	9
10	1.040212	1.040626	1.041041	1.041456	1.041872	10
11	1.044370	1.044788	1.045206	1.045625	1.046043	11
12	1.048559	1.048980	1.049401	1.049822	1.050243	12
13	1.052778	1.053202	1.053626	1.054050	1.054475	13
14	1.057029	1.057455	1.057882	1.058310	1.058737	14
15	1.061308	1.061738	1.062168	1.062598	1.063029	15
16	1.065621	1.066054	1.066487	1.066921	1.067355	16
17	1.069964	1.070400	1.070836	1.071273	1.071710	17
18	1.074338	1.074777	1.075217	1.075657	1.076097	18
19	1.078744	1.079187	1.079629	1.080072	1.080515	19
20	1.083182	1.083628	1.084074	1.084520	1.084967	20
21	1.087652	1.088101	1.088550	1.089000	1.089450	21
22	1.092155	1.092607	1.093060	1.093513	1.093966	22
23	1.096691	1.097147	1.097603	1.098058	1.098514	23
24	1.101259	1.101718	1.102177	1.102637	1.103097	24
25	1.105862	1.106324	1.106786	1.107248	1.107711	25
26	1.110497	1.110963	1.111429	1.111895	1.112361	26
27	1.115166	1.115635	1.116104	1.116572	1.117042	27
28	1.119867	1.120339	1.120812	1.121284	1.121757	28
29	1.124603	1.125079	1.125555	1.126030	1.126507	29
30	1.129374	1.129853	1.130332	1.130812	1.131292	30
31	1.134180	1.134663	1.135146	1.135628	1.136112	31
32	1.139020	1.139506	1.139993	1.140479	1.140966	32
33	1.143894	1.144384	1.144874	1.145363	1.145854	33
34	1.148805	1.149298	1.149792	1.150286	1.150780	34
35	1.153752	1.154249	1.154746	1.155242	1.155740	35
36	1.158733	1.159233	1.159733	1.160233	1.160734	36
37	1.163748	1.164252	1.164756	1.165259	1.165764	37
38	1.168800	1.169307	1.169815	1.170322	1.170831	38
39	1.173889	1.174400	1.174911	1.175423	1.175935	39
40	1.179014	1.179527	1.180044	1.180560	1.181076	40
41	1.184178	1.184696	1.185215	1.185734	1.186253	41
42	1.189379	1.189901	1.190423	1.190946	1.191469	42
43	1.194616	1.195141	1.195667	1.196193	1.196720	43
44	1.199890	1.200420	1.200950	1.201480	1.202010	44
45	1.205200	1.205733	1.206266	1.206801	1.207335	45
46	1.210549	1.211086	1.211623	1.212162	1.212700	46
47	1.215936	1.216476	1.217017	1.217559	1.218101	47
48	1.221360	1.221904	1.222449	1.222995	1.223540	48
49	1.226823	1.227371	1.227919	1.228469	1.229018	49

TABLE A2.—DENSITIES OF SOLUTIONS OF CANE SUGAR AT 20° C.—*Continued*

Per cent sugar	Tenths of Per Cent					Per cent sugar
	0	1	2	3	4	
50	1.229567	1.230117	1.230668	1.231219	1.231770	50
51	1.235085	1.235639	1.236194	1.236748	1.237303	51
52	1.240641	1.241198	1.241757	1.242315	1.242873	52
53	1.246234	1.246795	1.247358	1.247920	1.248482	53
54	1.251866	1.252431	1.252997	1.253563	1.254129	54
55	1.257535	1.258104	1.258674	1.259244	1.259815	55
56	1.263243	1.263816	1.264390	1.264963	1.265537	56
57	1.268989	1.269565	1.270143	1.270720	1.271299	57
58	1.274774	1.275354	1.275936	1.276517	1.277098	58
59	1.280595	1.281179	1.281764	1.282349	1.282935	59
60	1.286456	1.287044	1.287633	1.288222	1.288811	60
61	1.292354	1.292946	1.293539	1.294131	1.294725	61
62	1.298291	1.298886	1.299483	1.300079	1.300677	62
63	1.304267	1.304867	1.305467	1.306068	1.306669	63
64	1.310282	1.310885	1.311489	1.312093	1.312699	64
65	1.316334	1.316941	1.317549	1.318157	1.318766	65
66	1.322425	1.323036	1.323648	1.324259	1.324872	66
67	1.328554	1.329170	1.329785	1.330401	1.331017	67
68	1.334722	1.335342	1.335961	1.336581	1.337200	68
69	1.340928	1.341551	1.342174	1.342798	1.343421	69
70	1.347174	1.347801	1.348427	1.349055	1.349682	70
71	1.353456	1.354087	1.354717	1.355349	1.355980	71
72	1.359778	1.360413	1.361047	1.361682	1.362317	72
73	1.366139	1.366777	1.367415	1.368054	1.368693	73
74	1.372536	1.373178	1.373820	1.374463	1.375105	74
75	1.378971	1.379617	1.380262	1.380909	1.381555	75
76	1.385446	1.386096	1.386745	1.387396	1.388045	76
77	1.391956	1.392610	1.393263	1.393917	1.394571	77
78	1.398505	1.399162	1.399819	1.400477	1.401134	78
79	1.405091	1.405752	1.406412	1.407074	1.407735	79
80	1.411715	1.412380	1.413044	1.413709	1.414374	80
81	1.418374	1.419043	1.419711	1.420380	1.421049	81
82	1.425072	1.425744	1.426416	1.427089	1.427761	82
83	1.431807	1.432483	1.433158	1.433835	1.434511	83
84	1.438579	1.439259	1.439938	1.440619	1.441299	84
85	1.445388	1.446071	1.446754	1.447438	1.448121	85
86	1.452232	1.452919	1.453605	1.454292	1.454980	86
87	1.459114	1.459805	1.460495	1.461186	1.461877	87
88	1.466032	1.466726	1.467420	1.468115	1.468810	88
89	1.472986	1.473684	1.474381	1.475080	1.475779	89
90	1.479976	1.480677	1.481378	1.482080	1.482782	90
91	1.487002	1.487707	1.488411	1.489117	1.489823	91
92	1.494063	1.494771	1.495479	1.496188	1.496897	92
93	1.501158	1.501870	1.502582	1.503293	1.504006	93
94	1.508289	1.509004	1.509720	1.510435	1.511151	94
95	1.515455	1.516174	1.516893	1.517612	1.518332	95
96	1.522656	1.523378	1.524100	1.524823	1.525546	96
97	1.529891	1.530616	1.531342	1.532068	1.532794	97
98	1.537161	1.537889	1.538618	1.539347	1.540076	98
99	1.544462	1.545194	1.545926	1.546659	1.547392	99
100	1.551800					100

TABLE A2.—DENSITIES OF SOLUTIONS OF CANE SUGAR AT 20° C.—*Continued*

Per cent sugar	Tenths of Per Cent					Per cent sugar
	5	6	7	8	9	
50	1.232322	1.232874	1.233426	1.233979	1.234532	50
51	1.237859	1.238414	1.238970	1.239527	1.240084	51
52	1.243433	1.243992	1.244552	1.245113	1.245673	52
53	1.249046	1.249609	1.250172	1.250737	1.251301	53
54	1.254697	1.255264	1.255831	1.256400	1.256967	54
55	1.260385	1.260955	1.261527	1.262099	1.262671	55
56	1.266112	1.266686	1.267261	1.267837	1.268413	56
57	1.271877	1.272455	1.273035	1.273614	1.274194	57
58	1.277680	1.278262	1.278844	1.279428	1.280011	58
59	1.283521	1.284107	1.284694	1.285281	1.285869	59
60	1.289401	1.289991	1.290581	1.291172	1.291763	60
61	1.295318	1.295911	1.296506	1.297100	1.297696	61
62	1.301274	1.301871	1.302470	1.303068	1.303668	62
63	1.307271	1.307872	1.308475	1.309077	1.309680	63
64	1.313304	1.313909	1.314515	1.315121	1.315728	64
65	1.319374	1.319983	1.320593	1.321203	1.321814	65
66	1.325484	1.326097	1.326711	1.327325	1.327940	66
67	1.331633	1.332250	1.332868	1.333485	1.334103	67
68	1.337821	1.338441	1.339063	1.339684	1.340306	68
69	1.344046	1.344671	1.345296	1.345922	1.346547	69
70	1.350311	1.350939	1.351568	1.352197	1.352827	70
71	1.356612	1.357245	1.357877	1.358511	1.359144	71
72	1.362953	1.363590	1.364226	1.364864	1.365501	72
73	1.369333	1.369973	1.370613	1.371254	1.371894	73
74	1.375749	1.376392	1.377036	1.377680	1.378326	74
75	1.382203	1.382851	1.383499	1.384148	1.384796	75
76	1.388696	1.389347	1.389999	1.390651	1.391303	76
77	1.395226	1.395881	1.396536	1.397192	1.397848	77
78	1.401793	1.402452	1.403111	1.403771	1.404430	78
79	1.408398	1.409061	1.409723	1.410387	1.411051	79
80	1.415040	1.415706	1.416373	1.417039	1.417707	80
81	1.421719	1.422390	1.423059	1.423730	1.424400	81
82	1.428435	1.429109	1.429782	1.430457	1.431131	82
83	1.435188	1.435866	1.436543	1.437222	1.437900	83
84	1.441980	1.442661	1.443342	1.444024	1.444705	84
85	1.448806	1.449491	1.450175	1.450860	1.451545	85
86	1.455668	1.456357	1.457045	1.457735	1.458424	86
87	1.462568	1.463260	1.463953	1.464645	1.465338	87
88	1.469504	1.470200	1.470896	1.471592	1.472289	88
89	1.476477	1.477176	1.477876	1.478575	1.479275	89
90	1.483484	1.484187	1.484890	1.485593	1.486297	90
91	1.490528	1.491234	1.491941	1.492647	1.493355	91
92	1.497606	1.498316	1.499026	1.499736	1.500447	92
93	1.504719	1.505432	1.506146	1.506859	1.507574	93
94	1.511868	1.512585	1.513302	1.514019	1.514737	94
95	1.519051	1.519771	1.520492	1.521212	1.521934	95
96	1.526269	1.526993	1.527717	1.528441	1.529166	96
97	1.533521	1.534248	1.534976	1.535704	1.536432	97
98	1.540806	1.541536	1.542267	1.542998	1.543730	98
99	1.548127	1.548861	1.549595	1.550329	1.551064	99
100						100

TABLE A3.—RELATION BETWEEN SPECIFIC GRAVITY AT 15.56/15.56° C. AND ALCOHOL CONTENT OF MIXTURES OF ETHYL ALCOHOL AND WATER *

Sp. gr. 15.56°/15.56°	Per cent by volume	Per cent by weight	Grams per 100 cc.	Sp. gr. 15.56°/15.56°	Per cent by volume	Per cent by weight	Grams per 100 cc.	Sp. gr. 15.56°/15.56°	Per cent by volume	Per cent by weight	Grams per 100 cc.	Sp. gr. 15.56°/15.56°	Per cent by volume	Per cent by weight	Grams per 100 cc.
1.00000	0.0	0.00	0.00	0.99085	6.5	5.21	5.16	0.98319	13.0	10.50	10.31	0.97645	19.5	15.85	15.47
0.99985	0.1	0.08	0.08	0.99073	6.6	5.29	5.23	0.98308	13.1	10.58	10.39	0.97635	19.6	15.94	15.55
0.99970	0.2	0.16	0.16	0.99060	6.7	5.37	5.31	0.98297	13.2	10.66	10.47	0.97625	19.7	16.02	15.62
0.99955	0.3	0.24	0.24	0.99047	6.8	5.45	5.39	0.98286	13.3	10.74	10.55	0.97615	19.8	16.10	15.70
0.99940	0.4	0.32	0.32	0.99035	6.9	5.53	5.47	0.98275	13.4	10.82	10.63	0.97605	19.9	16.19	15.78
0.99925	0.5	0.40	0.40	0.99022	7.0	5.61	5.55	0.98264	13.5	10.91	10.71	0.97596	20.0	16.27	15.86
0.99910	0.6	0.48	0.48	0.99009	7.1	5.69	5.63	0.98254	13.6	10.99	10.79	0.97586	20.1	16.35	15.94
0.99895	0.7	0.56	0.56	0.98997	7.2	5.77	5.71	0.98243	13.7	11.07	10.87	0.97576	20.2	16.44	16.02
0.99880	0.8	0.64	0.63	0.98984	7.3	5.85	5.79	0.98232	13.8	11.15	10.95	0.97566	20.3	16.52	16.10
0.99865	0.9	0.72	0.71	0.98972	7.4	5.94	5.87	0.98221	13.9	11.23	11.02	0.97556	20.4	16.60	16.18
0.99850	1.0	0.80	0.79	0.98960	7.5	6.02	5.95	0.98210	14.0	11.32	11.10	0.97546	20.5	16.68	16.26
0.99835	1.1	0.87	0.87	0.98947	7.6	6.10	6.03	0.98200	14.1	11.40	11.18	0.97536	20.6	16.77	16.34
0.99820	1.2	0.95	0.95	0.98935	7.7	6.18	6.11	0.98190	14.2	11.48	11.26	0.97526	20.7	16.85	16.42
0.99806	1.3	1.03	1.03	0.98923	7.8	6.26	6.19	0.98179	14.3	11.56	11.34	0.97516	20.8	16.93	16.50
0.99791	1.4	1.11	1.11	0.98911	7.9	6.34	6.27	0.98168	14.4	11.65	11.42	0.97506	20.9	17.02	16.58
0.99776	1.5	1.19	1.19	0.98899	8.0	6.42	6.35	0.98157	14.5	11.73	11.50	0.97496	21.0	17.10	16.66
0.99761	1.6	1.27	1.27	0.98887	8.1	6.50	6.42	0.98147	14.6	11.81	11.58	0.97486	21.1	17.18	16.74
0.99747	1.7	1.35	1.35	0.98875	8.2	6.58	6.50	0.98136	14.7	11.89	11.66	0.97476	21.2	17.27	16.81
0.99732	1.8	1.43	1.43	0.98863	8.3	6.67	6.58	0.98125	14.8	11.97	11.74	0.97466	21.3	17.35	16.89
0.99717	1.9	1.51	1.51	0.98851	8.4	6.75	6.66	0.98115	14.9	12.06	11.82	0.97456	21.4	17.43	16.97
0.99703	2.0	1.59	1.59	0.98838	8.5	6.83	6.74	0.98104	15.0	12.14	11.90	0.97446	21.5	17.52	17.05
0.99688	2.1	1.67	1.67	0.98826	8.6	6.91	6.82	0.98093	15.1	12.22	11.98	0.97436	21.6	17.60	17.13
0.99674	2.2	1.75	1.74	0.98814	8.7	6.99	6.90	0.98083	15.2	12.30	12.06	0.97425	21.7	17.68	17.21
0.99659	2.3	1.83	1.82	0.98803	8.8	7.07	6.98	0.98072	15.3	12.39	12.13	0.97415	21.8	17.77	17.29
0.99645	2.4	1.91	1.90	0.98791	8.9	7.15	7.06	0.98062	15.4	12.47	12.21	0.97405	21.9	17.85	17.37
0.99630	2.5	1.99	1.98	0.98779	9.0	7.23	7.14	0.98051	15.5	12.55	12.29	0.97395	22.0	17.93	17.45
0.99616	2.6	2.07	2.06	0.98767	9.1	7.31	7.22	0.98040	15.6	12.63	12.37	0.97385	22.1	18.02	17.53
0.99602	2.7	2.15	2.14	0.98755	9.2	7.40	7.30	0.98030	15.7	12.71	12.45	0.97375	22.2	18.10	17.61
0.99587	2.8	2.23	2.22	0.98743	9.3	7.48	7.38	0.98019	15.8	12.80	12.53	0.97365	22.3	18.18	17.69
0.99573	2.9	2.31	2.30	0.98731	9.4	7.56	7.46	0.98009	15.9	12.88	12.61	0.97354	22.4	18.27	17.77

0.99559	3.0	*2.39*	2.38	0.98720	7.53	*7.64*	9.5	0.97998	16.0	*12.96*	12.69	0.97344	22.5	*18.35*	17.85
0.99545	3.1	*2.47*	2.46	0.98708	7.61	*7.72*	9.6	0.97988	16.1	*13.04*	12.77	0.97334	22.6	*18.43*	17.92
0.99531	3.2	*2.55*	2.54	0.98696	7.69	*7.80*	9.7	0.97977	16.2	*13.13*	12.85	0.97324	22.7	*18.52*	18.00
0.99516	3.3	*2.63*	2.62	0.98684	7.77	*7.88*	9.8	0.97967	16.3	*13.21*	12.93	0.97313	22.8	*18.60*	18.08
0.99502	3.4	*2.71*	2.70	0.98672	7.85	*7.97*	9.9	0.97956	16.4	*13.29*	13.01	0.97303	22.9	*18.68*	18.16
0.99488	3.5	*2.79*	2.78	0.98661	7.93	*8.05*	10.0	0.97946	16.5	*13.37*	13.09	0.97293	23.0	*18.77*	18.24
0.99474	3.6	*2.87*	2.86	0.98649	8.01	*8.13*	10.1	0.97936	16.6	*13.46*	13.17	0.97283	23.1	*18.85*	18.32
0.99460	3.7	*2.95*	2.93	0.98637	8.09	*8.21*	10.2	0.97925	16.7	*13.54*	13.25	0.97272	23.2	*18.93*	18.40
0.99446	3.8	*3.03*	3.01	0.98625	8.17	*8.29*	10.3	0.97915	16.8	*13.62*	13.32	0.97262	23.3	*19.02*	18.48
0.99432	3.9	*3.11*	3.09	0.98614	8.25	*8.37*	10.4	0.97905	16.9	*13.70*	13.40	0.97252	23.4	*19.10*	18.56
0.99419	4.0	*3.19*	3.17	0.98602	8.33	*8.45*	10.5	0.97895	17.0	*13.79*	13.48	0.97241	23.5	*19.19*	18.64
0.99405	4.1	*3.27*	3.25	0.98590	8.41	*8.54*	10.6	0.97885	17.1	*13.87*	13.56	0.97231	23.6	*19.27*	18.72
0.99391	4.2	*3.35*	3.33	0.98577	8.49	*8.62*	10.7	0.97875	17.2	*13.95*	13.64	0.97221	23.7	*19.35*	18.80
0.99378	4.3	*3.44*	3.41	0.98567	8.57	*8.70*	10.8	0.97864	17.3	*14.03*	13.72	0.97210	23.8	*19.44*	18.88
0.99364	4.4	*3.52*	3.49	0.98556	8.65	*8.78*	10.9	0.97854	17.4	*14.12*	13.80	0.97200	23.9	*19.52*	18.96
0.99350	4.5	*3.60*	3.57	0.98544	8.72	*8.86*	11.0	0.97844	17.5	*14.20*	13.88	0.97189	24.0	*19.60*	19.04
0.99336	4.6	*3.68*	3.65	0.98532	8.80	*8.94*	11.1	0.97834	17.6	*14.28*	13.96	0.97179	24.1	*19.69*	19.11
0.99323	4.7	*3.76*	3.73	0.98521	8.88	*9.03*	11.2	0.97824	17.7	*14.36*	14.04	0.97168	24.2	*19.77*	19.19
0.99309	4.8	*3.84*	3.81	0.98509	8.96	*9.11*	11.3	0.97814	17.8	*14.45*	14.12	0.97158	24.3	*19.86*	19.27
0.99296	4.9	*3.92*	3.89	0.98498	9.04	*9.19*	11.4	0.97804	17.9	*14.53*	14.20	0.97147	24.4	*19.94*	19.35
0.99282	5.0	*4.00*	3.97	0.98487	9.12	*9.27*	11.5	0.97794	18.0	*14.61*	14.28	0.97137	24.5	*20.02*	19.43
0.99269	5.1	*4.08*	4.04	0.98475	9.20	*9.35*	11.6	0.97784	18.1	*14.70*	14.36	0.97126	24.6	*20.11*	19.51
0.99255	5.2	*4.16*	4.12	0.98464	9.28	*9.43*	11.7	0.97774	18.2	*14.78*	14.43	0.97116	24.7	*20.19*	19.59
0.99242	5.3	*4.24*	4.20	0.98452	9.36	*9.52*	11.8	0.97764	18.3	*14.86*	14.51	0.97105	24.8	*20.28*	19.67
0.99228	5.4	*4.32*	4.28	0.98441	9.44	*9.60*	11.9	0.97754	18.4	*14.94*	14.59	0.97095	24.9	*20.36*	19.75
0.99215	5.5	*4.40*	4.36	0.98430	9.52	*9.68*	12.0	0.97744	18.5	*15.03*	14.67	0.97084	25.0	*20.44*	19.83
0.99202	5.6	*4.48*	4.44	0.98419	9.60	*9.76*	12.1	0.97734	18.6	*15.11*	14.75				
0.99189	5.7	*4.56*	4.52	0.98408	9.68	*9.84*	12.2	0.97724	18.7	*15.19*	14.83				
0.99176	5.8	*4.64*	4.60	0.98396	9.76	*9.92*	12.3	0.97714	18.8	*15.27*	14.91				
0.99163	5.9	*4.72*	4.68	0.98385	9.83	*10.01*	12.4	0.97704	18.9	*15.36*	14.99				
0.99150	6.0	*4.80*	4.76	0.98374	9.91	*10.09*	12.5	0.97694	19.0	*15.44*	15.07				
0.99137	6.1	*4.88*	4.84	0.98363	9.99	*10.17*	12.6	0.97684	19.1	*15.52*	15.15				
0.99124	6.2	*4.97*	4.92	0.98352	10.07	*10.25*	12.7	0.97674	19.2	*15.61*	15.23				
0.99111	6.3	*5.05*	5.00	0.98341	10.15	*10.33*	12.8	0.97664	19.3	*15.69*	15.31				
0.99098	6.4	*5.13*	5.08	0.98330	10.23	*10.42*	12.9	0.97654	19.4	*15.77*	15.39				

* Compiled at the U. S. Bur. of Standards. This table is based upon true specific gravity instead of apparent. For concentrations of 25% alcohol by volume the apparent specific gravity is 0.97080, the true specific gravity is 0.97084. For lower concentrations the difference between the two specific gravities is less; for higher concentrations the difference is greater.

TABLE A4.—RELATION BETWEEN SPECIFIC GRAVITY AT 20/20° C. AND ALCOHOL CONTENT OF MIXTURES OF ETHYL ALCOHOL AND WATER *

Sp. gr. 20°/20°	Per cent by volume	Per cent by weight	Grams per 100 cc.	Sp. gr. 20°/20°	Per cent by volume	Per cent by weight	Grams per 100 cc.	Sp. gr. 20°/20°	Per cent by volume	Per cent by weight	Grams per 100 cc.	Sp. gr. 20°/20°	Per cent by volume	Per cent by weight	Grams per 100 cc.
1.00000	0.0	0.00	0.00	0.99083	6.5	5.19	5.13	0.98296	13.0	10.46	10.26	0.97584	19.5	15.80	15.39
0.99985	0.1	0.08	0.08	0.99070	6.6	5.27	5.21	0.98285	13.1	10.54	10.34	0.97574	19.6	15.88	15.47
0.99970	0.2	0.16	0.16	0.99057	6.7	5.35	5.29	0.98274	13.2	10.62	10.42	0.97563	19.7	15.97	15.55
0.99955	0.3	0.24	0.24	0.99045	6.8	5.43	5.37	0.98263	13.3	10.70	10.50	0.97553	19.8	16.05	15.63
0.99939	0.4	0.32	0.32	0.99032	6.9	5.51	5.45	0.98251	13.4	10.78	10.58	0.97542	19.9	16.13	15.71
0.99924	0.5	0.40	0.40	0.99020	7.0	5.59	5.53	0.98239	13.5	10.86	10.66	0.97531	20.0	16.21	15.79
0.99910	0.6	0.47	0.47	0.99007	7.1	5.67	5.60	0.98227	13.6	10.95	10.74	0.97521	20.1	16.30	15.87
0.99895	0.7	0.55	0.55	0.98994	7.2	5.75	5.68	0.98216	13.7	11.03	10.81	0.97511	20.2	16.38	15.95
0.99880	0.8	0.63	0.63	0.98981	7.3	5.83	5.76	0.98204	13.8	11.11	10.89	0.97500	20.3	16.46	16.02
0.99866	0.9	0.71	0.71	0.98969	7.4	5.91	5.84	0.98193	13.9	11.19	10.97	0.97489	20.4	16.55	16.10
0.99851	1.0	0.79	0.79	0.98956	7.5	5.99	5.92	0.98182	14.0	11.28	11.05	0.97478	20.5	16.63	16.18
0.99836	1.1	0.87	0.87	0.98944	7.6	6.07	6.00	0.98171	14.1	11.36	11.13	0.97467	20.6	16.71	16.26
0.99821	1.2	0.95	0.95	0.98931	7.7	6.15	6.08	0.98159	14.2	11.44	11.21	0.97456	20.7	16.80	16.34
0.99807	1.3	1.03	1.03	0.98919	7.8	6.24	6.16	0.98148	14.3	11.52	11.29	0.97445	20.8	16.88	16.42
0.99792	1.4	1.11	1.11	0.98906	7.9	6.32	6.24	0.98137	14.4	11.60	11.37	0.97435	20.9	16.96	16.50
0.99777	1.5	1.19	1.19	0.98893	8.0	6.40	6.32	0.98126	14.5	11.68	11.44	0.97424	21.0	17.04	16.58
0.99763	1.6	1.27	1.26	0.98881	8.1	6.48	6.39	0.98115	14.6	11.77	11.52	0.97414	21.1	17.13	16.66
0.99748	1.7	1.35	1.34	0.98869	8.2	6.56	6.47	0.98103	14.7	11.85	11.60	0.97404	21.2	17.21	16.73
0.99733	1.8	1.43	1.42	0.98857	8.3	6.64	6.55	0.98092	14.8	11.93	11.68	0.97393	21.3	17.29	16.81
0.99719	1.9	1.51	1.50	0.98845	8.4	6.72	6.63	0.98081	14.9	12.01	11.76	0.97382	21.4	17.38	16.89
0.99704	2.0	1.59	1.58	0.98833	8.5	6.80	6.71	0.98070	15.0	12.09	11.84	0.97371	21.5	17.46	16.97
0.99689	2.1	1.67	1.66	0.98820	8.6	6.88	6.79	0.98058	15.1	12.18	11.92	0.97360	21.6	17.54	17.05
0.99675	2.2	1.75	1.74	0.98807	8.7	6.96	6.87	0.98047	15.2	12.26	12.00	0.97350	21.7	17.63	17.13
0.99661	2.3	1.82	1.81	0.98794	8.8	7.04	6.95	0.98036	15.3	12.34	12.08	0.97339	21.8	17.71	17.21
0.99646	2.4	1.90	1.89	0.98782	8.9	7.12	7.03	0.98025	15.4	12.42	12.16	0.97328	21.9	17.79	17.29
0.99632	2.5	1.98	1.97	0.98770	9.0	7.20	7.10	0.98014	15.5	12.50	12.23	0.97317	22.0	17.88	17.37
0.99618	2.6	2.06	2.05	0.98758	9.1	7.29	7.18	0.98003	15.6	12.59	12.31	0.97306	22.1	17.96	17.45
0.99603	2.7	2.14	2.13	0.98746	9.2	7.37	7.26	0.97992	15.7	12.67	12.39	0.97295	22.2	18.04	17.52
0.99589	2.8	2.22	2.21	0.98734	9.3	7.45	7.34	0.97981	15.8	12.75	12.47	0.97285	22.3	18.13	17.60
0.99574	2.9	2.30	2.29	0.98722	9.4	7.53	7.42	0.97970	15.9	12.83	12.55	0.97274	22.4	18.21	17.68

Sp. gr.	% vol.	*wt.*	wt.
0.99560	3.0	*2.38*	2.37
0.99546	3.1	*2.46*	2.45
0.99531	3.2	*2.54*	2.53
0.99517	3.3	*2.62*	2.60
0.99503	3.4	*2.70*	2.68
0.99489	3.5	*2.78*	2.76
0.99475	3.6	*2.86*	2.84
0.99461	3.7	*2.94*	2.92
0.99447	3.8	*3.02*	3.00
0.99433	3.9	*3.10*	3.08
0.99419	4.0	*3.18*	3.16
0.99405	4.1	*3.26*	3.24
0.99391	4.2	*3.34*	3.32
0.99377	4.3	*3.42*	3.39
0.99363	4.4	*3.50*	3.47
0.99349	4.5	*3.58*	3.55
0.99336	4.6	*3.66*	3.63
0.99322	4.7	*3.74*	3.71
0.99308	4.8	*3.82*	3.79
0.99295	4.9	*3.90*	3.87
0.99281	5.0	*3.98*	3.95
0.99268	5.1	*4.06*	4.03
0.99255	5.2	*4.14*	4.10
0.99241	5.3	*4.22*	4.18
0.99228	5.4	*4.30*	4.26
0.99215	5.5	*4.38*	4.34
0.99201	5.6	*4.46*	4.42
0.99188	5.7	*4.54*	4.50
0.99174	5.8	*4.62*	4.58
0.99161	5.9	*4.70*	4.66
0.99148	6.0	*4.78*	4.74
0.99135	6.1	*4.87*	4.82
0.99122	6.2	*4.95*	4.89
0.99109	6.3	*5.03*	4.97
0.99096	6.4	*5.11*	5.05

Sp. gr.	% vol.	*wt.*	wt.
0.98710	9.5	*7.61*	7.50
0.98698	9.6	*7.69*	7.58
0.98686	9.7	*7.77*	7.66
0.98674	9.8	*7.85*	7.73
0.98662	9.9	*7.93*	7.81
0.98650	10.0	*8.02*	7.89
0.98637	10.1	*8.10*	7.97
0.98626	10.2	*8.18*	8.05
0.98614	10.3	*8.26*	8.13
0.98602	10.4	*8.34*	8.21
0.98590	10.5	*8.42*	8.29
0.98578	10.6	*8.50*	8.37
0.98566	10.7	*8.58*	8.45
0.98554	10.8	*8.66*	8.52
0.98542	10.9	*8.75*	8.60
0.98530	11.0	*8.83*	8.68
0.98518	11.1	*8.91*	8.76
0.98506	11.2	*8.99*	8.84
0.98494	11.3	*9.07*	8.92
0.98482	11.4	*9.15*	9.00
0.98470	11.5	*9.23*	9.08
0.98459	11.6	*9.32*	9.16
0.98447	11.7	*9.40*	9.24
0.98435	11.8	*9.48*	9.31
0.98424	11.9	*9.56*	9.39
0.98412	12.0	*9.64*	9.47
0.98400	12.1	*9.72*	9.55
0.98388	12.2	*9.80*	9.63
0.98377	12.3	*9.89*	9.71
0.98365	12.4	*9.97*	9.79
0.98354	12.5	*10.05*	9.87
0.98342	12.6	*10.13*	9.95
0.98330	12.7	*10.21*	10.03
0.98318	12.8	*10.29*	10.10
0.98307	12.9	*10.38*	10.18

Sp. gr.	% vol.	*wt.*	wt.
0.97959	16.0	*12.92*	12.63
0.97948	16.1	*13.00*	12.71
0.97937	16.2	*13.08*	12.79
0.97926	16.3	*13.16*	12.87
0.97915	16.4	*13.24*	12.95
0.97905	16.5	*13.33*	13.02
0.97894	16.6	*13.41*	13.10
0.97883	16.7	*13.49*	13.18
0.97872	16.8	*13.57*	13.26
0.97862	16.9	*13.66*	13.34
0.97851	17.0	*13.74*	13.42
0.97840	17.1	*13.82*	13.50
0.97829	17.2	*13.90*	13.58
0.97818	17.3	*13.98*	13.66
0.97807	17.4	*14.07*	13.74
0.97797	17.5	*14.15*	13.81
0.97786	17.6	*14.23*	13.89
0.97775	17.7	*14.31*	13.97
0.97764	17.8	*14.40*	14.05
0.97754	17.9	*14.48*	14.13
0.97743	18.0	*14.56*	14.21
0.97732	18.1	*14.64*	14.29
0.97721	18.2	*14.73*	14.37
0.97711	18.3	*14.81*	14.45
0.97700	18.4	*14.89*	14.52
0.97690	18.5	*14.97*	14.60
0.97679	18.6	*15.06*	14.68
0.97668	18.7	*15.14*	14.76
0.97657	18.8	*15.22*	14.84
0.97646	18.9	*15.30*	14.92
0.97636	19.0	*15.39*	15.00
0.97626	19.1	*15.47*	15.08
0.97616	19.2	*15.55*	15.16
0.97605	19.3	*15.64*	15.23
0.97595	19.4	*15.72*	15.31

Sp. gr.	% vol.	*wt.*	wt.
0.97263	22.5	*18.29*	17.76
0.97252	22.6	*18.38*	17.84
0.97241	22.7	*18.46*	17.92
0.97230	22.8	*18.54*	18.00
0.97219	22.9	*18.63*	18.08
0.97208	23.0	*18.71*	18.16
0.97197	23.1	*18.79*	18.24
0.97185	23.2	*18.88*	18.31
0.97174	23.3	*18.96*	18.39
0.97163	23.4	*19.04*	18.47
0.97152	23.5	*19.13*	18.55
0.97141	23.6	*19.21*	18.63
0.97130	23.7	*19.29*	18.71
0.97118	23.8	*19.38*	18.79
0.97107	23.9	*19.46*	18.87
0.97096	24.0	*19.55*	18.94
0.97084	24.1	*19.63*	19.02
0.97073	24.2	*19.71*	19.10
0.97062	24.3	*19.80*	19.18
0.97051	24.4	*19.88*	19.26
0.97040	24.5	*19.96*	19.34
0.97028	24.6	*20.05*	19.42
0.97017	24.7	*20.13*	19.50
0.97006	24.8	*20.22*	19.58
0.96994	24.9	*20.30*	19.66
0.96983	25.0	*20.38*	19.73

* Compiled at the U. S. Bureau of Standards. This table is based upon true specific gravity instead of apparent. In a solution containing 25% alcohol by volume, the apparent specific gravity is 0.96980, the true specific gravity 0.96983. For lower concentrations the difference between the two specific gravities is less; for higher concentrations the difference is greater.

Table A5.—Relation between Specific Gravity at 25/25° C. and Alcohol Content of Mixtures of Ethyl Alcohol and Water*

Sp. gr. 25°/25°	Percent by volume	Percent by weight	Grams per 100 cc.	Sp. gr. 25°/25°	Percent by volume	Percent by weight	Grams per 100 cc.	Sp. gr. 25°/25°	Percent by volume	Percent by weight	Grams per 100 cc.	Sp. gr. 25°/25°	Percent by volume	Percent by weight	Grams per 100 cc.
1.00000	0.0	0.00	0.00	0.99080	6.5	5.17	5.10	0.98269	13.0	10.42	10.21	0.97518	19.5	15.74	15.31
0.99985	0.1	0.08	0.08	0.99067	6.6	5.25	5.18	0.98257	13.1	10.50	10.28	0.97507	19.6	15.83	15.39
0.99970	0.2	0.16	0.16	0.99054	6.7	5.33	5.26	0.98245	13.2	10.58	10.36	0.97496	19.7	15.91	15.47
0.99955	0.3	0.24	0.24	0.99041	6.8	5.41	5.34	0.98233	13.3	10.66	10.44	0.97485	19.8	15.99	15.54
0.99940	0.4	0.31	0.31	0.99027	6.9	5.49	5.42	0.98221	13.4	10.74	10.52	0.97473	19.9	16.07	15.62
0.99925	0.5	0.39	0.39	0.99014	7.0	5.57	5.50	0.98209	13.5	10.82	10.60	0.97462	20.0	16.16	15.70
0.99910	0.6	0.47	0.47	0.99001	7.1	5.65	5.57	0.98197	13.6	10.90	10.68	0.97451	20.1	16.24	15.78
0.99895	0.7	0.55	0.55	0.98988	7.2	5.73	5.65	0.98185	13.7	10.99	10.76	0.97440	20.2	16.32	15.86
0.99881	0.8	0.63	0.63	0.98975	7.3	5.81	5.73	0.98174	13.8	11.07	10.83	0.97428	20.3	16.41	15.94
0.99866	0.9	0.71	0.71	0.98962	7.4	5.89	5.81	0.98162	13.9	11.15	10.91	0.97417	20.4	16.49	16.02
0.99851	1.0	0.79	0.79	0.98950	7.5	5.97	5.89	0.98151	14.0	11.23	10.99	0.97406	20.5	16.57	16.09
0.99836	1.1	0.87	0.86	0.98937	7.6	6.05	5.97	0.98139	14.1	11.31	11.07	0.97394	20.6	16.65	16.17
0.99822	1.2	0.95	0.94	0.98925	7.7	6.13	6.04	0.98127	14.2	11.39	11.15	0.97383	20.7	16.74	16.25
0.99807	1.3	1.03	1.02	0.98912	7.8	6.21	6.12	0.98115	14.3	11.48	11.23	0.97371	20.8	16.82	16.33
0.99792	1.4	1.10	1.10	0.98899	7.9	6.29	6.20	0.98103	14.4	11.56	11.30	0.97360	20.9	16.90	16.41
0.99777	1.5	1.18	1.18	0.98887	8.0	6.37	6.28	0.98092	14.5	11.64	11.38	0.97349	21.0	16.98	16.49
0.99763	1.6	1.26	1.26	0.98874	8.1	6.45	6.36	0.98080	14.6	11.72	11.46	0.97337	21.1	17.07	16.56
0.99748	1.7	1.34	1.33	0.98861	8.2	6.53	6.44	0.98068	14.7	11.80	11.54	0.97326	21.2	17.15	16.64
0.99733	1.8	1.42	1.41	0.98849	8.3	6.61	6.52	0.98057	14.8	11.88	11.62	0.97314	21.3	17.23	16.72
0.99718	1.9	1.50	1.49	0.98836	8.4	6.69	6.59	0.98045	14.9	11.97	11.70	0.97303	21.4	17.32	16.80
0.99704	2.0	1.58	1.57	0.98824	8.5	6.77	6.67	0.98033	15.0	12.05	11.78	0.97291	21.5	17.40	16.88
0.99689	2.1	1.66	1.65	0.98811	8.6	6.85	6.75	0.98021	15.1	12.13	11.85	0.97280	21.6	17.48	16.96
0.99675	2.2	1.74	1.73	0.98799	8.7	6.93	6.83	0.98009	15.2	12.21	11.93	0.97268	21.7	17.57	17.04
0.99660	2.3	1.82	1.81	0.98787	8.8	7.01	6.91	0.97997	15.3	12.29	12.01	0.97256	21.8	17.65	17.11
0.99646	2.4	1.90	1.88	0.98774	8.9	7.09	6.99	0.97986	15.4	12.37	12.09	0.97245	21.9	17.73	17.19
0.99631	2.5	1.98	1.96	0.98762	9.0	7.18	7.07	0.97974	15.5	12.46	12.17	0.97233	22.0	17.81	17.27
0.99617	2.6	2.06	2.04	0.98749	9.1	7.26	7.14	0.97962	15.6	12.54	12.25	0.97222	22.1	17.90	17.35
0.99603	2.7	2.13	2.12	0.98737	9.2	7.34	7.22	0.97951	15.7	12.62	12.33	0.97210	22.2	17.98	17.43
0.99589	2.8	2.21	2.20	0.98724	9.3	7.42	7.30	0.97939	15.8	12.70	12.40	0.97198	22.3	18.06	17.51
0.99574	2.9	2.29	2.28	0.98712	9.4	7.50	7.38	0.97928	15.9	12.78	12.48	0.97187	22.4	18.15	17.59

Sp. gr.	Per cent by volume		
0.99560	3.0	2.37	2.36
0.99546	3.1	2.45	2.43
0.99531	3.2	2.53	2.51
0.99517	3.3	2.61	2.59
0.99503	3.4	2.69	2.67
0.99488	3.5	2.77	2.75
0.99474	3.6	2.85	2.83
0.99460	3.7	2.93	2.90
0.99446	3.8	3.01	2.98
0.99432	3.9	3.09	3.06
0.99418	4.0	3.17	3.14
0.99404	4.1	3.25	3.22
0.99390	4.2	3.33	3.30
0.99376	4.3	3.41	3.38
0.99362	4.4	3.49	3.45
0.99348	4.5	3.57	3.53
0.99335	4.6	3.65	3.61
0.99321	4.7	3.73	3.69
0.99307	4.8	3.81	3.77
0.99293	4.9	3.89	3.85
0.99280	5.0	3.97	3.93
0.99266	5.1	4.05	4.00
0.99253	5.2	4.13	4.08
0.99239	5.3	4.20	4.16
0.99226	5.4	4.28	4.24
0.99212	5.5	4.36	4.32
0.99199	5.6	4.44	4.40
0.99185	5.7	4.52	4.47
0.99172	5.8	4.60	4.55
0.99159	5.9	4.68	4.63
0.99145	6.0	4.76	4.71
0.99132	6.1	4.84	4.79
0.99119	6.2	4.93	4.87
0.99106	6.3	5.01	4.95
0.99093	6.4	5.09	5.02

Sp. gr.	Per cent by volume		
0.98699	9.5	7.58	7.46
0.98686	9.6	7.66	7.54
0.98674	9.7	7.74	7.62
0.98661	9.8	7.82	7.69
0.98649	9.9	7.90	7.77
0.98637	10.0	7.98	7.85
0.98624	10.1	8.06	7.93
0.98612	10.2	8.14	8.01
0.98600	10.3	8.22	8.09
0.98587	10.4	8.31	8.16
0.98575	10.5	8.39	8.24
0.98563	10.6	8.47	8.32
0.98550	10.7	8.55	8.40
0.98538	10.8	8.63	8.48
0.98525	10.9	8.71	8.56
0.98513	11.0	8.79	8.64
0.98501	11.1	8.87	8.71
0.98488	11.2	8.95	8.79
0.98476	11.3	9.03	8.87
0.98464	11.4	9.12	8.95
0.98451	11.5	9.20	9.03
0.98439	11.6	9.28	9.11
0.98427	11.7	9.36	9.19
0.98415	11.8	9.44	9.26
0.98402	11.9	9.52	9.34
0.98390	12.0	9.60	9.42
0.98378	12.1	9.68	9.50
0.98366	12.2	9.77	9.58
0.98354	12.3	9.85	9.66
0.98341	12.4	9.93	9.73
0.98329	12.5	10.01	9.81
0.98317	12.6	10.09	9.89
0.98305	12.7	10.17	9.97
0.98293	12.8	10.25	10.05
0.98281	12.9	10.33	10.13

Sp. gr.	Per cent by volume		
0.97916	16.0	12.87	12.56
0.97905	16.1	12.95	12.64
0.97893	16.2	13.03	12.72
0.97882	16.3	13.11	12.80
0.97870	16.4	13.19	12.87
0.97858	16.5	13.28	12.95
0.97847	16.6	13.36	13.03
0.97835	16.7	13.44	13.11
0.97824	16.8	13.52	13.19
0.97812	16.9	13.60	13.27
0.97801	17.0	13.69	13.35
0.97789	17.1	13.77	13.42
0.97778	17.2	13.85	13.50
0.97767	17.3	13.93	13.58
0.97755	17.4	14.01	13.66
0.97744	17.5	14.10	13.74
0.97733	17.6	14.18	13.82
0.97722	17.7	14.26	13.90
0.97711	17.8	14.34	13.97
0.97699	17.9	14.43	14.05
0.97687	18.0	14.51	14.13
0.97676	18.1	14.59	14.21
0.97665	18.2	14.67	14.29
0.97653	18.3	14.75	14.37
0.97642	18.4	14.84	14.45
0.97630	18.5	14.92	14.52
0.97619	18.6	15.00	14.60
0.97608	18.7	15.08	14.68
0.97597	18.8	15.17	14.76
0.97585	18.9	15.25	14.84
0.97574	19.0	15.33	14.92
0.97563	19.1	15.41	14.99
0.97551	19.2	15.50	15.07
0.97540	19.3	15.58	15.15
0.97529	19.4	15.66	15.23

Sp. gr.	Per cent by volume		
0.97175	22.5	18.23	17.66
0.97163	22.6	18.31	17.74
0.97151	22.7	18.40	17.82
0.97140	22.8	18.48	17.90
0.97128	22.9	18.56	17.98
0.97116	23.0	18.65	18.06
0.97104	23.1	18.73	18.13
0.97092	23.2	18.81	18.21
0.97080	23.3	18.90	18.29
0.97069	23.4	18.98	18.37
0.97057	23.5	19.06	18.45
0.97045	23.6	19.15	18.53
0.97033	23.7	19.23	18.61
0.97021	23.8	19.31	18.68
0.97009	23.9	19.40	18.76
0.96997	24.0	19.48	18.84
0.96985	24.1	19.57	18.92
0.96973	24.2	19.65	19.00
0.96961	24.3	19.73	19.08
0.96949	24.4	19.82	19.16
0.96937	24.5	19.90	19.23
0.96925	24.6	19.98	19.31
0.96913	24.7	20.07	19.39
0.96901	24.8	20.15	19.47
0.96889	24.9	20.23	19.55
0.96876	25.0	20.32	19.63

* Compiled at the U. S. Bureau of Standards. This table is based upon true specific gravity instead of apparent. For concentration of 25% by volume the apparent specific gravity is 0.96872, the true specific gravity 0.96876. For lower concentration the difference between the two specific gravities is less; for higher concentration it is relatively greater.

TABLE A6.—ALCOHOL TABLE FOR CALCULATING THE PERCENTAGES OF ALCOHOL IN MIXTURES OF ETHYL ALCOHOL AND WATER FROM THEIR ZEISS IMMERSION REFRACTOMETER READINGS AT 17.5–25° C.*

Scale reading	17.5°C Per cent by volume	17.5°C Per cent by weight	18°C vol	18°C wt	19°C vol	19°C wt	20°C vol	20°C wt	21°C vol	21°C wt	22°C vol	22°C wt	23°C vol	23°C wt	24°C vol	24°C wt	25°C vol	25°C wt	Scale reading
13.2																	0.00	0.00	13.2
13.4																	0.18	0.14	13.4
13.6															0.14	0.11	0.35	0.28	13.6
13.8													0.10	0.08	0.31	0.25	0.53	0.42	13.8
14.0									0.04	0.03	0.08	0.06	0.28	0.22	0.49	0.30	0.70	0.56	14.0
14.2									0.21	0.17	0.24	0.19	0.45	0.36	0.67	0.53	0.88	0.70	14.2
14.4									0.38	0.30	0.41	0.33	0.63	0.50	0.84	0.67	1.06	0.85	14.4
14.6							0.16	0.13	0.55	0.44	0.59	0.47	0.80	0.64	1.02	0.81	1.24	0.99	14.6
14.8					0.14	0.11	0.34	0.27	0.73	0.58	0.77	0.61	0.98	0.78	1.19	0.95	1.40	1.11	14.8
15.0	0.00	0.00	0.10	0.08	0.31	0.24	0.52	0.41	0.91	0.72	0.94	0.75	1.16	0.92	1.36	1.08	1.55	1.23	15.0
15.2	0.17	0.13	0.27	0.21	0.48	0.38	0.69	0.55	1.07	0.85	1.12	0.89	1.32	1.05	1.51	1.20	1.71	1.36	15.2
15.4	0.34	0.27	0.44	0.35	0.65	0.51	0.85	0.68	1.24	0.99	1.29	1.02	1.47	1.17	1.66	1.32	1.86	1.48	15.4
15.6	0.51	0.40	0.60	0.48	0.82	0.65	1.03	0.82	1.40	1.11	1.44	1.15	1.62	1.29	1.82	1.44	2.01	1.60	15.6
15.8	0.68	0.54	0.78	0.62	0.99	0.79	1.21	0.96	1.55	1.23	1.60	1.27	1.77	1.41	1.97	1.57	2.17	1.72	15.8
16.0	0.84	0.67	0.94	0.75	1.17	0.93	1.36	1.08	1.70	1.35	1.75	1.30	1.92	1.53	2.12	1.60	2.33	1.85	16.0
16.2	1.02	0.81	1.12	0.89	1.32	1.05	1.51	1.20	1.85	1.47	1.90	1.51	2.08	1.65	2.27	1.81	2.48	1.97	16.2
16.4	1.18	0.94	1.29	1.02	1.47	1.17	1.66	1.32	2.00	1.59	2.05	1.63	2.24	1.78	2.43	1.93	2.62	2.00	16.4
16.6	1.34	1.06	1.43	1.13	1.62	1.29	1.81	1.44	2.15	1.71	2.20	1.75	2.39	1.90	2.57	2.05	2.77	2.21	16.6
16.8	1.49	1.18	1.57	1.25	1.77	1.41	1.96	1.56	2.30	1.83	2.35	1.87	2.53	2.02	2.72	2.17	2.92	2.33	16.8
17.0	1.63	1.30	1.72	1.37	1.92	1.53	2.11	1.68	2.45	1.95	2.50	1.99	2.69	2.14	2.87	2.29	3.06	2.44	17.0
17.2	1.77	1.41	1.87	1.40	2.06	1.64	2.26	1.80	2.59	2.07	2.65	2.11	2.82	2.25	3.02	2.41	3.21	2.56	17.2
17.4	1.92	1.53	2.01	1.60	2.21	1.76	2.41	1.92	2.74	2.18	2.79	2.23	2.97	2.37	3.17	2.53	3.36	2.68	17.4
17.6	2.07	1.65	2.16	1.72	2.36	1.88	2.56	2.04	2.89	2.30	2.94	2.34	3.12	2.49	3.32	2.65	3.51	2.80	17.6
17.8	2.21	1.76	2.31	1.84	2.51	2.00	2.70	2.15			3.09	2.46	3.27	2.61	3.46	2.76	3.66	2.92	17.8

18.0	3.04	3.81	2.88	3.61	2.73	3.42	2.58	3.23	2.42	3.04	2.27	2.85	2.12	2.66	1.95	2.45	1.88	2.36	18.0
18.2	3.16	3.96	3.00	3.76	2.85	3.57	2.69	3.37	2.54	3.19	2.39	3.00	2.24	2.81	2.07	2.60	2.00	2.50	18.2
18.4	3.28	4.11	3.12	3.91	2.96	3.71	2.81	3.52	2.66	3.34	2.51	3.15	2.36	2.96	2.19	2.75	2.11	2.65	18.4
18.6	3.40	4.26	3.24	4.06	3.08	3.86	2.92	3.66	2.78	3.48	2.63	3.30	2.47	3.10	2.31	2.90	2.23	2.80	18.6
18.8	3.52	4.41	3.36	4.21	3.20	4.01	3.04	3.81	2.89	3.63	2.75	3.45	2.59	3.25	2.43	3.05	2.35	2.95	18.8
19.0	3.64	4.56	3.48	4.36	3.32	4.16	3.16	3.96	3.01	3.77	2.86	3.59	2.71	3.40	2.54	3.19	2.47	3.10	19.0
19.2	3.76	4.70	3.60	4.51	3.44	4.31	3.28	4.11	3.13	3.92	2.98	3.73	2.83	3.55	2.66	3.34	2.58	3.25	19.2
19.4	3.88	4.85	3.72	4.65	3.56	4.46	3.40	4.26	3.25	4.07	3.10	3.88	2.95	3.70	2.78	3.48	2.70	3.39	19.4
19.6	4.00	5.00	3.84	4.80	3.68	4.61	3.52	4.41	3.37	4.22	3.22	4.03	3.06	3.84	2.90	3.63	2.82	3.53	19.6
19.8	4.12	5.15	3.96	4.95	3.80	4.75	3.64	4.56	3.49	4.37	3.33	4.17	3.18	3.98	3.02	3.78	2.94	3.68	19.8
20.0	4.23	5.29	4.08	5.10	3.92	4.90	3.77	4.72	3.61	4.52	3.45	4.32	3.30	4.13	3.13	3.93	3.06	3.83	20.0
20.2	4.35	5.44	4.19	5.24	4.04	5.05	3.89	4.87	3.73	4.66	3.57	4.47	3.41	4.27	3.25	4.07	3.17	3.97	20.2
20.4	4.47	5.58	4.31	5.38	4.16	5.20	4.01	5.01	3.85	4.83	3.68	4.61	3.53	4.42	3.37	4.22	3.29	4.12	20.4
20.6	4.58	5.72	4.42	5.52	4.27	5.34	4.12	5.15	3.97	4.96	3.80	4.75	3.65	4.56	3.49	4.36	3.41	4.26	20.6
20.8	4.70	5.87	4.54	5.67	4.39	5.48	4.24	5.29	4.08	5.10	3.92	4.90	3.76	4.70	3.60	4.51	3.52	4.41	20.8
21.0	4.81	6.02	4.66	5.82	4.50	5.62	4.35	5.44	4.20	5.24	4.03	5.04	3.88	4.85	3.72	4.65	3.64	4.56	21.0
21.2	4.93	6.16	4.77	5.96	4.62	5.77	4.47	5.58	4.31	5.39	4.15	5.19	3.99	4.99	3.84	4.80	3.76	4.70	21.2
21.4	5.05	6.30	4.89	6.11	4.73	5.91	4.58	5.72	4.43	5.53	4.26	5.33	4.11	5.14	3.95	4.94	3.87	4.84	21.4
21.6	5.16	6.44	5.01	6.25	4.85	6.06	4.70	5.87	4.54	5.67	4.38	5.47	4.22	5.28	4.07	5.09	3.99	4.99	21.6
21.8	5.28	6.59	5.12	6.39	4.97	6.20	4.81	6.01	4.66	5.82	4.49	5.61	4.34	5.43	4.18	5.23	4.10	5.13	21.8
22.0	5.40	6.73	5.24	6.54	5.08	6.34	4.93	6.15	4.77	5.96	4.61	5.76	4.45	5.57	4.30	5.37	4.22	5.27	22.0
22.2	5.51	6.87	5.35	6.68	5.20	6.49	5.04	6.29	4.89	6.11	4.72	5.90	4.57	5.71	4.41	5.51	4.33	5.41	22.2
22.4	5.63	7.01	5.47	6.82	5.31	6.63	5.15	6.43	5.00	6.25	4.84	6.05	4.68	5.85	4.53	5.65	4.45	5.56	22.4
22.6	5.74	7.16	5.58	6.96	5.43	6.77	5.27	6.57	5.11	6.39	4.95	6.19	4.80	6.00	4.64	5.80	4.56	5.70	22.6
22.8	5.86	7.31	5.70	7.10	5.54	6.91	5.38	6.71	5.23	6.53	5.07	6.33	4.91	6.14	4.75	5.94	4.68	5.85	22.8
23.0	5.99	7.45	5.81	7.24	5.66	7.06	5.50	6.86	5.34	6.67	5.18	6.47	5.03	6.28	4.87	6.08	4.79	5.99	23.0
23.2	6.13	7.59	5.93	7.39	5.77	7.20	5.61	7.00	5.46	6.81	5.30	6.61	5.14	6.42	4.98	6.22	4.91	6.13	23.2
23.4	6.27	7.73	6.04	7.53	5.89	7.34	5.73	7.14	5.57	6.95	5.41	6.75	5.26	6.56	5.10	6.36	5.02	6.27	23.4
23.6	6.41	7.87	6.15	7.67	6.00	7.48	5.84	7.28	5.69	7.09	5.53	6.90	5.37	6.70	5.21	6.50	5.14	6.41	23.6
23.8	6.55	8.00	6.27	7.81	6.11	7.62	5.96	7.42	5.80	7.23	5.64	7.04	5.49	6.85	5.32	6.64	5.25	6.55	23.8

* Calculated and arranged by B. H. St. John from data of Doroshevskii and Dvorzhanchik.

TABLE A6.—(Continued)

Scale reading	17.5°C Per cent by volume	17.5°C Per cent by weight	18°C Per cent by volume	18°C Per cent by weight	19°C Per cent by volume	19°C Per cent by weight	20°C Per cent by volume	20°C Per cent by weight	21°C Per cent by volume	21°C Per cent by weight	22°C Per cent by volume	22°C Per cent by weight	23°C Per cent by volume	23°C Per cent by weight	24°C Per cent by volume	24°C Per cent by weight	25°C Per cent by volume	25°C Per cent by weight	Scale reading
24.0	6.69	5.36	6.78	5.44	6.99	5.60	7.18	5.76	7.38	5.92	7.56	6.07	7.76	6.22	7.95	6.38	8.14	6.53	24.0
24.2	6.83	5.47	6.92	5.55	7.13	5.71	7.32	5.87	7.52	6.03	7.70	6.18	7.90	6.34	8.09	6.49	8.28	6.65	24.2
24.4	6.97	5.59	7.06	5.66	7.27	5.83	7.46	5.99	7.66	6.14	7.84	6.29	8.04	6.45	8.23	6.60	8.42	6.76	24.4
24.6	7.11	5.70	7.20	5.78	7.41	5.94	7.60	6.10	7.80	6.25	7.98	6.41	8.17	6.56	8.37	6.72	8.55	6.87	24.6
24.8	7.25	5.81	7.35	5.89	7.55	6.06	7.74	6.21	7.93	6.36	8.12	6.52	8.31	6.67	8.51	6.83	8.69	6.98	24.8
25.0	7.39	5.03	7.49	6.01	7.68	6.16	7.88	6.32	8.06	6.47	8.26	6.63	8.45	6.79	8.64	6.94	8.84	7.10	25.0
25.2	7.53	6.04	7.63	6.12	7.82	6.27	8.01	6.43	8.20	6.50	8.40	6.75	8.59	6.90	8.78	7.06	8.98	7.21	25.2
25.4	7.66	6.15	7.76	6.23	7.95	6.38	8.14	6.54	8.34	6.70	8.54	6.86	8.73	7.01	8.92	7.17	9.12	7.33	25.4
25.6	7.80	6.26	7.90	6.34	8.09	6.49	8.28	6.65	8.48	6.81	8.68	6.97	8.86	7.12	9.06	7.28	9.26	7.44	25.6
25.8	7.94	6.37	8.03	6.44	8.22	6.60	8.42	6.76	8.62	6.92	8.82	7.08	9.00	7.23	9.20	7.40	9.39	7.55	25.8
26.0	8.07	6.48	8.16	6.55	8.36	6.71	8.55	6.87	8.75	7.03	8.95	7.10	9.14	7.35	9.34	7.51	9.53	7.67	26.0
26.2	8.21	6.59	8.30	6.66	8.50	6.82	8.69	6.98	8.89	7.14	9.09	7.30	9.28	7.46	9.48	7.62	9.67	7.78	26.2
26.4	8.34	6.70	8.44	6.78	8.63	6.93	8.82	7.09	9.03	7.25	9.22	7.41	9.42	7.57	9.61	7.73	9.81	7.90	26.4
26.6	8.48	6.81	8.57	6.88	8.77	7.04	8.96	7.20	9.16	7.36	9.36	7.52	9.55	7.68	9.75	7.85	9.95	8.01	26.6
26.8	8.62	6.92	8.71	6.99	8.91	7.15	9.10	7.31	9.30	7.47	9.49	7.63	9.69	7.79	9.89	7.96	10.09	8.12	26.8
27.0	8.75	7.03	8.85	7.11	9.05	7.27	9.23	7.42	9.44	7.59	9.63	7.74	9.83	7.91	10.03	8.07	10.23	8.24	27.0
27.2	8.89	7.14	8.98	7.22	9.18	7.38	9.37	7.54	9.58	7.70	9.76	7.85	9.97	8.02	10.17	8.18	10.37	8.35	27.2
27.4	9.02	7.25	9.12	7.33	9.32	7.49	9.51	7.65	9.71	7.81	9.90	7.96	10.10	8.13	10.31	8.30	10.51	8.46	27.4
27.6	9.16	7.36	9.26	7.44	9.45	7.60	9.65	7.76	9.85	7.92	10.03	8.07	10.24	8.24	10.45	8.41	10.65	8.57	27.6
27.8	9.29	7.47	9.39	7.55	9.59	7.71	9.79	7.87	9.98	8.03	10.17	8.18	10.38	8.35	10.58	8.52	10.79	8.69	27.8
28.0	9.43	7.58	9.53	7.66	9.72	7.82	9.92	7.98	10.12	8.14	10.31	8.30	10.51	8.46	10.72	8.63	10.93	8.80	28.0
28.2	9.57	7.69	9.66	7.77	9.86	7.93	10.06	8.09	10.25	8.25	10.45	8.41	10.65	8.58	10.86	8.74	11.06	8.91	28.2
28.4	9.70	7.80	9.80	7.88	9.99	8.04	10.19	8.20	10.39	8.36	10.59	8.52	10.79	8.69	11.00	8.86	11.20	9.02	28.4
28.6	9.84	7.91	9.93	7.99	10.13	8.15	10.32	8.31	10.52	8.47	10.72	8.63	10.93	8.80	11.13	8.97	11.33	9.13	28.6
28.8	9.97	8.02	10.07	8.10	10.26	8.26	10.46	8.42	10.66	8.58	10.86	8.74	11.06	8.91	11.27	9.08	11.47	9.24	28.8

29.0	9.36	11.61	9.19	11.40	9.02	11.20	8.86	11.00	8.69	10.79	8.53	10.59	8.36	10.40	8.20	10.19	8.13	10.10
29.2	9.47	11.75	9.30	11.54	9.13	11.33	8.97	11.13	8.80	10.93	8.64	10.73	8.47	10.52	8.31	10.33	8.24	10.24
29.4	9.58	11.88	9.41	11.67	9.24	11.47	9.08	11.27	8.91	11.06	8.74	10.86	8.58	10.66	8.42	10.46	8.34	10.36
29.6	9.69	12.01	9.52	11.81	9.35	11.60	9.18	11.39	9.02	11.20	8.85	10.99	8.69	10.79	8.53	10.59	8.45	10.50
29.8	9.80	12.15	9.63	11.94	9.46	11.74	9.29	11.53	9.13	11.33	8.96	11.12	8.80	10.93	8.63	10.72	8.56	10.63
30.0	9.91	12.29	9.74	12.08	9.57	11.87	9.40	11.66	9.24	11.46	9.07	11.26	8.91	11.05	8.74	10.86	8.66	10.76
30.2	10.02	12.42	9.85	12.21	9.68	12.00	9.51	11.79	9.34	11.59	9.18	11.38	9.02	11.18	8.85	10.99	8.77	10.89
30.4	10.13	12.56	9.96	12.34	9.79	12.13	9.62	11.93	9.45	11.72	9.28	11.51	9.12	11.31	8.96	11.12	8.88	11.02
30.6	10.24	12.70	10.07	12.48	9.90	12.27	9.73	12.06	9.56	11.85	9.39	11.64	9.23	11.44	9.06	11.25	8.98	11.15
30.8	10.36	12.84	10.18	12.61	10.01	12.40	9.84	12.19	9.67	11.98	9.50	11.78	9.34	11.58	9.17	11.38	9.09	11.28
31.0	10.47	12.97	10.29	12.75	10.12	12.54	9.95	12.32	9.77	12.12	9.60	11.91	9.44	11.71	9.28	11.51	9.20	11.41
31.2	10.58	13.11	10.40	12.89	10.23	12.67	10.05	12.46	9.88	12.25	9.71	12.04	9.55	11.84	9.38	11.64	9.30	11.54
31.4	10.69	13.24	10.51	13.02	10.34	12.81	10.16	12.59	9.99	12.38	9.82	12.17	9.65	11.97	9.49	11.77	9.40	11.66
31.6	10.80	13.37	10.62	13.15	10.45	12.94	10.27	12.72	10.10	12.51	9.93	12.30	9.76	12.10	9.59	11.90	9.51	11.79
31.8	10.91	13.51	10.73	13.29	10.55	13.07	10.38	12.85	10.21	12.64	10.03	12.43	9.87	12.23	9.70	12.03	9.62	11.92
32.0	11.02	13.64	10.84	13.42	10.66	13.20	10.49	12.99	10.31	12.78	10.14	12.57	9.97	12.36	9.80	12.15	9.72	12.05
32.2	11.13	13.77	10.95	13.55	10.77	13.34	10.59	13.12	10.42	12.91	10.25	12.70	10.08	12.49	9.91	12.28	9.83	12.18
32.4	11.24	13.91	11.06	13.69	10.88	13.47	10.70	13.25	10.53	13.04	10.35	12.83	10.19	12.62	10.02	12.40	9.93	12.31
32.6	11.35	14.04	11.17	13.82	10.99	13.60	10.81	13.38	10.64	13.17	10.46	12.96	10.29	12.75	10.12	12.54	10.04	12.43
32.8	11.46	14.17	11.28	13.95	11.10	13.73	10.92	13.51	10.75	13.30	10.57	13.09	10.40	12.88	10.23	12.67	10.14	12.56
33.0	11.57	14.31	11.39	14.09	11.21	13.86	11.03	13.64	10.85	13.43	10.68	13.22	10.50	13.01	10.33	12.79	10.24	12.69
33.2	11.68	14.44	11.50	14.22	11.31	13.99	11.13	13.78	10.96	13.56	10.79	13.35	10.61	13.13	10.43	12.92	10.35	12.82
33.4	11.79	14.58	11.61	14.35	11.42	14.13	11.24	13.91	11.07	13.69	10.89	13.48	10.71	13.26	10.54	13.05	10.45	12.95
33.6	11.90	14.71	11.72	14.48	11.53	14.26	11.35	14.04	11.17	13.82	11.00	13.61	10.82	13.39	10.64	13.18	10.56	13.08
33.8	12.01	14.85	11.83	14.62	11.64	14.39	11.45	14.17	11.28	13.95	11.10	13.74	10.93	13.52	10.75	13.30	10.66	13.20
34.0	12.12	14.98	11.93	14.75	11.75	14.52	11.56	14.30	11.38	14.08	11.21	13.86	11.03	13.64	10.85	13.43	10.77	13.33
34.2	12.23	15.11	12.04	14.88	11.85	14.65	11.67	14.43	11.49	14.21	11.31	13.99	11.13	13.77	10.96	13.56	10.87	13.45
34.4	12.34	15.25	12.15	15.01	11.96	14.78	11.78	14.57	11.59	14.34	11.41	14.12	11.24	13.90	11.06	13.68	10.97	13.58
34.6	12.45	15.38	12.26	15.14	12.07	14.91	11.88	14.70	11.70	14.47	11.52	14.25	11.34	14.02	11.16	13.81	11.08	13.70
34.8	12.56	15.51	12.36	15.28	12.18	15.05	11.99	14.83	11.81	14.59	11.62	14.37	11.44	14.14	11.27	13.94	11.18	13.83

Table A6.—(Continued)

Scale reading	17.5°C Per cent by volume	17.5°C Per cent by weight	18°C Per cent by volume	18°C Per cent by weight	19°C Per cent by volume	19°C Per cent by weight	20°C Per cent by volume	20°C Per cent by weight	21°C Per cent by volume	21°C Per cent by weight	22°C Per cent by volume	22°C Per cent by weight	23°C Per cent by volume	23°C Per cent by weight	24°C Per cent by volume	24°C Per cent by weight	25°C Per cent by volume	25°C Per cent by weight	Scale reading
35.0	13.96	11.28	14.06	11.37	14.27	11.55	14.50	11.73	14.72	11.91	14.96	12.10	15.18	12.28	15.41	12.47	15.65	12.67	35.0
35.2	14.08	11.38	14.19	11.47	14.39	11.65	14.62	11.83	14.85	12.02	15.09	12.20	15.31	12.39	15.54	12.58	15.78	12.78	35.2
35.4	14.21	11.49	14.31	11.57	14.52	11.75	14.75	11.93	14.97	12.12	15.22	12.31	15.44	12.50	15.67	12.69	15.91	12.89	35.4
35.6	14.33	11.59	14.44	11.68	14.65	11.86	14.87	12.04	15.10	12.23	15.34	12.41	15.56	12.60	15.80	12.80	16.05	13.00	35.6
35.8	14.46	11.69	14.56	11.78	14.78	11.96	15.00	12.14	15.23	12.33	15.47	12.52	15.69	12.71	15.93	12.91	16.18	13.11	35.8
36.0	14.58	11.79	14.69	11.88	14.90	12.06	15.13	12.24	15.35	12.43	15.59	12.62	15.82	12.82	16.06	13.02	16.31	13.21	36.0
36.2	14.71	11.90	14.81	11.99	15.03	12.16	15.25	12.35	15.48	12.54	15.72	12.73	15.95	12.92	16.19	13.12	16.44	13.32	36.2
36.4	14.83	12.00	14.94	12.09	15.16	12.27	15.38	12.45	15.61	12.64	15.85	12.84	16.08	13.03	16.32	13.23	16.56	13.43	36.4
36.6	14.96	12.10	15.06	12.19	15.28	12.37	15.51	12.56	15.73	12.75	15.97	12.94	16.21	13.14	16.45	13.34	16.69	13.53	36.6
36.8	15.08	12.20	15.19	12.29	15.41	12.47	15.63	12.66	15.86	12.85	16.10	13.05	16.34	13.24	16.58	13.44	16.82	13.64	36.8
37.0	15.20	12.30	15.31	12.40	15.53	12.58	15.76	12.77	15.99	12.96	16.23	13.15	16.47	13.35	16.71	13.55	16.95	13.75	37.0
37.2	15.33	12.41	15.44	12.50	15.66	12.68	15.89	12.87	16.11	13.06	16.35	13.26	16.60	13.45	16.84	13.65	17.08	13.86	37.2
37.4	15.45	12.51	15.56	12.60	15.79	12.78	16.01	12.97	16.24	13.16	16.48	13.36	16.72	13.56	16.97	13.76	17.21	13.96	37.4
37.6	15.57	12.61	15.69	12.70	15.91	12.89	16.14	13.08	16.37	13.27	16.61	13.46	16.85	13.66	17.09	13.87	17.34	14.07	37.6
37.8	15.70	12.71	15.81	12.81	16.04	12.99	16.26	13.18	16.49	13.37	16.73	13.57	16.98	13.77	17.22	13.97	17.46	14.17	37.8
38.0	15.82	12.81	15.94	12.91	16.16	13.09	16.39	13.28	16.62	13.47	16.86	13.67	17.11	13.87	17.35	14.08	17.59	14.28	38.0
38.2	15.94	12.91	16.06	13.01	16.29	13.19	16.51	13.38	16.75	13.58	16.99	13.78	17.23	13.98	17.47	14.18	17.72	14.38	38.2
38.4	16.07	13.02	16.18	13.11	16.41	13.30	16.64	13.49	16.87	13.68	17.11	13.88	17.36	14.08	17.60	14.29	17.85	14.49	38.4
38.6	16.19	13.12	16.31	13.21	16.53	13.40	16.76	13.59	17.00	13.79	17.24	13.98	17.48	14.19	17.73	14.39	17.97	14.59	38.6
38.8	16.31	13.22	16.43	13.31	16.66	13.50	16.89	13.69	17.13	13.89	17.36	14.09	17.61	14.29	17.85	14.49	18.10	14.70	38.8
39.0	16.44	13.32	16.55	13.42	16.78	13.61	17.01	13.79	17.25	14.00	17.49	14.19	17.74	14.40	17.98	14.60	18.23	14.80	39.0
39.2	16.56	13.42	16.67	13.52	16.91	13.71	17.14	13.90	17.38	14.10	17.62	14.30	17.86	14.50	18.11	14.70	18.35	14.91	39.2
39.4	16.68	13.52	16.80	13.62	17.03	13.81	17.26	14.00	17.50	14.20	17.74	14.40	17.99	14.60	18.23	14.81	18.48	15.01	39.4
39.6	16.80	13.62	16.92	13.72	17.15	13.91	17.39	14.10	17.63	14.30	17.87	14.50	18.11	14.71	18.36	14.91	18.61	15.12	39.6
39.8	16.93	13.73	17.04	13.82	17.28	14.02	17.51	14.21	17.75	14.41	17.99	14.61	18.24	14.81	18.48	15.02	18.73	15.22	39.8

40.0	15.32	18.86	15.12	18.61	14.92	18.36	14.71	18.12	14.51	17.88	14.31	17.63	14.12	17.40	13.02	17.16	13.83	17.05
40.2	15.43	18.99	15.22	18.74	15.02	18.49	14.82	18.24	14.61	18.00	14.41	17.76	14.22	17.53	14.03	17.29	13.93	17.17
40.4	15.53	19.11	15.33	18.86	15.12	18.61	14.92	18.37	14.72	18.12	14.51	17.88	14.32	17.64	14.13	17.41	14.03	17.29
40.6	15.64	19.24	15.43	18.99	15.23	18.74	15.03	18.49	14.82	18.25	14.62	18.01	14.42	17.77	14.23	17.53	14.13	17.41
40.8	15.74	19.37	15.53	19.11	15.33	18.86	15.13	18.61	14.92	18.37	14.72	18.13	14.52	17.89	14.33	17.65	14.23	17.54
41.0	15.85	19.49	15.64	19.24	15.43	18.99	15.23	18.74	15.03	18.49	14.82	18.25	14.62	18.01	14.43	17.77	14.33	17.66
41.2	15.95	19.62	15.74	19.36	15.53	19.11	15.33	18.86	15.13	18.62	14.92	18.37	14.73	18.13	14.53	17.90	14.43	17.78
41.4	16.06	19.75	15.84	19.49	15.64	19.24	15.43	18.99	15.23	18.74	15.03	18.50	14.83	18.26	14.63	18.03	14.53	17.90
41.6	16.16	19.87	15.95	19.61	15.74	19.36	15.53	19.11	15.33	18.86	15.13	18.62	14.93	18.38	14.73	18.14	14.63	18.02
41.8	16.27	20.00	16.05	19.74	15.84	19.48	15.63	19.23	15.43	18.99	15.23	18.74	15.03	18.50	14.83	18.26	14.73	18.14
42.0	16.37	20.13	16.16	19.86	15.94	19.61	15.74	19.36	15.53	19.11	15.33	18.87	15.13	18.62	14.93	18.38	14.83	18.27
42.2	16.48	20.25	16.26	19.99	16.05	19.73	15.84	19.48	15.63	19.23	15.43	18.99	15.23	18.74	15.03	18.50	14.93	18.39
42.4	16.58	20.38	16.36	20.11	16.15	19.86	15.94	19.60	15.74	19.36	15.53	19.11	15.33	18.87	15.13	18.62	15.03	18.51
42.6	16.69	20.50	16.47	20.24	16.25	19.98	16.04	19.72	15.84	19.48	15.63	19.23	15.43	18.99	15.23	18.75	15.13	18.63
42.8	16.79	20.63	16.57	20.36	16.35	20.10	16.14	19.85	15.94	19.60	15.74	19.36	15.53	19.11	15.33	18.87	15.23	18.75
43.0	16.90	20.75	16.67	20.49	16.46	20.23	16.24	19.97	16.04	19.72	15.84	19.48	15.63	19.23	15.43	18.99	15.33	18.87
43.2	17.00	20.88	16.78	20.61	16.56	20.35	16.35	20.09	16.14	19.85	15.94	19.60	15.73	19.35	15.53	19.11	15.43	18.99
43.4	17.10	21.01	16.88	20.74	16.66	20.47	16.45	20.21	16.24	19.97	16.04	19.72	15.83	19.47	15.63	19.23	15.53	19.11
43.6	17.21	21.13	16.98	20.86	16.76	20.60	16.55	20.34	16.34	20.09	16.14	19.85	15.93	19.59	15.73	19.35	15.63	19.23
43.8	17.31	21.25	17.08	20.99	16.87	20.72	16.65	20.46	16.45	20.21	16.24	19.97	16.03	19.72	15.83	19.47	15.73	19.35
44.0	17.41	21.38	17.19	21.11	16.97	20.84	16.76	20.58	16.55	20.34	16.34	20.09	16.13	19.84	15.93	19.59	15.83	19.46
44.2	17.52	21.50	17.29	21.23	17.07	20.96	16.86	20.71	16.65	20.46	16.44	20.21	16.23	19.96	16.03	19.71	15.93	19.58
44.4	17.62	21.63	17.40	21.36	17.17	21.09	16.96	20.83	16.75	20.58	16.55	20.33	16.33	20.08	16.13	19.83	16.02	19.70
44.6	17.73	21.75	17.50	21.48	17.28	21.21	17.06	20.95	16.85	20.70	16.65	20.45	16.43	20.20	16.23	19.95	16.12	19.82
44.8	17.83	21.88	17.60	21.60	17.38	21.33	17.16	21.07	16.95	20.82	16.75	20.58	16.53	20.32	16.32	20.07	16.22	19.94
45.0	17.93	22.00	17.71	21.73	17.48	21.45	17.26	21.19	17.06	20.95	16.85	20.70	16.63	20.44	16.42	20.18	16.32	20.06
45.2	18.04	22.13	17.81	21.85	17.58	21.58	17.36	21.31	17.16	21.07	16.95	20.82	16.73	20.56	16.52	20.30	16.41	20.18
45.4	18.14	22.25	17.91	21.98	17.68	21.70	17.46	21.43	17.26	21.19	17.05	20.94	16.83	20.68	16.62	20.42	16.51	20.29
45.6	18.25	22.38	18.02	22.10	17.79	21.82	17.56	21.55	17.36	21.31	17.15	21.06	16.93	20.80	16.72	20.54	16.61	20.41
45.8	18.36	22.51	18.12	22.23	17.89	21.94	17.66	21.67	17.46	21.43	17.25	21.18	17.03	20.92	16.81	20.66	16.71	20.53

TABLE A6.—(Continued)

Scale reading	17.5° C		18° C		19° C		20° C		21° C		22° C		23° C		24° C		25° C		Scale reading
	Per cent by volume	Per cent by weight	Per cent by volume	Per cent by weight	Per cent by volume	Per cent by weight	Per cent by volume	Per cent by weight	Per cent by volume	Per cent by weight	Per cent by volume	Per cent by weight	Per cent by volume	Per cent by weight	Per cent by volume	Per cent by weight	Per cent by volume	Per cent by weight	
46.0	20.65	*16.80*	20.78	*16.91*	21.04	*17.13*	21.30	*17.35*	21.54	*17.56*	21.79	*17.76*	22.07	*17.99*	22.35	*18.23*	22.64	*18.47*	46.0
46.2	20.76	*16.90*	20.89	*17.01*	21.16	*17.23*	21.42	*17.45*	21.66	*17.66*	21.91	*17.86*	22.19	*18.09*	22.48	*18.33*	22.76	*18.57*	46.2
46.4	20.88	*17.00*	21.01	*17.11*	21.28	*17.33*	21.54	*17.55*	21.78	*17.76*	22.03	*17.96*	22.32	*18.20*	22.61	*18.44*	22.89	*18.68*	46.4
46.6	21.00	*17.10*	21.13	*17.21*	21.40	*17.43*	21.66	*17.65*	21.90	*17.86*	22.16	*18.06*	22.44	*18.30*	22.73	*18.55*	23.02	*18.79*	46.6
46.8	21.12	*17.20*	21.25	*17.31*	21.52	*17.53*	21.78	*17.75*	22.02	*17.96*	22.28	*18.17*	22.57	*18.41*	22.86	*18.65*	23.15	*18.90*	46.8
47.0	21.24	*17.30*	21.37	*17.41*	21.64	*17.63*	21.90	*17.85*	22.15	*18.06*	22.41	*18.27*	22.69	*18.51*	22.99	*18.76*	23.28	*19.00*	47.0
47.2	21.36	*17.40*	21.49	*17.51*	21.76	*17.73*	22.02	*17.95*	22.27	*18.16*	22.53	*18.38*	22.82	*18.62*	23.12	*18.86*	23.41	*19.11*	47.2
47.4	21.48	*17.50*	21.61	*17.61*	21.88	*17.83*	22.15	*18.06*	22.39	*18.26*	22.66	*18.48*	22.94	*18.72*	23.24	*18.97*	23.54	*19.22*	47.4
47.6	21.60	*17.60*	21.73	*17.71*	22.00	*17.94*	22.27	*18.16*	22.51	*18.36*	22.78	*18.58*	23.07	*18.83*	23.37	*19.08*	23.67	*19.33*	47.6
47.8	21.72	*17.70*	21.85	*17.81*	22.12	*18.04*	22.39	*18.26*	22.64	*18.47*	22.91	*18.69*	23.20	*18.93*	23.50	*19.18*	23.80	*19.43*	47.8
48.0	21.84	*17.80*	21.97	*17.91*	22.24	*18.14*	22.51	*18.36*	22.76	*18.57*	23.03	*18.79*	23.32	*19.04*	23.63	*19.29*	23.93	*19.54*	48.0
48.2	21.96	*17.90*	22.09	*18.01*	22.36	*18.24*	22.63	*18.46*	22.88	*18.67*	23.16	*18.90*	23.45	*19.14*	23.76	*19.40*	24.06	*19.65*	48.2
48.4	22.08	*18.00*	22.21	*18.11*	22.48	*18.34*	22.75	*18.56*	23.01	*18.77*	23.28	*19.00*	23.58	*19.25*	23.89	*19.51*	24.19	*19.76*	48.4
48.6	22.20	*18.10*	22.33	*18.21*	22.60	*18.44*	22.87	*18.66*	23.13	*18.88*	23.41	*19.11*	23.71	*19.36*	24.02	*19.61*	24.32	*19.87*	48.6
48.8	22.32	*18.20*	22.45	*18.31*	22.72	*18.54*	22.99	*18.76*	23.26	*18.98*	23.54	*19.21*	23.83	*19.47*	24.14	*19.72*	24.45	*19.98*	48.8
49.0	22.44	*18.30*	22.57	*18.41*	22.84	*18.64*	23.12	*18.86*	23.38	*19.08*	23.66	*19.32*	23.96	*19.57*	24.27	*19.83*	24.59	*20.09*	49.0
49.2	22.56	*18.40*	22.69	*18.51*	22.96	*18.74*	23.24	*18.96*	23.51	*19.19*	23.79	*19.43*	24.09	*19.68*	24.40	*19.94*	24.72	*20.20*	49.2
49.4	22.68	*18.50*	22.81	*18.61*	23.08	*18.84*	23.36	*19.07*	23.63	*19.29*	23.92	*19.54*	24.22	*19.79*	24.53	*20.04*	24.85	*20.31*	49.4
49.6	22.80	*18.60*	22.93	*18.71*	23.21	*18.94*	23.48	*19.17*	23.76	*19.40*	24.04	*19.64*	24.35	*19.89*	24.66	*20.15*	24.98	*20.42*	49.6
49.8	22.92	*18.70*	23.05	*18.81*	23.33	*19.04*	23.61	*19.27*	23.88	*19.51*	24.17	*19.75*	24.48	*20.00*	24.79	*20.27*	25.11	*20.54*	49.8
50.0	23.04	*18.80*	23.17	*18.91*	23.45	*19.14*	23.73	*19.38*	24.01	*19.61*	24.30	*19.86*	24.61	*20.11*	24.92	*20.38*	25.25	*20.65*	50.0
50.2	23.16	*18.90*	23.30	*19.02*	23.57	*19.24*	23.85	*19.48*	24.13	*19.72*	24.43	*19.96*	24.74	*20.22*	25.05	*20.49*	25.38	*20.76*	50.2
50.4	23.28	*19.00*	23.42	*19.12*	23.69	*19.35*	23.98	*19.58*	24.26	*19.82*	24.56	*20.07*	24.86	*20.33*	25.18	*20.60*	25.51	*20.87*	50.4
50.6	23.40	*19.10*	23.54	*19.22*	23.81	*19.45*	24.10	*19.69*	24.38	*19.93*	24.69	*20.18*	24.99	*20.44*	25.32	*20.71*	25.65	*20.98*	50.6
50.8	23.51	*19.20*	23.66	*19.32*	23.93	*19.55*	24.22	*19.79*	24.51	*20.03*	24.81	*20.29*	25.12	*20.55*	25.45	*20.82*	25.78	*21.10*	50.8

51.0	21.21	25.91	20.03	25.58	20.66	25.25	20.30	24.94	20.14	24.64	19.80	24.35	19.65	24.05	19.42	23.78	19.30	23.63	51.0
51.2	21.32	26.05	21.04	25.71	20.77	25.38	20.50	25.07	20.24	24.76	20.00	24.47	19.75	24.18	19.52	23.90	19.40	23.75	51.2
51.4	21.44	26.18	21.15	25.84	20.87	25.51	20.61	25.20	20.35	24.89	20.10	24.59	19.85	24.30	19.62	24.02	19.50	23.87	51.4
51.6	21.55	26.32	21.26	25.97	20.08	25.64	20.72	25.33	20.46	25.01	20.20	24.72	19.95	24.42	19.72	24.14	19.60	23.99	51.6
51.8	21.66	26.45	21.37	26.11	21.09	25.77	20.82	25.46	20.56	25.14	20.31	24.84	20.06	24.54	19.82	24.26	19.70	24.11	51.8
52.0	21.78	26.59	21.40	26.24	21.20	25.90	20.93	25.58	20.67	25.27	20.41	24.96	20.16	24.66	19.02	24.38	19.80	24.23	52.0
52.2	21.89	26.72	21.60	26.37	21.31	26.03	21.04	25.71	20.77	25.39	20.52	25.09	20.26	24.79	20.02	24.50	19.00	24.36	52.2
52.4	22.01	26.86	21.71	26.51	21.42	26.16	21.15	25.84	20.88	25.52	20.62	25.21	20.37	24.91	20.12	24.62	20.00	24.48	52.4
52.6	22.12	26.99	21.82	26.64	21.53	26.29	21.26	25.97	20.08	25.65	20.72	25.34	20.47	25.03	20.22	24.74	20.10	24.60	52.6
52.8	22.24	27.13	21.93	26.77	21.64	26.42	21.36	26.10	21.09	25.77	20.83	25.46	20.57	25.15	20.33	24.86	20.20	24.72	52.8
53.0	22.35	27.27	22.05	26.91	21.75	26.56	21.47	26.23	21.20	25.90	20.93	25.59	20.68	25.28	20.43	24.98	20.30	24.84	53.0
53.2	22.47	27.40	22.16	27.04	21.86	26.69	21.58	26.35	21.31	26.03	21.04	25.71	20.78	25.40	20.53	25.10	20.41	24.96	53.2
53.4	22.58	27.54	22.27	27.17	21.07	26.82	21.69	26.48	21.42	26.15	21.14	25.84	20.88	25.52	20.63	25.23	20.51	25.08	53.4
53.6	22.70	27.67	22.30	27.31	22.08	26.95	21.80	26.61	21.52	26.28	21.25	25.96	20.08	25.65	20.74	25.35	20.61	25.20	53.6
53.8	22.81	27.81	22.50	27.44	22.20	27.08	21.01	26.74	21.63	26.41	21.36	26.09	21.09	25.77	20.84	25.47	20.71	25.32	53.8
54.0	22.03	27.95	22.61	27.58	22.31	27.21	22.02	26.87	21.74	26.54	21.47	26.22	21.19	25.90	20.94	25.59	20.81	25.44	54.0
54.2	23.04	28.08	22.73	27.71	22.42	27.35	22.13	27.00	21.85	26.67	21.57	26.34	21.30	26.02	21.04	25.71	20.91	25.56	54.2
54.4	23.16	28.22	22.84	27.85	22.53	27.48	22.24	27.13	21.06	26.79	21.68	26.47	21.40	26.14	21.14	25.84	21.02	25.68	54.4
54.6	23.27	28.36	22.84	27.98	22.65	27.61	22.35	27.26	22.06	26.92	21.79	26.59	21.51	26.27	21.25	25.96	21.12	25.81	54.6
54.8	23.39	28.49	23.07	28.11	22.76	27.75	22.46	27.39	22.17	27.05	21.90	26.72	21.62	26.39	21.35	26.08	21.22	25.93	54.8
55.0	23.51	28.63	23.18	28.25	22.87	27.88	22.57	27.52	22.28	27.18	22.00	26.85	21.72	26.52	21.45	26.20	21.32	26.05	55.0
55.2	23.62	28.77	23.30	28.38	22.08	28.01	22.68	27.65	22.30	27.31	22.11	26.97	21.83	26.64	21.56	26.32	21.43	26.17	55.2
55.4	23.74	28.90	23.41	28.52	23.10	28.15	22.79	27.78	22.40	27.43	22.21	27.10	21.03	26.76	21.66	26.45	21.53	26.29	55.4
55.6	23.86	29.04	23.53	28.65	23.21	28.28	22.90	27.92	22.60	27.55	22.32	27.23	22.04	26.89	21.76	26.57	21.63	26.41	55.6
55.8	23.07	29.18	23.64	28.78	23.32	28.41	23.01	28.05	22.71	27.69	22.42	27.35	22.14	27.01	21.87	26.09	21.73	26.53	55.8
56.0	24.09	29.31	23.75	28.92	23.43	28.54	23.12	28.18	22.82	27.82	22.53	27.48	22.24	27.14	21.07	26.81	21.84	26.65	56.0
56.2	24.20	29.45	23.87	29.05	23.54	28.68	23.23	28.31	22.02	27.94	22.64	27.60	22.35	27.26	22.07	26.93	21.04	26.78	56.2
56.4	24.32	29.58	23.08	29.19	23.66	28.81	23.34	28.44	23.03	28.07	22.74	27.73	22.45	27.38	22.18	27.05	22.04	26.90	56.4
56.6	24.43	29.72	24.10	29.33	23.77	28.94	23.45	28.56	23.14	28.20	22.85	27.85	22.56	27.51	22.28	27.18	22.14	27.02	56.6
56.8	24.55	29.86	24.21	29.46	23.88	29.07	23.56	28.69	23.25	28.33	22.95	27.98	22.66	27.63	22.38	27.30	22.25	27.14	56.8

TABLE A6.—(Continued)

Scale reading	17.5°C Per cent by volume	17.5°C Per cent by weight	18°C Per cent by volume	18°C Per cent by weight	19°C Per cent by volume	19°C Per cent by weight	20°C Per cent by volume	20°C Per cent by weight	21°C Per cent by volume	21°C Per cent by weight	22°C Per cent by volume	22°C Per cent by weight	23°C Per cent by volume	23°C Per cent by weight	24°C Per cent by volume	24°C Per cent by weight	25°C Per cent by volume	25°C Per cent by weight	Scale reading
57.0	27.26	*22.35*	27.42	*22.48*	27.75	*22.77*	28.10	*23.06*	28.46	*23.36*	28.82	*23.67*	29.20	*23.99*	29.59	*24.32*	29.99	*24.66*	57.0
57.2	27.38	*22.45*	27.54	*22.58*	27.88	*22.87*	28.23	*23.17*	28.59	*23.47*	28.95	*23.78*	29.34	*24.11*	29.73	*24.44*	30.13	*24.78*	57.2
57.4	27.50	*22.55*	27.66	*22.69*	28.00	*22.97*	28.35	*23.27*	28.72	*23.58*	29.08	*23.90*	29.47	*24.22*	29.86	*24.55*	30.27	*24.90*	57.4
57.6	27.62	*22.66*	27.79	*22.79*	28.13	*23.08*	28.48	*23.38*	28.85	*23.69*	29.21	*24.01*	29.60	*24.33*	30.00	*24.66*	30.41	*25.01*	57.6
57.8	27.75	*22.76*	27.91	*22.90*	28.25	*23.19*	28.60	*23.49*	28.97	*23.80*	29.34	*24.12*	29.73	*24.44*	30.14	*24.78*	30.55	*25.13*	57.8
58.0	27.87	*22.86*	28.03	*23.00*	28.38	*23.29*	28.73	*23.59*	29.10	*23.91*	29.47	*24.23*	29.87	*24.55*	30.27	*24.89*	30.69	*25.25*	58.0
58.2	27.99	*22.96*	28.15	*23.10*	28.50	*23.40*	28.86	*23.70*	29.23	*24.02*	29.60	*24.34*	29.99	*24.66*	30.41	*25.01*	30.83	*25.37*	58.2
58.4	28.11	*23.07*	28.28	*23.21*	28.62	*23.50*	28.98	*23.81*	29.36	*24.13*	29.73	*24.45*	30.13	*24.78*	30.54	*25.13*	30.97	*25.49*	58.4
58.6	28.23	*23.17*	28.40	*23.31*	28.75	*23.61*	29.11	*23.91*	29.48	*24.23*	29.86	*24.56*	30.26	*24.89*	30.68	*25.23*	31.11	*25.61*	58.6
58.8	28.35	*23.27*	28.52	*23.41*	28.88	*23.71*	29.23	*24.02*	29.61	*24.34*	29.99	*24.67*	30.40	*25.00*	30.82	*25.35*	31.25	*25.73*	58.8
59.0	28.47	*23.37*	28.64	*23.52*	29.00	*23.82*	29.36	*24.13*	29.74	*24.45*	30.13	*24.78*	30.53	*25.12*	30.95	*25.47*	31.40	*25.86*	59.0
59.2	28.59	*23.48*	28.77	*23.62*	29.12	*23.93*	29.49	*24.24*	29.87	*24.56*	30.26	*24.89*	30.67	*25.24*	31.09	*25.59*	31.54	*25.98*	59.2
59.4	28.71	*23.58*	28.89	*23.73*	29.25	*24.03*	29.61	*24.34*	29.99	*24.67*	30.39	*25.00*	30.81	*25.36*	31.23	*25.71*	31.68	*26.10*	59.4
59.6	28.84	*23.68*	29.01	*23.83*	29.37	*24.14*	29.74	*24.45*	30.13	*24.78*	30.53	*25.11*	30.94	*25.47*	31.38	*25.83*	31.83	*26.23*	59.6
59.8	28.96	*23.79*	29.13	*23.93*	29.50	*24.24*	29.87	*24.56*	30.26	*24.89*	30.66	*25.23*	31.08	*25.59*	31.52	*25.95*	31.97	*26.35*	59.8
60.0	29.08	*23.89*	29.26	*24.04*	29.62	*24.35*	29.99	*24.67*	30.39	*24.99*	30.79	*25.34*	31.22	*25.71*	31.66	*26.08*	32.12	*26.48*	60.0
60.2	29.20	*23.99*	29.38	*24.14*	29.74	*24.46*	30.12	*24.77*	30.52	*25.11*	30.93	*25.46*	31.36	*25.83*	31.80	*26.20*	32.27	*26.61*	60.2
60.4	29.32	*24.10*	29.50	*24.25*	29.87	*24.56*	30.25	*24.88*	30.65	*25.22*	31.06	*25.57*	31.50	*25.95*	31.94	*26.33*	32.41	*26.73*	60.4
60.6	29.45	*24.20*	29.63	*24.35*	29.99	*24.67*	30.38	*24.99*	30.78	*25.34*	31.20	*25.69*	31.64	*26.07*	32.09	*26.45*	32.56	*26.86*	60.6
60.8	29.57	*24.30*	29.75	*24.46*	30.12	*24.77*	30.51	*25.10*	30.91	*25.45*	31.33	*25.80*	31.78	*26.19*	32.23	*26.58*	32.71	*26.99*	60.8
61.0	29.69	*24.41*	29.87	*24.56*	30.25	*24.88*	30.64	*25.21*	31.05	*25.56*	31.47	*25.92*	31.92	*26.31*	32.38	*26.70*	32.86	*27.12*	61.0
61.2	29.81	*24.51*	29.99	*24.66*	30.38	*24.98*	30.77	*25.32*	31.18	*25.68*	31.61	*26.04*	32.06	*26.43*	32.52	*26.83*	33.01	*27.24*	61.2
61.4	29.93	*24.61*	30.12	*24.77*	30.50	*25.09*	30.90	*25.44*	31.32	*25.79*	31.74	*26.16*	32.20	*26.55*	32.67	*26.95*	33.16	*27.37*	61.4
61.6	30.06	*24.72*	30.25	*24.87*	30.63	*25.20*	31.03	*25.55*	31.45	*25.90*	31.88	*26.28*	32.34	*26.67*	32.81	*27.08*	33.31	*27.50*	61.6
61.8	30.18	*24.82*	30.37	*24.98*	30.76	*25.31*	31.16	*25.66*	31.59	*26.02*	32.01	*26.40*	32.49	*26.79*	32.96	*27.20*	33.46	*27.63*	61.8

62.0	27.76	33.60	27.33	33.10	26.90	32.63	26.51	32.16	26.14	31.72	25.77	31.29	25.43	30.89	25.09	30.50	24.93	30.31
62.2	27.88	33.75	27.46	33.25	27.04	32.77	26.63	32.32	26.25	31.86	25.88	31.43	25.54	31.01	25.20	30.63	25.03	30.43
62.4	28.01	33.90	27.58	33.40	27.16	32.91	26.75	32.44	26.37	31.99	25.99	31.56	25.65	31.14	25.31	30.75	25.14	30.56
62.6	28.15	34.05	27.71	33.55	27.29	33.06	26.87	32.58	26.49	32.13	26.11	31.69	25.76	31.28	25.42	30.88	25.25	30.69
62.8	28.28	34.21	27.84	33.70	27.41	33.20	26.99	32.72	26.61	32.27	26.23	31.83	25.87	31.41	25.53	31.01	25.36	30.81
63.0	28.42	34.36	27.97	33.84	27.54	33.35	27.12	32.87	26.73	32.41	26.35	31.96	25.98	31.54	25.64	31.14	25.47	30.94
63.2	28.55	34.52	28.10	33.99	27.66	33.50	27.24	33.01	26.85	32.55	26.46	32.10	26.09	31.67	25.75	31.26	25.58	31.06
63.4	28.69	34.67	28.23	34.15	27.79	33.64	27.37	33.15	26.98	32.69	26.58	32.23	26.21	31.80	25.86	31.39	25.69	31.19
63.6	28.82	34.83	28.36	34.30	27.91	33.79	27.50	33.30	27.09	32.83	26.70	32.37	26.32	31.93	25.97	31.52	25.80	31.32
63.8	28.96	34.98	28.49	34.45	28.04	33.93	27.62	33.44	27.21	32.97	26.82	32.51	26.44	32.07	26.08	31.65	25.91	31.45
64.0	29.10	35.15	28.63	34.61	28.17	34.08	27.74	33.59	27.33	33.11	26.94	32.65	26.55	32.20	26.19	31.78	26.02	31.58
64.2	29.24	35.31	28.76	34.76	28.30	34.23	27.87	33.73	27.45	33.25	27.05	32.79	26.67	32.34	26.30	31.91	26.13	31.70
64.4	29.38	35.48	28.90	34.92	28.43	34.39	28.00	33.88	27.57	33.39	27.17	32.92	26.78	32.47	26.41	32.04	26.24	31.83
64.6	29.52	35.64	29.03	35.07	28.57	34.54	28.12	34.02	27.70	33.53	27.29	33.06	26.90	32.60	26.53	32.17	26.35	31.96
64.8	29.67	35.80	29.17	35.23	28.70	34.69	28.25	34.17	27.82	33.67	27.41	33.20	27.01	32.74	26.64	32.30	26.46	32.09
65.0	29.81	35.97	29.31	35.39	28.83	34.84	28.38	34.32	27.94	33.82	27.53	33.34	27.13	32.87	26.75	32.43	26.57	32.22
65.2	29.95	36.13	29.44	35.55	28.96	34.99	28.51	34.47	28.06	33.96	27.65	33.48	27.25	33.01	26.86	32.57	26.68	32.35
65.4	30.09	36.30	29.58	35.71	29.10	35.15	28.63	34.61	28.19	34.10	27.77	33.62	27.37	33.15	26.97	32.70	26.79	32.48
65.6	30.24	36.46	29.72	35.87	29.23	35.30	28.76	34.76	28.32	34.25	27.89	33.76	27.49	33.28	27.09	32.83	26.90	32.61
65.8	30.39	36.63	29.86	36.02	29.37	35.46	28.89	34.91	28.45	34.40	28.01	33.90	27.60	33.42	27.21	32.96	27.01	32.75
66.0	30.54	36.79	30.00	36.19	29.51	35.62	29.02	35.06	28.57	34.54	28.13	34.04	27.72	33.56	27.32	33.10	27.13	32.88
66.2	30.68	36.96	30.15	36.35	29.64	35.77	29.16	35.22	28.70	34.69	28.26	34.18	27.84	33.70	27.44	33.23	27.25	33.01
66.4	30.83	37.13	30.29	36.52	29.78	35.93	29.29	35.38	28.83	34.84	28.38	34.33	27.96	33.84	27.56	33.37	27.36	33.14
66.6	30.98	37.30	30.43	36.68	29.92	36.09	29.43	35.53	28.96	34.99	28.51	34.47	28.08	33.98	27.68	33.51	27.48	33.28
66.8	31.13	37.48	30.58	36.84	30.06	36.25	29.57	35.69	29.09	35.14	28.64	34.62	28.20	34.12	27.80	33.65	27.60	33.41
67.0	31.28	37.65	30.72	37.01	30.20	36.41	29.71	35.84	29.22	35.29	28.76	34.76	28.33	34.26	27.91	33.79	27.71	33.55
67.2	31.44	37.83	30.87	37.18	30.34	36.57	29.84	36.00	29.35	35.44	28.89	34.91	28.45	34.41	28.03	33.92	27.83	33.69
67.4	31.59	38.00	31.02	37.35	30.49	36.73	29.98	36.16	29.49	35.60	29.01	35.05	28.58	34.55	28.15	34.06	27.95	33.82
67.6	31.74	38.18	31.17	37.52	30.63	36.90	30.12	36.32	29.62	35.75	29.14	35.20	28.70	34.69	28.27	34.20	28.06	33.96
67.8	31.89	38.35	31.32	37.69	30.77	37.06	30.26	36.48	29.75	35.90	29.27	35.35	28.83	34.84	28.40	34.34	28.18	34.09

TABLE A6.—(Continued)

Scale reading	17.5°C Per cent by volume	17.5°C Per cent by weight	18°C Per cent by volume	18°C Per cent by weight	19°C Per cent by volume	19°C Per cent by weight	20°C Per cent by volume	20°C Per cent by weight	21°C Per cent by volume	21°C Per cent by weight	22°C Per cent by volume	22°C Per cent by weight	23°C Per cent by volume	23°C Per cent by weight	24°C Per cent by volume	24°C Per cent by weight	25°C Per cent by volume	25°C Per cent by weight	Scale reading
68.0	34.23	28.30	34.48	28.52	34.98	28.95	35.50	29.41	36.05	29.89	36.63	30.40	37.23	30.91	37.86	31.47	38.53	32.05	68.0
68.2	34.36	28.42	34.62	28.64	35.13	29.08	35.65	29.54	36.21	30.02	36.79	30.54	37.39	31.06	38.03	31.62	38.70	32.21	68.2
68.4	34.50	28.54	34.76	28.76	35.27	29.21	35.80	29.67	36.37	30.16	36.95	30.68	37.56	31.21	38.21	31.77	38.88	32.37	68.4
68.6	34.64	28.65	34.90	28.88	35.42	29.33	35.95	29.80	36.52	30.30	37.12	30.82	37.73	31.35	38.38	31.92	39.06	32.53	68.6
68.8	34.77	28.77	35.04	29.01	35.57	29.46	36.10	29.93	36.68	30.43	37.28	30.96	37.90	31.50	38.56	32.07	39.24	32.69	68.8
69.0	34.91	28.89	35.19	29.13	35.71	29.59	36.25	30.06	36.84	30.57	37.45	31.10	38.07	31.65	38.73	32.23	39.43	32.86	69.0
69.2	35.04	29.01	35.33	29.26	35.86	29.72	36.41	30.20	36.99	30.71	37.61	31.25	38.24	31.79	38.90	32.39	39.61	33.02	69.2
69.4	35.19	29.14	35.47	29.38	36.01	29.85	36.56	30.33	37.15	30.85	37.78	31.39	38.41	31.94	39.08	32.55	39.80	33.18	69.4
69.6	35.34	29.26	35.62	29.51	36.16	29.97	36.72	30.47	37.32	30.99	37.94	31.54	38.58	32.09	39.26	32.71	39.98	33.34	69.6
69.8	35.49	29.39	35.76	29.64	36.31	30.11	36.87	30.61	37.48	31.13	38.11	31.68	38.75	32.25	39.45	32.86	40.17	33.51	69.8
70.0	35.64	29.52	35.91	29.76	36.46	30.24	37.02	30.74	37.64	31.27	38.28	31.83	38.92	32.41	39.63	33.02	40.35	33.67	70.0
70.2	35.78	29.65	36.05	29.89	36.61	30.38	37.19	30.88	37.80	31.42	38.45	31.97	39.10	32.57	39.81	33.19	40.53	33.84	70.2
70.4	35.93	29.79	36.20	30.01	36.76	30.51	37.35	31.01	37.97	31.56	38.61	32.12	39.28	32.72	39.99	33.35	40.72	34.00	70.4
70.6	36.08	29.91	36.35	30.15	36.92	30.64	37.51	31.16	38.13	31.70	38.78	32.28	39.46	32.88	40.17	33.51	40.90	34.17	70.6
70.8	36.23	30.04	36.50	30.28	37.07	30.78	37.67	31.30	38.30	31.85	38.95	32.43	39.64	33.04	40.35	33.68	41.08	34.33	70.8
71.0	36.38	30.17	36.65	30.41	37.23	30.91	37.83	31.44	38.47	31.99	39.12	32.59	39.82	33.20	40.54	33.84	41.27	34.50	71.0
71.2	36.53	30.30	36.80	30.55	37.39	31.05	37.99	31.59	38.63	32.15	39.30	32.74	40.00	33.36	40.72	34.00	41.46	34.67	71.2
71.4	36.68	30.44	36.95	30.68	37.55	31.19	38.16	31.73	38.80	32.30	39.48	32.90	40.18	33.52	40.90	34.17	41.64	34.83	71.4
71.6	36.83	30.57	37.11	30.81	37.71	31.33	38.32	31.87	38.97	32.45	39.65	33.05	40.36	33.68	41.08	34.33	41.83	35.00	71.6
71.8	36.98	30.70	37.27	30.95	37.87	31.47	38.49	32.01	39.14	32.60	39.83	33.21	40.54	33.84	41.27	34.50	42.02	35.17	71.8
72.0	37.13	30.84	37.42	31.08	38.02	31.61	38.65	32.17	39.31	32.76	40.01	33.37	40.72	34.00	41.45	34.66	42.21	35.34	72.0
72.2	37.29	30.97	37.58	31.22	38.19	31.75	38.82	32.32	39.49	32.91	40.18	33.52	40.90	34.16	41.64	34.83	42.40	35.51	72.2
72.4	37.44	31.10	37.73	31.36	38.35	31.89	38.98	32.47	39.66	33.06	40.36	33.68	41.08	34.33	41.82	34.99	42.58	35.68	72.4
72.6	37.60	31.24	37.89	31.49	38.51	32.04	39.16	32.62	39.83	33.22	40.54	33.84	41.26	34.49	42.01	35.16	42.77	35.85	72.6
72.8	37.75	31.37	38.05	31.63	38.67	32.18	39.33	32.77	40.01	33.37	40.71	33.99	41.45	34.65	42.19	35.33	42.96	36.02	72.8

73.0	37.91	31.51	38.21	31.77	38.84	32.33	39.50	32.02	40.18	33.52	40.88	34.15	34.81	42.38	35.40	43.15	36.18
73.2	38.06	31.65	38.37	31.90	39.00	32.48	39.67	32.21	40.36	33.68	41.06	34.31	34.98	42.56	35.66	43.33	36.35
73.4	38.22	31.78	38.53	32.04	39.17	32.62	39.84	32.40	40.53	33.83	41.24	34.47	35.14	42.75	35.83	43.52	36.52
73.6	38.38	31.92	38.69	32.19	39.34	32.77	40.02	32.59	40.71	33.98	41.42	34.63	35.31	42.93	35.99	43.70	36.68
73.8	38.54	32.06	38.85	32.34	39.50	32.92	40.19	32.78	40.88	34.14	41.60	34.79	35.47	43.12	36.16	43.89	36.85
74.0	38.70	32.20	39.01	32.48	39.67	33.07	40.36	32.97	41.05	34.30	41.78	34.95	35.64	43.31	36.33	44.08	37.02
74.2	38.86	32.35	39.18	32.63	39.84	33.22	40.53	33.16	41.23	34.46	41.96	35.12	35.80	43.49	36.50	44.28	37.20
74.4	39.02	32.49	39.34	32.77	40.01	33.37	40.71	33.35	41.41	34.62	42.15	35.28	35.97	43.68	36.66	44.48	37.38
74.6	39.18	32.63	39.51	32.92	40.18	33.53	40.88	33.54	41.59	34.78	42.33	35.45	36.13	43.86	36.83	44.67	37.56
74.8	39.35	32.78	39.68	33.07	40.35	33.68	41.05	33.73	41.77	34.94	42.51	35.61	36.30	44.05	36.99	44.87	37.75
75.0	39.51	32.92	39.84	33.22	40.53	33.83	41.23	33.92	41.95	35.10	42.70	35.78	36.47	44.25	37.17	45.07	37.93
75.2	39.68	33.07	40.01	33.37	40.70	33.98	41.41	34.11	42.13	35.26	42.88	35.95	36.64	44.44	37.35	45.29	38.12
75.4	39.84	33.22	40.18	33.53	40.87	34.14	41.58	34.30	42.31	35.43	43.07	36.11	36.81	44.63	37.53	45.50	38.31
75.6	40.01	33.37	40.35	33.68	41.04	34.30	41.76	34.49	42.49	35.59	43.25	36.28	36.97	44.83	37.71	45.71	38.50
75.8	40.18	33.53	40.53	33.83	41.22	34.45	41.94	34.68	42.67	35.75	43.44	36.45	37.15	45.03	37.89	45.92	38.69
76.0	40.35	33.68	40.70	33.98	41.40	34.61	42.12	34.87	42.85	35.92	43.63	36.62	37.33	45.24	38.08	46.12	38.88
76.2	40.53	33.83	40.87	34.14	41.57	34.77	42.30	35.06	43.04	36.08	43.81	36.79	37.50	45.44	38.27	46.34	39.08
76.4	40.70	33.98	41.04	34.29	41.75	34.92	42.48	35.25	43.22	36.25	44.00	36.96	37.68	45.65	38.46	46.56	39.29
76.6	40.87	34.14	41.22	34.45	41.92	35.08	42.66	35.44	43.41	36.41	44.19	37.13	37.86	45.86	38.65	46.78	39.49
76.8	41.04	34.29	41.40	34.60	42.10	35.24	42.84	35.63	43.60	36.59	44.38	37.30	38.04	46.07	38.84	47.00	39.69
77.0	41.22	34.45	41.57	34.76	42.28	35.40	43.02	35.82	43.79	36.76	44.57	37.47	38.23	46.29	39.03	47.23	39.90
77.2	41.39	34.60	41.74	34.91	42.46	35.56	43.20	36.01	43.97	36.93	44.76	37.65	38.42	46.51	39.23	47.45	40.11
77.4	41.57	34.75	41.91	35.07	42.63	35.72	43.39	36.20	44.16	37.10	44.95	37.82	38.60	46.73	39.44	47.68	40.32
77.6	41.75	34.91	42.09	35.23	42.81	35.88	43.57	36.39	44.35	37.28	45.15	37.99	38.79	46.95	39.64	47.91	40.54
77.8	41.92	35.06	42.26	35.38	42.99	36.04	43.76	36.58	44.54	37.45	45.35	38.18	38.98	47.17	39.85	48.14	40.75
78.0	42.09	35.22	42.43	35.54	43.17	36.21	43.94	36.77	44.73	37.62	45.56	38.37	39.18	47.40	40.05	48.37	40.97
78.2	42.26	35.38	42.61	35.70	43.36	36.38	44.13	36.96	44.92	37.79	45.76	38.56	39.39	47.63	40.27	48.60	41.18
78.4	42.44	35.53	42.78	35.85	43.54	36.54	44.32	37.15	45.12	37.97	45.96	38.75	39.59	47.85	40.48	48.84	41.40
78.6	42.61	35.69	42.96	36.01	43.72	36.71	44.51	37.34	45.32	38.15	46.17	38.94	39.80	48.08	40.69	49.07	41.62
78.8	42.78	35.84	43.14	36.17	43.91	36.88	44.70	37.53	45.52	38.33	46.39	39.13	40.00	48.31	40.90	49.31	41.84
79.0	42.95	36.00	43.32	36.34	44.09	37.04	44.89	37.72	45.72	38.52	46.61	39.33	40.21	48.53	41.12	49.54	42.05
79.2	43.13	36.16	43.50	36.50	44.28	37.21	45.08	37.91	45.92	38.70	46.83	39.54	40.42	48.76	41.33	49.77	42.27
79.4	43.31	36.33	43.68	36.67	44.47	37.38	45.28	38.10	46.13	38.89	47.04	39.74	40.63	48.99	41.54	50.01	42.49
79.6	43.49	36.49	43.86	36.83	44.65	37.56	45.48	38.29	46.34	39.08	47.26	39.94	40.84	49.22	41.76	50.24	42.71
79.8	43.67	36.66	44.05	37.00	44.84	37.73	45.68	38.48	46.56	39.28	47.48	40.14	41.05	49.45	41.97	50.48	42.93
80.0	43.85	36.82	44.24	37.17	45.04	37.90	45.88	38.67	46.77	39.48	47.70	40.35	41.26	49.68	42.18	50.71	43.15

TABLE A7.—MUNSON AND WALKER'S TABLE * FOR CALCULATING DEXTROSE, INVERT SUGAR ALONE, INVERT SUGAR IN THE PRESENCE OF SUCROSE (0.4 GRAM AND 2 GRAMS TOTAL SUGAR), LACTOSE, LACTOSE AND SUCROSE (2 MIXTURES), AND MALTOSE (CRYSTALLIZED)

(Expressed in milligrams)

Cuprous oxide (Cu₂O)	Copper (Cu)	Dextrose (d-glucose)	Invert sugar	Invert Sugar and Sucrose		Lactose	Lactose and Sucrose		Maltose	Cuprous oxide (Cu₂O)
				0.4 gram total sugar	2 grams total sugar	$C_{12}H_{22}O_{11}+H_2O$	1 lactose, 4 sucrose	1 lactose, 12 sucrose	$C_{12}H_{22}O_{11}+H_2O$	
10	8.9	4.0	4.5	1.6	....	6.3	6.1	....	6.2	10
12	10.7	4.9	5.4	2.5	....	7.5	7.3	....	7.9	12
14	12.4	5.7	6.3	3.4	....	8.8	8.5	....	9.5	14
16	14.2	6.6	7.2	4.3	....	10.0	9.7	....	11.2	16
18	16.0	7.5	8.1	5.2	....	11.3	10.9	....	12.9	18
20	17.8	8.3	8.9	6.1	....	12.5	12.1	....	14.6	20
22	19.5	9.2	9.8	7.0	....	13.8	13.3	....	16.2	22
24	21.3	10.0	10.7	7.9	....	15.0	14.5	....	17.9	24
26	23.1	10.9	11.6	8.8	....	16.3	15.8	....	19.6	26
28	24.9	11.8	12.5	9.7	....	17.6	17.0	....	21.2	28
30	26.6	12.6	13.4	10.7	4.3	18.8	18.2	....	22.9	30
32	28.4	13.5	14.3	11.6	5.2	20.1	19.4	....	24.6	32
34	30.2	14.3	15.2	12.5	6.1	21.4	20.7	....	26.2	34
36	32.0	15.2	16.1	13.4	7.0	22.8	22.0	....	27.9	36
38	33.8	16.1	16.9	14.3	7.9	24.2	23.3	....	29.6	38
40	35.5	16.9	17.8	15.2	8.8	25.5	24.7	....	31.3	40
42	37.3	17.8	18.7	16.1	9.7	29.6	26.0	....	32.9	42
44	39.1	18.7	19.6	17.0	10.7	28.3	27.3	....	34.6	44
46	40.9	19.6	20.5	17.9	11.6	29.6	28.6	....	36.3	46
48	42.6	20.4	21.4	18.8	12.5	31.0	30.0	....	37.9	48
50	44.4	21.3	22.3	19.7	13.4	32.3	31.3	....	39.6	50
52	46.2	22.2	23.2	20.7	14.3	33.7	32.6	....	41.3	52
54	48.0	23.0	24.1	21.6	15.2	35.1	34.0	....	42.9	54
56	49.7	23.9	25.0	22.5	16.2	36.4	35.3	....	44.6	56
58	51.5	24.8	25.9	23.4	17.1	37.8	36.6	....	46.3	58
60	53.3	25.6	26.8	24.3	18.0	39.2	37.9	....	48.0	60
62	55.1	26.5	27.7	25.2	18.9	40.5	39.3	....	49.6	62
64	56.8	27.4	28.6	26.2	19.8	41.9	40.6	....	51.3	64
66	58.6	28.3	29.5	27.1	20.8	43.3	41.9	....	53.0	66
68	60.4	29.2	30.4	28.0	21.7	44.7	43.3	40.7	54.6	68
70	62.2	30.0	31.3	28.9	22.6	46.0	44.6	41.9	56.3	70
72	64.0	30.9	32.3	29.8	23.5	47.4	45.9	43.1	58.0	72
74	65.7	31.8	33.2	30.8	24.5	48.8	47.3	44.2	59.6	74
76	67.5	32.7	34.1	31.7	25.4	50.1	48.6	45.4	61.3	76
78	69.3	33.6	35.0	32.6	26.3	51.5	49.9	46.6	63.0	78
80	71.1	34.4	35.9	33.5	27.3	52.9	51.3	47.8	64.6	80
82	72.8	35.3	36.8	34.5	28.2	54.2	52.6	49.0	66.3	82
84	74.6	36.2	37.7	35.4	29.1	55.6	53.9	50.1	68.0	84
86	76.4	37.1	38.6	36.3	30.0	57.0	55.3	51.3	69.7	86
88	78.2	38.0	39.5	37.2	31.0	58.4	56.6	52.5	71.3	88

*U. S. Bur. Standards Circ. 44, p. 321. The columns headed "Lactose" and "Lactose and Sucrose" are taken from "Methods of Sugar Analysis and Allied Determinations" by Arthur Given.

TABLE A7.—MUNSON AND WALKER'S TABLE.—(*Continued*)

(Expressed in milligrams)

Cuprous oxide (Cu₂O)	Copper (Cu)	Dextrose (*d*-glucose)	Invert sugar	Invert Sugar and Sucrose		Lactose	Lactose and Sucrose		Maltose	Cuprous oxide (Cu₂O)
				0.4 gram total sugar	2 grams total sugar	$C_{12}H_{22}O_{11}+H_2O$	1 lactose, 4 sucrose	1 lactose, 12 sucrose	$C_{12}H_{22}O_{11}+H_2O$	
90	79.9	38.9	40.4	38.2	31.9	59.7	57.9	53.7	73.0	90
92	81.7	39.8	41.4	39.1	32.8	61.1	59.3	54.9	74.7	92
94	83.5	40.6	42.3	40.0	33.8	62.5	60.6	56.0	76.3	94
96	85.3	41.5	43.2	41.0	34.7	63.8	61.9	57.2	78.0	96
98	87.1	42.4	44.1	41.9	35.6	65.2	63.3	58.4	79.7	98
100	88.8	43.3	45.0	42.8	36.6	66.6	64.6	59.6	81.3	100
102	90.6	44.2	46.0	43.8	37.5	68.0	66.0	60.8	83.0	102
104	92.4	45.1	46.9	44.7	38.5	69.3	67.3	62.0	84.7	104
106	94.2	46.0	47.8	45.6	39.4	70.7	68.6	63.2	86.3	106
108	95.9	46.9	48.7	46.6	40.3	72.1	70.0	64.4	88.0	108
110	97.7	47.8	49.6	47.5	41.3	73.5	71.3	65.6	89.7	110
112	99.5	48.7	50.6	48.4	42.2	74.8	72.6	66.7	91.3	112
114	101.3	49.6	51.5	49.4	43.2	76.2	74.0	67.9	93.0	114
116	103.0	50.5	52.4	50.3	44.1	77.6	75.3	69.1	94.7	116
118	104.8	51.4	53.3	51.2	45.0	79.0	76.7	70.3	96.4	118
120	106.6	52.3	54.3	52.2	46.0	80.3	78.0	71.5	98.0	120
122	108.4	53.2	55.2	53.1	46.9	81.7	79.3	72.7	99.7	122
124	110.1	54.1	56.1	54.1	47.9	83.1	80.7	73.9	101.4	124
126	111.9	55.0	57.0	55.0	48.8	84.5	82.0	75.1	103.0	126
128	113.7	55.9	58.0	55.9	49.8	85.8	83.4	76.3	104.7	128
130	115.5	56.8	58.9	56.9	50.7	87.2	84.7	77.5	106.4	130
132	117.3	57.7	59.8	57.8	51.7	88.6	86.0	78.7	108.0	132
134	119.0	58.6	60.8	58.8	52.6	90.0	87.4	79.7	109.7	134
136	120.8	59.5	61.7	59.7	53.6	91.3	88.7	81.1	111.4	136
138	122.6	60.4	62.6	60.7	54.5	92.7	90.1	82.3	113.0	138
140	124.4	61.3	63.6	61.6	55.5	94.1	91.4	83.5	114.7	140
142	126.1	62.2	64.5	62.6	56.4	95.5	92.8	84.7	116.4	142
144	127.9	63.1	65.4	63.5	57.4	96.8	94.1	85.9	118.0	144
146	129.7	64.0	66.4	64.5	58.3	98.2	95.4	87.1	119.7	146
148	131.5	65.0	67.3	65.4	59.3	99.6	96.8	88.3	121.4	148
150	133.2	65.9	68.3	66.4	60.2	101.0	98.1	89.5	123.0	150
152	135.0	66.8	69.2	67.3	61.2	102.3	99.5	90.8	124.7	152
154	136.8	67.7	70.1	68.3	62.1	103.7	100.8	92.0	126.4	154
156	138.6	68.6	71.1	69.2	63.1	105.1	102.2	93.2	128.0	156
158	140.3	69.5	72.0	70.2	64.1	106.5	103.5	94.4	129.7	158
160	142.1	70.4	73.0	71.2	65.0	107.9	104.8	95.6	131.4	160
162	143.9	71.4	73.9	72.1	66.0	109.2	106.2	96.8	133.0	162
164	145.7	72.3	74.9	73.1	66.9	110.6	107.5	98.0	134.7	164
166	147.5	73.2	75.8	74.0	67.9	112.0	108.9	99.2	136.4	166
168	149.2	74.1	76.8	75.0	68.9	113.4	110.2	100.4	138.0	168

TABLE A7.—MUNSON AND WALKER'S TABLE.—(Continued)

(Expressed in milligrams)

Cuprous oxide (Cu₂O)	Copper (Cu)	Dextrose (d-glucose)	Invert sugar	Invert Sugar and Sucrose		Lactose	Lactose and Sucrose		Maltose	Cuprous oxide (Cu₂O)
				0.4 gram total sugar	2 grams total sugar	$C_{12}H_{22}O_{11}+H_2O$	1 lactose, 4 sucrose	1 lactose, 12 sucrose	$C_{12}H_{22}O_{11}+H_2O$	
170	151.0	75.1	77.7	76.0	69.8	114.8	111.6	101.6	139.7	170
172	152.8	76.0	78.7	76.9	70.8	116.1	112.9	102.8	141.4	172
174	154.6	76.9	79.6	77.9	71.7	117.5	114.3	104.1	143.0	174
176	156.3	77.8	80.6	78.8	72.7	118.9	115.6	105.3	144.7	176
178	158.1	78.8	81.5	79.8	73.7	120.3	117.0	106.5	146.4	178
180	159.9	79.7	82.5	80.8	74.6	121.6	118.3	107.7	148.0	180
182	161.7	80.6	83.4	81.7	75.6	123.1	119.7	108.9	149.7	182
184	163.4	81.5	84.4	82.7	76.6	124.3	121.0	110.1	151.4	184
186	165.2	82.5	85.3	83.7	77.6	125.8	122.4	111.3	153.0	186
188	167.0	83.4	86.3	84.6	78.5	127.2	123.7	112.5	154.7	188
190	168.8	84.3	87.2	85.6	79.5	128.5	125.1	113.8	156.4	190
192	170.5	85.3	88.2	86.6	80.5	129.9	126.4	115.0	158.0	192
194	172.3	86.2	89.2	87.6	81.4	131.3	127.8	116.2	159.7	194
196	174.1	87.1	90.1	88.5	82.4	132.7	129.2	117.4	161.4	196
198	175.9	88.1	91.1	89.5	83.4	134.1	130.5	118.6	163.0	198
200	177.7	89.0	92.0	90.5	84.4	135.4	131.9	119.8	164.7	200
202	179.4	89.9	93.0	91.4	85.3	136.8	133.2	121.0	166.4	202
204	181.2	90.9	94.0	92.4	86.3	138.2	134.6	122.3	168.0	204
206	183.0	91.8	94.9	93.4	87.3	139.6	135.9	123.5	169.7	206
208	184.8	92.8	95.9	94.4	88.3	141.0	137.3	124.7	171.4	208
210	186.5	93.7	96.9	95.4	89.2	142.3	138.6	126.0	173.0	210
212	188.3	94.6	97.8	96.3	90.2	143.7	140.0	127.2	174.7	212
214	190.1	95.6	98.8	97.3	91.2	145.1	141.4	128.4	176.4	214
216	191.9	96.5	99.8	98.3	92.2	146.5	142.7	129.6	178.0	216
218	193.6	97.5	100.8	99.3	93.2	147.9	144.1	130.9	179.7	218
220	195.4	98.4	101.7	100.3	94.2	149.3	145.4	132.1	181.4	220
222	197.2	99.4	102.7	101.2	95.1	150.7	146.8	133.3	183.0	222
224	199.0	100.3	103.7	102.2	96.1	152.0	148.1	134.5	184.7	224
226	200.7	101.3	104.6	103.2	97.1	153.4	149.5	135.8	186.4	226
228	202.5	102.2	105.6	104.2	98.1	154.8	150.8	137.0	188.0	228
230	204.3	103.2	106.6	105.2	99.1	156.2	152.2	138.2	189.7	230
232	206.1	104.1	107.6	106.2	100.1	157.6	153.6	139.4	191.3	232
234	207.9	105.1	108.6	107.2	101.1	159.0	154.9	140.7	193.0	234
236	209.6	106.0	109.5	108.2	102.1	160.3	156.3	141.9	194.7	236
238	211.4	107.0	110.5	109.2	103.1	161.7	157.6	143.2	196.3	238
240	213.2	108.0	111.5	110.1	104.0	163.1	159.0	144.4	198.0	240
242	215.0	108.9	112.5	111.1	105.0	164.5	160.3	145.6	199.7	242
244	216.7	109.9	113.5	112.1	106.0	165.9	161.7	146.9	201.3	244
246	218.5	110.8	114.5	113.1	107.0	167.3	163.1	148.1	203.0	246
248	220.3	111.8	115.4	114.1	108.0	168.7	164.4	149.3	204.7	248

TABLE A7.— MUNSON AND WALKER'S TABLE.—*(Continued)*

(Expressed in milligrams)

Cuprous oxide (Cu₂O)	Copper (Cu)	Dextrose (*d*-glucose)	Invert sugar	Invert Sugar and Sucrose		Lactose	Lactose and Sucrose		Maltose	Cuprous oxide (Cu₂O)
				0.4 gram total sugar	2 grams total sugar	$C_{12}H_{22}O_{11}+H_2O$	1 lactose, 4 sucrose	1 lactose, 12 sucrose	$C_{12}H_{22}O_{11}+H_2O$	
250	222.1	112.8	116.4	115.1	109.0	170.1	165.8	150.6	206.3	250
252	223.8	113.7	117.4	116.1	110.0	171.5	167.2	151.8	208.0	252
254	225.6	114.7	118.4	117.1	111.0	172.8	168.5	153.1	209.7	254
256	227.4	115.7	119.4	118.1	112.0	174.2	169.9	154.3	211.3	256
258	229.2	116.6	120.4	119.1	113.0	175.6	171.3	155.5	213.0	258
260	231.0	117.6	121.4	120.1	114.0	177.0	172.6	156.8	214.7	260
262	232.7	118.6	122.4	121.1	115.0	178.4	174.0	158.0	216.3	262
264	234.5	119.5	123.4	122.1	116.0	179.8	175.3	159.3	218.0	264
266	236.3	120.5	124.4	123.1	117.0	181.2	176.7	160.5	219.7	266
268	238.1	121.5	125.4	124.1	118.0	182.6	178.1	161.8	221.3	268
270	239.8	122.5	126.4	125.1	119.0	184.0	179.4	163.0	223.0	270
272	241.6	123.4	127.4	126.2	120.0	185.3	180.8	164.3	224.6	272
274	243.4	124.4	128.4	127.2	121.1	186.7	182.2	165.5	226.3	274
276	245.2	125.4	129.4	128.2	122.1	188.1	183.5	166.8	228.0	276
278	246.9	126.4	130.4	129.2	123.1	189.5	184.9	168.0	229.6	278
280	248.7	127.3	131.4	130.2	124.1	190.9	186.3	169.3	231.3	280
282	250.5	128.3	132.4	131.2	125.1	192.3	187.6	170.5	233.0	282
284	252.3	129.3	133.4	132.2	126.1	193.7	189.0	171.8	234.6	284
286	254.0	130.3	134.4	133.2	127.1	195.1	190.4	173.0	236.3	286
288	255.8	131.3	135.4	134.3	128.1	196.5	191.7	174.3	238.0	288
290	257.6	132.3	136.4	135.3	129.2	197.8	193.1	175.5	239.6	290
292	259.4	133.2	137.4	136.3	130.2	199.2	194.4	176.8	241.3	292
294	261.2	134.2	138.4	137.3	131.2	200.6	195.8	178.1	242.9	294
296	262.9	135.2	139.4	138.3	132.2	202.0	197.2	179.3	244.6	296
298	264.7	136.2	140.5	139.4	133.2	203.4	198.6	180.6	246.3	298
300	266.5	137.2	141.5	140.4	134.2	204.8	199.9	181.8	247.9	300
302	268.3	138.2	142.5	141.4	135.3	206.2	201.3	183.1	249.6	302
304	270.0	139.2	143.5	142.4	136.3	207.6	202.7	184.4	251.3	304
306	271.8	140.2	144.5	143.4	137.3	209.0	204.0	185.6	252.9	306
308	273.6	141.2	145.5	144.5	138.3	210.4	205.4	186.9	254.6	308
310	275.4	142.2	146.6	145.5	139.4	211.8	206.8	188.1	256.3	310
312	277.1	143.2	147.6	146.5	140.4	213.2	208.1	189.4	257.9	312
314	278.9	144.2	148.6	147.6	141.4	214.6	209.5	190.7	259.6	314
316	280.7	145.2	149.6	148.6	142.4	216.0	210.9	191.9	261.2	316
318	282.5	146.2	150.7	149.6	143.5	217.3	212.2	193.2	262.9	318
320	284.2	147.2	151.7	150.7	144.5	218.7	213.6	194.4	264.6	320
322	286.0	148.2	152.7	151.7	145.5	220.1	215.5	195.7	266.2	322
324	287.8	149.2	153.7	152.7	146.6	221.5	216.4	197.0	267.9	324
326	289.6	150.2	154.8	153.8	147.6	222.9	217.7	198.2	269.6	326
328	291.4	151.2	155.8	154.8	148.6	224.3	219.1	199.5	271.2	328

TABLE A7.—MUNSON AND WALKER'S TABLE.—(Continued)

(Expressed in milligrams)

Cuprous oxide (Cu$_2$O)	Copper (Cu)	Dextrose (d-glucose)	Invert sugar	Invert Sugar and Sucrose		Lactose	Lactose and Sucrose		Maltose	Cuprous oxide (Cu$_2$O)
				0.4 gram total sugar	2 grams total sugar	C$_{12}$H$_{22}$O$_{11}$+H$_2$O	1 lactose, 4 sucrose	1 lactose, 12 sucrose	C$_{12}$H$_{22}$O$_{11}$+H$_2$O	
330	293.1	152.2	156.8	155.8	149.7	225.7	220.5	200.8	272.9	330
332	294.9	153.2	157.9	156.9	150.7	227.1	221.8	202.0	274.6	332
334	296.7	154.2	158.9	157.9	151.7	228.5	223.2	203.3	276.2	334
336	298.5	155.2	159.9	159.0	152.8	229.9	224.6	204.6	277.9	336
338	300.2	156.3	161.0	160.0	153.8	231.3	226.0	205.9	279.5	338
340	302.0	157.3	162.0	161.0	154.8	232.7	227.4	207.1	281.2	340
342	303.8	158.3	163.1	162.1	155.9	234.1	228.7	208.4	282.9	342
344	305.6	159.3	164.1	163.1	156.9	235.5	230.1	209.7	284.5	344
346	307.3	160.3	165.1	164.2	158.0	236.9	231.5	211.0	286.2	346
348	309.1	161.4	166.2	165.2	159.0	238.3	232.9	212.2	287.9	348
350	310.9	162.4	167.2	166.3	160.1	239.7	234.3	213.5	289.5	350
352	312.7	163.4	168.3	167.3	161.1	241.1	235.6	214.8	291.2	352
354	314.4	164.4	169.3	168.4	162.2	242.5	237.0	216.1	292.8	354
356	316.2	165.4	170.4	169.4	163.2	243.9	238.4	217.3	294.5	356
358	318.0	166.5	171.4	170.5	164.3	245.3	239.8	218.6	296.2	358
360	319.8	167.5	172.5	171.5	165.3	246.7	241.2	219.9	297.8	360
362	321.6	168.5	173.5	172.6	166.4	248.1	242.5	221.2	299.5	362
364	323.3	169.6	174.6	173.7	167.4	249.5	243.9	222.5	301.2	364
366	325.1	170.6	175.6	174.7	168.5	250.9	245.3	223.7	302.8	366
368	326.9	171.6	176.7	175.8	169.5	252.3	246.7	225.0	304.5	368
370	328.7	172.7	177.7	176.8	170.6	253.7	248.1	226.3	306.1	370
372	330.4	173.7	178.8	177.9	171.6	255.1	249.5	227.6	307.8	372
374	332.2	174.7	179.8	179.0	172.7	256.5	250.9	228.9	309.5	374
376	334.0	175.8	180.9	180.0	173.7	257.9	252.2	230.2	311.1	376
378	335.8	176.8	182.0	181.1	174.8	259.3	253.6	231.5	312.8	378
380	337.5	177.9	183.0	182.1	175.9	260.7	255.0	232.8	314.5	380
382	339.3	178.9	184.1	183.2	176.9	262.1	256.4	234.1	316.1	382
384	341.1	180.0	185.2	184.3	178.0	263.5	257.8	235.4	317.8	384
386	342.9	181.0	186.2	185.4	179.1	264.9	259.2	236.6	319.4	386
388	344.6	182.0	187.3	186.4	180.1	266.5	260.5	237.9	321.1	388
390	346.4	183.1	188.4	187.5	181.2	267.7	261.9	239.2	322.8	390
392	348.2	184.1	189.4	188.6	182.3	269.1	263.3	240.5	324.4	392
394	350.0	185.2	190.5	189.7	183.3	270.5	264.7	241.8	326.1	394
396	351.8	186.2	191.6	190.7	184.4	271.9	266.1	243.1	327.7	396
398	353.5	187.3	192.7	191.8	185.5	273.3	267.5	244.4	329.4	398
400	355.3	188.4	193.7	192.9	186.5	274.7	268.9	245.7	331.1	400
402	357.1	189.4	194.8	194.0	187.6	276.1	270.3	247.0	332.7	402
404	358.9	190.5	195.9	195.0	188.7	277.5	271.7	248.3	334.4	404
406	360.6	191.5	197.0	196.1	189.8	278.9	273.0	249.6	336.0	406
408	362.4	192.6	198.1	197.2	190.8	280.3	274.4	251.0	337.7	408

TABLE A7.—MUNSON AND WALKER'S TABLE.—(Continued)

(Expressed in milligrams)

Cuprous oxide (Cu_2O)	Copper (Cu)	Dextrose (d-glucose)	Invert sugar	Invert Sugar and Sucrose		Lactose	Lactose and Sucrose		Maltose	Cuprous oxide (Cu_2O)
				0.4 gram total sugar	2 grams total sugar	$C_{12}H_{22}O_{11}+H_2O$	1 lactose, 4 sucrose	1 lactose, 12 sucrose	$C_{12}H_{22}O_{11}+H_2O$	
410	364.2	193.7	199.1	198.3	191.9	281.7	275.8	252.3	339.4	410
412	366.0	194.7	200.2	199.4	193.0	283.2	277.2	253.6	341.0	412
414	367.7	195.8	201.3	200.5	194.1	284.6	278.6	254.9	342.7	414
416	369.5	196.8	202.4	201.6	195.2	286.0	280.0	256.2	344.4	416
418	371.3	197.9	203.5	202.6	196.2	287.4	281.4	257.5	346.0	418
420	373.1	199.0	204.6	203.7	197.3	288.8	282.8	258.8	347.7	420
422	374.8	200.1	205.7	204.8	198.4	290.2	284.2	260.1	349.3	422
424	376.6	201.1	206.7	205.9	199.5	291.6	285.6	261.4	351.0	424
426	378.4	202.2	207.8	207.0	200.6	293.0	287.0	262.7	352.7	426
428	380.2	203.3	208.9	208.1	201.7	294.4	288.4	264.0	354.3	428
430	382.0	204.4	210.0	209.2	202.7	295.8	289.8	265.4	356.0	430
432	383.7	205.5	211.1	210.3	203.8	297.2	291.2	266.6	357.6	432
434	385.5	206.5	212.3	211.4	204.9	298.6	292.6	268.0	359.3	434
436	387.3	207.6	213.3	212.5	206.0	300.0	294.0	269.3	361.0	436
438	389.1	208.7	214.4	213.6	207.1	301.4	295.4	270.6	362.6	438
440	390.8	209.8	215.5	214.7	208.2	302.8	296.8	272.0	364.3	440
442	392.6	210.9	216.6	215.8	209.3	304.2	298.2	273.3	365.9	442
444	394.4	212.0	217.8	216.9	210.4	305.6	299.6	274.6	367.6	444
446	396.2	213.1	218.9	218.0	211.5	307.0	301.0	275.9	369.3	446
448	397.9	214.1	220.0	219.1	212.6	308.4	302.4	277.2	370.9	448
450	399.7	215.2	221.1	220.2	213.7	309.9	303.8	278.6	372.6	450
452	401.5	216.3	222.2	221.4	214.8	311.3	305.2	279.9	374.2	452
454	403.3	217.4	223.3	222.5	215.9	312.7	306.6	281.2	375.9	454
456	405.1	218.5	224.4	223.6	217.0	314.1	308.0	282.5	377.6	456
458	406.8	219.6	225.5	224.7	218.1	315.5	309.4	283.9	379.2	458
460	408.6	220.7	226.7	225.8	219.2	316.9	310.8	285.2	380.9	460
462	410.4	221.8	227.8	226.9	220.3	318.3	312.2	286.5	382.5	462
464	412.2	222.9	228.9	228.1	221.4	319.7	313.6	287.8	384.2	464
466	413.9	224.0	230.0	229.2	222.5	321.1	315.0	289.2	385.9	466
468	415.7	225.1	231.2	230.3	223.7	322.5	316.4	290.5	387.5	468
470	417.5	226.2	232.3	231.4	224.8	323.9	317.7	291.8	389.2	470
472	419.3	227.4	233.4	232.5	225.9	325.3	319.1	293.2	390.8	472
474	421.0	228.3	234.5	233.7	227.0	326.8	320.5	294.5	392.5	474
476	422.8	229.6	235.7	234.8	228.1	328.2	321.9	295.8	394.2	476
478	424.6	230.7	236.8	235.9	229.2	329.6	323.3	297.1	395.8	478
480	426.4	231.8	237.9	237.1	230.3	331.0	324.7	298.5	397.5	480
482	428.1	232.9	239.1	238.2	231.5	332.4	326.1	299.8	399.1	482
484	429.9	234.1	240.2	239.3	232.6	333.8	327.5	301.1	400.8	484
486	431.7	235.2	241.4	240.5	233.7	335.2	328.9	302.5	402.4	486
488	433.5	236.3	242.5	241.6	234.8	336.6	330.3	303.8	404.1	488
490	435.3	237.4	243.6	242.7	236.0	338.0	331.7	305.1	405.8	490

CHAPTER XV

STATISTICS OF THE LIQUOR INDUSTRY

The importance of reliable and ample statistics to modern business is now widely recognized and in the last two decades the Federal Government and various trade associations have placed a vast amount of data at the disposal of the corporation executive.

While the Federal Government has compiled statistics of the liquor production and trade for at least three decades, unfortunately the prohibition era has interfered; so that such figures as are available cannot present an accurate picture of the liquor trade as we have known it to exist for the 1918 to 1932 period.

There are presented here a selection of statistical tables taken mainly from government publications and compilations obtained therefrom; which indicate with some accuracy the trends in the United States production and consumption of wines and liquors. The data are presented with a reservation that their accuracy, at best, is qualitative.

Size of the industry. That the industry was formerly both large and widespread is shown by the fact that in 1901 there were 3,736 distilleries in operation, of which 1,258 used grain and 2,478 used fruit as raw material.

By 1914 there were only 352 grain and 368 fruit distilleries operating, or a total of 720. Since production increased during this fourteen year period, we have evidence of merger into fewer and larger units.

According to recent data (Dec., 1933) there are registered 6 grain, 2 molasses and 25 fruit distilleries. All but five of the fruit distilleries are operating. In addition there are 43 industrial alcohol plants operating. See Table 27 for complete data.

In 1907, and again in 1912, over $156,000,000 in taxes were collected on withdrawals of distilled spirits. The Government collected $186,947,000 in 1917, $308,429,000 in 1918 and $353,737,000 in 1919. Taxes on wine ranged from $2,307,000 in 1915 to $11,474,000 in 1919. See Table 28.

In 1901 there remained aging in bonded warehouses after the annual withdrawals 154,438,407 gallons of distilled spirits. By 1914 the bonded warehouses held, after annual withdrawals, 286,900,000 gallons; of which 278,108,000 were whiskey, 1,217,000 rum, 216,000 gin, 4,865,000 brandy and 2,495,000 alcohol. See Table 29. In 1913 the industry used 4,252,583 bushels of malt and 5,828,450 bushels of rye in addition to other cereals. See Table 30.

Production of distilled spirits.—Production of distilled spirits exclusive of alcohol averaged 89,900,000 gallons from 1901 to 1914, reaching a maximum of 114,634,000 gallons in 1913.

Production of whiskey increased from 89,700,000 gallons in 1901 to an average of 99,400,000 for the three year period 1911 to 1913 inclusive.

Production of rum increased from 1,724,000 gallons in 1901 to 3,026,000 gallons in 1914.

Production of gin increased from 1,636,000 gallons in 1901 to 4,012,543 gallons in 1914.

Production of brandy increased from 4,047,000 gallons in 1901 to 7,307,800 gallons in 1914.

Production of alcohol in 1901 was 41,458,000 gallons and 81,101,000 gallons in 1915. Practically a 100 per cent increase in 15 years. By 1926 it had more than doubled to 202,270,000 gallons. It jumped from 79,900,000 in 1922 to 122,400,000 in 1923. Prior to that time 10,000,000 gallons was the highest jump, war years excepted. See Table 31 for more complete data.

Withdrawals of distilled spirits.—Withdrawals of distilled spirits exclusive of alcohol increased from 60,585,000 gallons in 1901 to 83,577,000 gallons in 1913 and to 93,210,000 in 1917, a war year. They declined abruptly in 1920 to 6,394,000 and reached a low of 1,000,000 gallons in 1932. See Table 32.

Foreign trade in distilled spirits.—Exports of distilled spirits averaged 1,400,000 gallons for the 15 year period 1901 to 1915 inclusive.

Exports of rum held fairly level for the 8 year period 1910 to 1917 and averaged 1,250,000 gallons.

Exports of whiskey declined from 685,729 gallons in 1901 to 155,880 gallons in 1918. They spurted suddenly to 3,315,861 in 1920, preceding prohibition. See Table 33.

Germany, Mexico, the Philippine Islands, South America, England, Central America, Canada and Bermuda have been our best customers for whiskey. Following prohibition, Bahama Islands, Bermuda, Canada, Cuba, England and Mexico all received large shipments, most of which probably found its way back to us via the bootleggers. See Table 34.

Imports of distilled liquors remained fairly level during the period 1901 to 1918 and averaged 3,450,000 gallons a year.

Imports of whiskey remained fairly level during the period 1910 to 1917 and averaged 1,420,000 gallons per year.

Imports of brandy averaged 494,000 gallons for the period 1901 to 1917.

Imports of gin averaged 955,000 gallons for the period 1910 to 1916.

Imports of cordials averaged 472,000 gallons for the period 1912 to 1916. See Table 35.

Consumption of distilled spirits.—Theoretically, the apparent consumption of distilled spirits may be calculated as follows. Withdrawals from bonded warehouses, plus imports, minus exports should give the apparent consumption. Unfortunately, in the last seven or eight years bootleggers succeeded in diverting large amounts of industrial alcohol to alcoholic beverages. Since this diversion was considered to amount to a large volume, any estimate of consumption during prohibition must include an allowance for industrial alcohol and must be arbitrary at best. By reference to Table 31 it will be observed that there was a tremendous expansion in industrial alcohol production for the period 1922 to 1932. However, in the period 1901 to 1915 production increased 100 per cent so that large expansion in this industry is not unknown. On the other hand, in the period 1905 to 1914 production averaged 71,570,000 gallons and in the period 1919 to 1922, 90,350,000 gallons. Production then increased rapidly, more than doubling (from 90,000,000 to 200,000,000) in the next four years, 1922 to 1926, and maintained a high rate thereafter.

If increased industrial consumption during 1922 to 1926 accounted for one-half of the increased production then the remaining half probably represents the amount diverted by bootleggers. In other words, 25 per cent of any increase over the 1919 to 1922 average.

On this basis an estimate of consumption has been arrived at for the period 1901 to 1932 and is shown in detail in Table 36. In period 1901 to 1919 consumption averaged 70,120,000 gallons. In period 1925 to 1932 it averaged 41,050,000 gallons with a high of 52,292,000.

It is impossible to determine whether prohibition really did cut consumption to this extent or whether too small credit has been given to the bootleg industry. However, by another method of the consumption it might be pointed out that in the 16 years 1901 to 1916 the gross increase of withdrawals was roughly 28 per cent. On this basis, if the normal increase for the next 16 years has been at the same rate then withdrawals in 1932 would have amounted to 98,630,000 gallons so that consumption would have been something slightly over 100,000,000 gallons allowing for excess of imports over exports.

Wine consumption.—Statistics on wine are divided into two classifications, namely, still wines, and champagne and other sparkling wines.

Production (total) of wines was fairly constant for the period 1912 to 1919 averaging 45,600,000 gallons and reaching a high of 55,756,000 gallons in 1919. Thereafter it dropped sharply, averaging 6,800,000 gallons for the period 1922 to 1932. See Table 37.

Exports of wine for the period 1901 to 1915 averaged 786,000 gallons. From 1916 to 1920 exports increased from 1,133,000 gallons to 4,573,000 gallons. Thereafter there was a sharp drop to 26,000 gallons in 1921 and in the period 1926 to 1932 exports were too negligible to record. See Table 38.

Imports of wine for the period 1901 to 1918 averaged

5,969,000 gallons of still wines; and 913,000 gallons of champagne and other sparkling wines. It is interesting to note that the period 1903 to 1910 marked the largest consumption. Champagne imports then averaged well over 1,000,000 gallons annually, and still wines also jumped to well over 7,000,000 gallons a year and even to 9,500,000 in 1910. Following enactment of prohibition there was a sharp drop to an average of 33,700 gallons of still wines and champagnes for the period 1924 to 1931. See Table 39.

By combining the data in Tables 37–39 apparent consumption of wine is obtained. For the period 1912 to 1919 it averaged 49,233,000 gallons.

Following prohibition it dropped to an average of 6,033,000 gallons for the period 1924 to 1932, with a high of 11,403,000 for the boom year 1929. See Table 40.

The liquor industry as a consumer of agricultural products. —The advocates for repeal of prohibition have made much of the large outlet for farmers' crops which the liquor industry will present. It is not within the scope of this book to consider the crops and amounts consumed by the beer brewing industry.

Reference to Table 30 shows that in 1913 the distilled liquor industry used 4,252,000 bushels of malt and 5,828,000 bushels of unmalted rye. In addition some portion of the corn and molasses consumption of 23,800,000 bushels and 64,600,000 gallons respectively must also be allocated to the liquor industry although larger proportions went into the manufacture of alcohol.

In 1925 the liquor industry used 96,170,000 pounds of raisins and 682,000 pounds of rice besides other materials.

Another factor to be taken into consideration and which, so far, has not had a great deal of publicity is the potential outlet for fruit surpluses. For example, applejack from apples; cordials and liqueurs from peaches, pears, oranges, apricots, plums, cherries, prunes, herbs and seeds, etc.

The possible future of the industry can be visualized by references to Table 41 which shows that in 1930 France produced 1,109,933,000 gallons, Italy 551,737,000 gallons, Spain 481,-585,000 and other countries similar large amounts of wine.

TABLE 27.—DISTILLERIES REGISTERED AND OPERATED AND INDUSTRIAL-ALCOHOL PLANTS OPERATED, FISCAL YEARS 1901 TO 1932, INCLUSIVE

Fiscal year	Grain		Molasses *		Fruit		Industrial alcohol plants operated †	Total operated *
	Registered	Operated	Registered	Operated	Registered	Operated		
1901	1,506	1,258	9	9	2,515	2,478	....	3,745
1902	1,372	1,089	11	11	1,869	1,838	....	2,938
1903	1,275	1,103	13	12	1,378	1,326	....	2,441
1904	1,131	744	15	15	1,453	1,413	....	2,172
1905	896	728	14	13	1,108	1,031	....	1,772
1906	912	740	14	13	1,215	1,132	....	1,885
1907	878	710	15	15	919	862	....	1,587
1908	799	579	15	14	667	607	....	1,200
1909	690	466	17	16	840	810	....	1,292
1910	633	444	18	16	470	446	....	906
1911	588	432	18	17	504	474	....	923
1912	575	417	19	18	401	386	....	821
1913	528	398	23	22	469	450	....	870
1914	475	352	25	23	380	368	....	743
1915	438	249	23	23	386	363	....	635
1916	338	279	25	25	337	301	....	605
1917	301	198	27	25	297	284	....	507
1918	154	72	27	27	192	137	....	236
1919	15	13	25	23	38	38	....	74
1920	8	6	21	21	21	20	44	91
1921	2	2	2	2	32	29	69	102
1922	2	2	2	2	31	30	68	102
1923	....	....	2	2	31	31	65	98
1924	....	....	2	2	23	23	70	95
1925	....	....	2	2	24	24	71	97
1926	....	....	2	2	27	27	64	93
1927	....	....	2	2	22	22	62	86
1928	....	....	2	2	25	25	55	82
1929	....	....	2	2	32	32	52	86
1930	7	7	2	2	25	25	50	84
1931	7	7	2	2	20	20	46	75
1932	6	6	2	2	25	20	43	71

* Rum only manufactured since 1921.
† Industrial-alcohol plants use both molasses and grain in manufacture of alcohol.

TABLE 28.—TAX COLLECTED ON DISTILLED SPIRITS, WINES AND FERMENTED LIQUORS,
FISCAL YEARS 1901 TO 1932, INCLUSIVE

Year	Distilled spirits	Wines	Fermented liquors	Total
1901	$116,027,979.56	....	$75,669,907.65	$191,697,887.21
1902	121,138,013.13	....	71,988,902.39	193,126,915.52
1903	131,953,472.39	....	47,547,856.08	179,501,328.47
1904	135,810,015.42	....	49,083,458.77	184,893,474.19
1905	135,958,513.12	....	50,360,553.18	186,319,066.30
1906	143,394,055.12	....	55,641,858.56	199,035,913.68
1907	156,336,901.89	....	59,567,818.18	215,904,720.07
1908	140,158,807.15	....	59,807,616.81	199,966,423.96
1909	134,868,034.12	....	57,456,411.42	192,324,445.54
1910	148,029,311.54	....	60,572,288.54	208,601,600.08
1911	155,279,858.25	....	64,367,777.65	219,647,635.90
1912	156,391,435.77	$52.00	63,268,770.51	219,660,258.28
1913	163,879,276.54 .	66.00	66,266,989.60	230,146,332.14
1914	159,098,177.31	....	67,081,512.45	226,179,689.76
1915	142,312,397.40	2,307,301.97	79,328,946.72	223,948,646.09
1916	156,050,909.55	2,631,529.98	88,771,103.99	247,453,543.52
1917	186,947,243.78	5,164,075.03	91,897,193.81	284,008,512.62
1918	308,429,318.77	9,124,368.56	126,285,857.65	443,839,544.98
1919	353,737,044.77	11,474,207.49	117,839,602.21	483,050,854.47
1920	93,161,205.60	4,744,070.11	41,965,874.09	139,871,149.80
1921	80,281,612.55	2,316,452.46	25,363.82	82,623,428.83
1922	44,257,100.75	1,306,249.72	46,086.00	45,609,436.47
1923	28,822,015.50	1,531,991.38	4,078.75	30,358,085.63
1924	26,126,317.76	1,454,062.88	5,327.73	27,585,708.37
1925	24,307,331.65	1,595,488.63	1,954.44	25,904,774.72
1926	24,756,900.06	1,679,434.38	15,694.19	26,452,028.63
1927	20,399,065.88	795,602.83	883.25	21,195,551.96
1928	14,414,088.04	893,408.41	300.00	15,307,796.45
1929	12,484,078.53	292,549.93	100.00	12,776,728.46
1930	11,455,883.99	239,383.68	....	11,695,267.67
1931	10,203,569.43	228,495.06	....	10,432,064.49
1932	8,517,399.98	186,563.29	....	8,703,963.27

TABLE 29.—SPIRITS REMAINING IN BONDED WAREHOUSES JUNE 30, 1901, TO 1932
BY KINDS

[Statement in tax gallons]

Year	Whiskey	Rum	Gin	Brandy	Alcohol	Aggregate
1901	150,652,832.5	679,302.7	268,105.7	1,705,269.7	1,132,897.1	154,438,407.7
1902	164,388,547.8	949,430.1	246,526.8	2,077,254.1	3,157,925.8	170,819,684.6
1903	183,930,488.3	1,229,162.2	172,118.6	2,757,382.8	3,019,009.0	191,108,160.9
1904	191,320,875.7	1,310,632.4	255,073.1	2,775,088.3	2,249,344.6	197,911,014.1
1905	210,780,752.6	1,195,443.9	320,568.9	3,177,271.9	3,260,558.2	218,734,595.5
1906	223,737,332.0	1,188,675.5	273,231.3	2,226,587.0	1,536,590.0	228,962,415.8
1907	242,319,516.7	1,222,581.1	242,370.8	2,153,250.4	1,654,347.4	247,592,066.4
1908	231,940,083.4	1,227,008.5	201,176.3	2,966,215.6	1,657,860.0	237,992,343.8
1909	226,096,519.0	1,108,327.9	181,479.0	3,679,936.7	1,755,108.1	232,821,370.7
1910	230,224,625.0	820,268.5	161,604.8	4,137,844.5	2,302,176.3	237,646,519.1
1911	246,203,020.4	983,387.6	214,794.0	4,519,762.1	1,878,144.6	253,799,108.7
1912	260,074,282.8	984,953.3	190,278.3	5,001,083.6	2,536,317.4	268,786,915.4
1913	272,504,285.5	1,086,063.4	180,458.0	5,784,226.8	3,013,733.1	282,568,766.8
1914	278,108,056.1	1,217,302.7	216,016.2	4,865,324.7	2,495,085.2	286,901,784.9
1915	249,714,721.4	1,218,392.7	234,965.4	6,143,372.3	2,500,261.8	259,811,713.6
1916	228,677,774.1	906,042.5	216,911.5	5,849,015.4	2,602,150.2	238,251,893.7
1917	189,675,854.7	966,644.5	533,065.0	4,244,404.8	3,657,118.4	199,257,087.4
1918	140,721,821.5	741,104.2	2,777,467.7	3,494,020.8	14,718,871.1	162,453,285.3
1919	63,942,931.5	460,709.6	1,551,101.8	1,260,344.9	6,403,408.2	73,618,496.0
1920	50,550,498.6	413,923.8	963,996.7	884,025.1	3,935,326.1	56,747,770.3
1921	39,961,943.8	399,419.1	885,912.9	641,558.1	8,643,577.4	50,532,411.3
1922	36,588,568.3	384,012.2	987,884.7	963,781.5	7,068,291.1	45,992,537.8
1923	33,151,029.0	366,244.2	878,597.2	1,269,206.5	7,364,040.9	43,029,117.8
1924	30,064,670.9	341,214.0	836,730.2	1,289,400.8	6,697,627.3	39,229,643.2
1925	26,840,953.5	327,379.1	819,599.3	1,229,141.7	9,641,420.9	38,852,494.5
1926	23,814,140.2	289,344.6	802,433.2	1,133,057.7	6,249,064.2	32,288,039.9
1927	20,904,071.2	252,320.5	819,443.7	1,029,598.0	9,263,241.6	32,268,684.0
1928	17,975,943.8	282,387.2	815,718.2	1,013,305.2	7,113,533.3	27,200,887.7
1929	15,127,390.8	226,830.6	794,447.5	906,702.8	12,075,898.9	29,131,270.6
1930	14,786,971.9	171,409.8	799,587.6	864,141.1	10,375,534.0	26,997,644.4
1931	15,179,327.9	188,648.2	781,191.8	971,828.5	16,346,160.8	33,467,157.2
1932	15,293,713.1	200,305.2	746,076.2	1,020,895.9	19,223,942.03	36,484,932.43

TABLE 30.—STATEMENT SHOWING THE QUANTITIES OF GRAIN AND OTHER MATERIALS USED IN THE PRODUCTION OF ALCOHOL AND OTHER DISTILLED SPIRITS, FISCAL YEARS 1901 TO 1932, INCLUSIVE

Year	Malt	Wheat	Barley	Rye	Corn	Oats	Other materials	Molasses	Dilute saccharine liquid	Liquids containing one-half of 1 per cent or more alcohol by volume
	Bushels	Bushels	Bushels	Bushels	Bushels	Bushels	Bushels	Gallons	Gallons	Gallons
1901	3,274,212	24,172	1,476	5,085,766	18,867,088	21,114	5,019	3,165,390		
1902	3,361,107	29,391	2,542	5,584,659	18,473,850	33,775	2,027	12,485,276		
1903	3,754,085	32,197	3,378	5,873,226	20,597,594	31,235	4,834	15,544,360		
1904	3,454,778	23,915	3,972	5,023,932	19,149,413	25,161	6,512	18,549,406		
1905	3,798,578	12,481	9,874	5,489,028	20,592,504	18,898	5,962	20,549,553		
1906	3,758,555	11,366	2,170	5,595,566	20,001,975	16,925	5,631	22,637,582		
1907	4,440,315	21,452	685	6,259,898	23,474,509	17,301	6,071	25,722,926		
1908	2,974,853	11,756	1,700	3,755,519	17,383,724	12,555	6,425	28,944,703		
1909	3,221,399	9,648	1,678	4,364,097	18,080,711	9,840	997	33,550,024		
1910	3,704,740	10,316	2,733	5,042,741	20,547,427	11,502	8,248	42,293,073		
1911	4,053,262	21,765	2,585	5,376,018	23,247,004	13,172	53,884	44,363,133		
1912	4,075,991	25,505	1,943	5,599,667	23,016,759	6,563	50,576	61,605,281		
1913	4,252,583	2,756	1,225	5,828,450	23,847,875	8,259	98,139	64,640,976		
1914	3,938,715	10,582	2,072	5,341,931	21,315,699	5,654	64,896	64,721,265		
1915	2,357,449	4,550	1,137	2,440,557	14,259,842	5,460	69,123	123,301,496		
1916	4,480,588	3,373	148	3,116,612	32,069,542	9,807	68,822	80,977,474	71,164,758	
1917	4,239,677			2,375,439	33,973,268	6,730	72,172	112,497,633	78,462,969	
1918	1,689,677			248,864	14,544,545		172,039	118,027,960	68,527,242	
1919	573,246			25,304	3,890,347		85,624	123,498,693	9,801,335	
1920	215,072			59,077	1,057,519		51,760	113,132,685	19,327,334	
1921	719,171			187,940	4,810,517		303,072	119,052,798	5,562,518	25,538,553

Year	Raisins (Pounds)	Rice (Pounds)	Hops (Pounds)	Meal, starch, sugar, and sirup (Pounds)	Hydrol acids, sulphuric acids, and ammonia (Pounds)	Chemicals (Pounds)	Mixed sulphate (Gallons)	Pineapple juice (Gallons)	Sirup (Gallons)	Wine and grape materials (Gallons)
1922	679,697			84,876	3,093,065		4,097,905	97,222,854		74,149,364
1923	443,845			5,108	3,105,963		387,598	148,711,458		32,542,885
1924	1,959,985			91,065	4,835,139		2,691,070	155,001,162		35,524,539
1925	1,056,397			96,793	7,200,818		145,071	203,270,135		37,824,066
1926	641,032			12,678	7,948,184		26,621	267,404,218		44,508,071
1927	496,969			6,787	8,383,041		25,701	211,518,647		40,941,512
1928	385,238			6,579	6,189,264		123,624	213,629,806		56,794,823
1929	358,658			8,475	9,801,852		10,891	268,044,822		53,884,275
1930	646,574	11,990	21,320	208,209	9,966,336		19,144	235,797,008		21,601,391
1931	643,350	28,379		6,385,365	2,454,445		1,961	187,790,358		18,378,443
1932	505,613	3,311,441		217,934	4,848,133		2,478	161,295,108		14,967,482

Year	Raisins (Pounds)	Rice (Pounds)	Hops (Pounds)	Meal, starch, sugar, and sirup (Pounds)	Hydrol acids, sulphuric acids, and ammonia (Pounds)	Chemicals (Pounds)	Mixed sulphate (Gallons)	Pineapple juice (Gallons)	Sirup (Gallons)	Wine and grape materials (Gallons)
1925	96,170,629	682,690	57,936	8,107,258	207,268					
1926	10,705,920	724,100	71,508	33,141,370	1,895,265					
1927			960	40,521,606	8,732,989					
1928			6,294	38,655,159	40,869,103					
1929					53,037,963					
1930					65,524,269					
1931 *						30,226,037	10,072,400	5,767,500	1,127	5,142,004
1932 †						20,103,526	11,146,057	4,235,380		4,216,865

* Year 1931 figures include 341,730,671 pounds of rye, 51,117,871 pounds of corn, 1,455,980 pounds of wheat, 5,060 pounds of malt, 30,008,713 pounds of hydrol acids, and 217,324 pounds of xylos, used at chemical plants producing butyl alcohol, acetone, and ethyl alcohol.

† Year 1932 figures include 211,084,675 pounds of corn, 19,865,419 pounds of wheat, 150,619 pounds of malt, and 20,103,526 pounds of hydrol acids, used at chemical plants producing butyl alcohol, acetone, and ethyl alcohol.

TABLE 31.—DISTILLED SPIRITS PRODUCED DURING FISCAL YEARS 1901-1932

[Production shown in tax gallons]

Year	Whiskey	Rum	Gin	Brandy	Alcohol	Total exclusive of alcohol
1901	79,701,170	1,724,582	1,636,299	4,047,602	41,458,547	89,109,653
1902	75,414,812	2,202,047	1,752,280	4,220,400	49,254,261	83,589,539
1903	70,673,931	2,247,906	1,913,404	6,430,673	66,940,959	81,265,914
1904	60,606,978	1,801,179	2,110,215	5,193,262	69,793,578	69,711,634
1905	71,083,421	1,791,987	2,187,709	5,448,584	72,747,676	80,511,701
1906	70,633,074	1,730,101	2,323,289	4,444,071	70,979,660	79,130,555
1907	86,552,027	2,022,407	2,947,687	6,138,304	77,051,166	97,661,049
1908	54,502,027	1,895,922	2,756,752	6,899,822	67,835,037	66,054,523
1909	70,152,174	1,952,374	2,483,743	6,440,857	58,862,463	81,029,248
1910	82,463,894	2,253,949	2,985,435	7,656,433	68,534,247	95,359,711
1911	100,647,155	2,631,059	3,345,370	7,953,131	68,778,809	104,576,715
1912	98,209,574	2,832,515	3,577,861	9,321,823	75,630,032	113,944,773
1913	99,615,828	2,750,846	4,014,600	8,252,874	78,972,108	114,634,148
1914	88,698,797	3,026,085	4,012,542	7,307,897	78,874,219	103,045,321
1915	44,552,489	2,844,313	3,636,285	8,521,951	81,101,063	....
1916	59,240,671	2,986,940	4,118,064	4,159,351	182,778,245	⎫
1917	57,651,834	2,842,921	5,756,666	8,251,097	211,582,744	⎬ War
1918	17,383,511	1,526,743	4,178,538	5,357,325	150,387,680	⎭
1919	....	815,794	....	1,802,422	98,160,323	....
1920	234,705	944,916	....	1,649,445	98,436,170	....
1921	753,374	543,507	....	1,530,792	85,068,776	....
1922	315,799	864,332	....	1,077,063	79,906,101	....
1923	....	805,322	....	1,417,461	122,402,849	....
1924	....	784,698	....	847,104	135,897,725	....
1925	....	784,986	....	547,727	166,165,517	....
1926	....	894,306	....	643,968	202,271,670	....
1927	....	810,449	....	338,430	184,323,016	....
1928	....	953,350	....	411,515	169,149,904	....
1929	....	1,227,413	....	1,194,292	200,832,051	....
1930	1,998,947	982,781	....	416,043	191,859,342	....
1931	2,435,631	1,123,977	....	820,278	166,014,346	....
1932	1,711,028	1,094,777	....	630,786	146,930,912	....
						....

Source: "Statistics concerning intoxicating liquors," Bureau of Industrial Alcohol, U. S. Treasury Department, Washington, D. C.

TABLE 32.—DISTILLED SPIRITS WITHDRAWN ON PAYMENT OF TAX, FISCAL YEARS 1901 TO 1932, INCLUSIVE

[Statement in tax gallons]

Year	Whiskey	Rum	Gin	Brandy	Alcohol	Aggregate	Distilled spirits excluding alcohol
1901	57,117,571.7	731,832.7	1,657,938.7	1,078,389.3	39,686,929.5	100,272,661.9	60,585,730
1902	54,948,215.3	798,038.9	1,768,472.6	1,096,718.1	45,790,578.2	104,402,023.1	58,611,443
1903	45,118,385.3	844,849.7	1,979,003.8	1,214,068.4	64,846,114.9	114,002,422.1	49,156,305
1904	45,611,673.3	872,209.9	2,023,842.3	1,254,540.0	67,526,449.7	117,288,715.2	49,762,264
1905	45,234,977.6	905,746.9	2,117,637.7	1,333,225.6	67,736,495.3	117,328,083.1	49,591,583
1906	49,543,257.7	894,951.6	2,364,890.8	1,513,062.8	69,815,214.6	124,131,377.5	54,316,160
1907	58,703,504.8	969,263.4	2,968,181.6	1,749,554.4	71,390,116.9	135,780,621.1	64,390,492
1908	56,099,838.0	630,431.9	2,793,529.7	1,472,676.9	60,179,879.6	121,176,356.1	60,996,474
1909	62,546,366.1	613,751.2	2,497,070.1	1,593,130.7	49,059,432.6	116,309,750.7	67,250,317
1910	67,290,394.7	690,188.6	2,999,476.5	2,014,420.7	55,404,831.3	128,399,311.8	72,994,478
1911	72,682,389.8	697,380.9	3,291,223.9	2,223,269.9	55,387,755.3	134,282,019.8	78,894,261
1912	72,355,460.6	715,701.1	3,593,502.6	2,284,825.6	56,594,728.7	135,544,218.6	78,949,488
1913	76,244,441.4	704,150.8	4,023,658.9	2,605,077.4	59,317,173.7	142,894,502.2	83,577,326
1914	72,866,983.3	654,920.3	3,972,939.7	2,570,420.1	58,775,333.5	138,840,596.9	80,065,262
1915	63,614,609.0	564,244.2	3,611,224.6	2,362,289.2	53,708,280.1	123,860,647.1	70,152,366
1916	69,468,144.9	646,900.4	4,131,348.9	2,830,144.1	58,779,676.7	135,856,215.0	77,076,536
1917	83,591,339.9	659,815.7	5,408,321.6	3,551,084.3	71,081,121.5	164,291,683.0	93,210,559
1918	56,222,591.6	331,634.5	1,907,120.8	2,299,885.5	29,326,590.9	90,087,823.3	60,761,230
1919	62,142,790.8	269,915.8	1,130,210.8	2,082,881.1	18,055,500.3	83,681,298.8	65,625,796
1920	6,187,984.3	21,229.9	76,198.8	109,486.0	22,639,355.7	29,034,254.7	6,394,897
1921	9,118,325.2	10,008.6	64,708.3	65,459.1	26,275,969.4	35,534,470.6	9,258,500
1922	2,676,022.3	8,914.0	6,850.4	26,748.5	16,391,489.6	19,109,724.8	2,718,534
1923	1,761,566.4	13,591.8	6,865.6	23,986.1	10,763,613.4	12,569,623.3	1,806,008
1924	1,822,869.0	12,452.9	3,084.1	26,054.1	9,382,302.4	11,246,762.5	1,864,459
1925	1,926,621.5	14,971.0	2,414.9	30,021.2	8,547,518.2	10,521,546.8	1,974,027
1926	1,892,003.1	19,881.6	1,956.6	36,979.0	8,801,398.9	10,752,219.2	1,950,819
1927	1,648,078.2	18,274.9	1,521.6	46,961.3	8,253,512.5	9,968,348.5	1,714,834
1928	1,545,618.2	17,323.0	3,144.6	50,147.6	8,675,717.2	10,291,950.6	1,616,232
1929	1,534,983.2	31708.7	2,372.8	48,082.4	8,892,513.2	10,509,660.3	1,617,145
1930	1,405,706.5	22,492.7	2,082.1	42,169.9	8,251,274.2	9,723,725.4	1,472,449
1931	1,202,510.5	15,987.7	1,911.3	43,714.4	7,398,386.8	8,662,990.7	1,264,122
1932	937,382.1	17,942.2	2,400.3	43,161.6	6,154,449.6	7,155,335.8	1,000,885

TABLE 33.—EXPORTS DISTILLED SPIRITS, 1901 TO 1932

In gallons

Year	Brandy	Rum	Whiskey B	Whiskey R	All other	Total
1901	15,323	1,076,711	525,372	160,357	23,562	1,801,325
1902	24,077	1,095,401	611,518	155,046	76,384	1,962,426
1903	18,117	1,096,719	169,396	104,236	48,014	1,536,482
1904	70,193	757,227	231,540	127,535	47,402	1,253,897
1905	21,171	911,371	212,001	106,893	83,771	1,335,207
1906	5,145	701,423	183,621	109,522	40,089	1,037,800
1907	14,172	914,074	190,067	134,110	19,779	1,272,202
1908	2,750	938,331	129,258	172,755	28,391	1,271,485
1909	14,718	926,049	331,909	121,320	11,204	1,405,200
1910	....	1,138,128	46,301	182,002	38,122	1,404,553
1911	....	1,129,578	58,459	133,450	42,246	1,363,733
1912	....	1,410,840	84,381	140,122	23,797	1,695,140
1913	....	1,268,054	60,252	177,341	29,271	1,534,918
1914	....	1,388,738	47,775	134,152	25,408	1,596,073
1915	....	1,240,804	34,823	86,564	30,152	1,392,343
1916	....	1,586,900	88,802	124,700	50,259	1,850,661
1917	....	1,394,796	59,611	139,619	515,113	2,109,139
1918	....	461,571	65,955	89,925	110,646	....
1919	....	120,519	247,583	842,942	247,238	....
1920	....	96,943	1,196,169	2,119,692	902,108	....
1921	....	....	....	....	....	264,000
1922	....	....	....	....	....	268,000
1923	....	....	....	....	....	303,000
1924	....	....	....	....	....	238,000
1925	....	....	....	....	....	118,000
1926	....	....	....	....	....	294,000
1927	....	....	....	....	....	34,000
1928	....	....	....	....	....	216,000
1929	....	....	....	....	....	383,000
1930	....	....	....	....	....	13,000
1931	....	....	....	....	....	31,000
1932	....	....	....	....	....	14,000

TABLE 34.—WHISKEY EXPORTED DURING THE FISCAL YEARS 1904 TO 1932, INCLUSIVE

[Statement in tax gallons]

Year	Africa	Asia	Australia	Austria-Hungary	Bahama Islands	Belgium	Bermuda	Canada	Central America	Cuba	Denmark
1904		31.6					4,633.8	2,296.8	5,520.1	118.2	212.2
1905		36.9				32.6	1,795.5	4,135.4	4,550.1	334.9	201.3
1906	38.8	116.8		75.0			3,690.8	5,047.0	3,002.3	741.9	36.9
1907					70.4		2,063.0	2,038.1	2,699.3	1,180.3	
1908	39.0				72.6	35.5	2,249.0	1,730.7	7,184.3	1,188.4	36.1
1909					35.0		2,182.2	5,494.4	8,465.4	1,702.6	105.9
1910		73.9	35.4	31.8	34.5		1,640.3	4,042.8	14,589.4	1,005.2	
1911					27.7		2,576.5	6,556.5	22,029.2	563.0	36.5
1912		426.6					2,805.4	12,496.6	14,299.2	776.0	
1913			392.8	35.3	21.6		1,272.0	41,411.0	18,699.4	625.7	
1914							3,227.7	24,244.9	19,488.9	462.6	173.0
1915		41.1	4,098.3				823.6	2,124.5	14,221.9	680.8	34.9
1916		411.8	195.5				46,201.8	1,154.1	10,549.3	1,362.3	
1917		189.0	27,870.9				417.6	1,670.9	5,085.6	1,315.8	
1918	706.3	727.1						2,324.8	3,018.7	401.2	11,420.5
1919	307.5	10,461.7			470,281.9		409.8	1,112.9	3,467.4	6,264.6	17,298.8
1920	1,251.1	18,377.3	682.4		14,992.8		109,410.2	135,372.7	7,903.3	324,581.4	
1921		3,680.3					343.9	145,475.9			
1922								89,720.7			
1923								61,470.8	8,787.7		
1924											
1925											
1926											
1927											
1928											
1929											
1930											
1931											
1932											

TABLE 34.—(Continued)

Year	East Indies	England	France	Germany	Greece	Holland	Ireland	Italy	Japan	Mexico
1904		2,643.6	273.4	122,902.7			35.4			20,530.4
1905		2,410.3	168.7	144,325.8		471.2	28.1		33.6	22,490.4
1906		2,324.5	149.2	138,929.7					32.6	17,266.9
1907		2,275.5	376.1	147,173.2			154.8	36.6		28,901.7
1908		1,349.1	319.7	134,926.7			33.1			22,377.5
1909		1,710.7	309.5	304,620.9				249.0		23,116.4
1910		568.2	312.4	50,402.6						32,427.1
1911		1,783.8	440.9	27,771.9						31,063.1
1912		1,421.1	218.0	31,731.4						31,949.5
1913		2,659.5	27.5	40,698.1		35.2				23,568.6
1914		1,231.0	34.9	1,332.5						19,125.4
1915	102.7	1,031.1						79.1	39.6	33,838.8
1916		63,409.8			4,866.9					24,346.5
1917	616.3		35.1							16,654.4
1918	646.4	63,184.1							1,081.4	21,368.8
1919	245.4	879,287.9	655,564.8	93,577.2		268.3			6,450.1	32,507.3
1920		7,124.4				12,280.5	230,822.6		17,707.5	209,156.4
1921	34.5	862.1			3,995.6					10,673.0
1922		62,511.2		2,996.8	6,238.0					
1923		9,159.0		28,742.8						
1924				60,163.4						
1925		8,502.9								
1926		2,595.2								
1927										
1928										
1929										
1930										
1931		300.0								
1932										

TABLE 34.—(Continued)

Year	Norway	Philippine Islands	Scotland	South America	Spain	Sweden	Switzerland	West Indies	Foreign, port not stated	Total
1904		414.9		194.5			64.9			159,872.5
1905		1,623.6				215.0				182,853.4
1906		108.2	220.8	32.4		73.6	69.9			171,887.4
1907	33.6	1,375.0	200.7	41.2	38.3		76.3			188,691.1
1908		2,515.6	115.3	111.2						174,325.1
1909		11,730.5		122.3						359,561.5
1910		14,048.0								119,457.2
1911		18,633.7		1,055.4						112,611.0
1912		15,198.8		669.2			39.2			111,640.9
1913		5,126.9		3,230.0				72.5		137,853.0
1914		8,870.9		564.8		6.4		37.4		79,149.8
1915		7,394.5		428.7				133.3		60,937.3
1916		6,585.7	9,886.3	427.7				375.1	5,117.4	178,930.6
1917	266.1	7,523.2	5,534.6	2,923.8				1,077.9	13,253.3	56,142.8
1918		11,648.4		827.5				162.6	2,697.5	73,451.5
1919		2,309.1		593.9					3,788.1	143,191.7
1920		3,750.3	529,637.8	6,558.5	1,652.8	43.0	168.0	492.3	250,659.6	3,976,763.7
1921			3,289.4							189,575.3
1922										93,614.1
1923			140,751.8							308,502.3
1924			42,477.7				17,461.2			111,800.1
1925			6,179.2							23,640.4
1926										8,502.9
1927										2,595.2
1928										
1929										
1930										
1931										
1932										300.0

TABLE 35.—IMPORTS OF DISTILLED SPIRITS AS SHOWN BY THE REPORTS OF THE BUREAU OF FOREIGN AND DOMESTIC COMMERCE, FISCAL YEARS 1901 TO 1932, INCLUSIVE

Year	Returned	Brandy	Whiskey	Gin	Cordials and liqueurs	All other	Total
	Gallons	Gallons	Gallons	Gallons	Gallons	Gallons	Gallons
1901	875,099	290,301	....	....	....	1,712,156	2,877,556
1902	805,212	316,222	....	....	....	1,909,887	3,031,321
1903	819,591	348,878	....	....	....	2,061,057	3,229,526
1904	471,596	390,988	....	....	....	2,238,842	3,101,426
1905	316,469	403,386	....	....	....	2,366,466	3,086,321
1906	177,499	470,433	....	....	....	2,639,680	3,287,612
1907	154,106	629,333	....	....	....	3,270,226	4,053,665
1908	148,298	592,384	....	....	....	3,216,228	3,956,910
1909	134,015	764,244	....	....	....	3,889,066	4,787,325
1910	119,646	716,259	1,060,300	1,240,662	....	1,245,200	4,382,067
1911	....	409,242	1,293,692	1,045,815	....	925,601	3,674,350
1912	....	509,286	1,373,010	824,694	532,151	411,595	3,650,736
1913	....	610,358	1,541,663	974,776	575,290	378,623	4,080,710
1914	....	602,563	1,571,870	1,055,885	515,575	414,950	4,160,843
1915	....	400,203	1,327,759	742,439	408,090	411,236	3,289,727
1916	....	536,342	1,742,197	805,749	330,452	538,759	3,953,499
1917	....	420,567	1,676,151	263,520	357,311	397,934	3,115,483
1918	....	234,912	796,267	112,649	76,120	154,148	1,374,096
1919	....	224	....	....	....	2,964	3,188
1920	....	28,919	167,310	23,283	58,497	4,665	282,674
1921	....	....	....	....	....	....	487
1922	....	....	....	....	....	....	64
1923	....	....	....	....	....	....	50
1924	....	....	....	....	....	....	53
1925	....	....	....	....	....	....	58
1926	....	....	....	....	....	....	72
1927	....	....	....	....	....	....	70
1928	....	....	....	....	....	....	79
1929	....	....	....	....	....	....	81
1930	....	....	....	....	....	....	43
1931	....	....	....	....	....	....	33
1932	....	....	....	....	....	....	39

TABLE 36.—APPARENT U. S. CONSUMPTION OF DISTILLED SPIRITS, 1901 TO 1932

[Statement in tax gallons]

Year	Withdrawals	Imports	Exports	Apparent consumption
1901	60,585,730	2,877,556	1,801,325	61,661,961
1902	58,611,443	3,031,321	1,962,426	59,680,338
1903	49,156,305	3,229,526	1,536,482	50,849,349
1904	49,762,264	3,101,426	1,253,897	51,609,793
1905	49,591,583	3,086,321	1,335,207	51,342,697
1906	54,316,160	3,287,612	1,039,800	56,563,972
1907	64,390,492	4,053,665	1,272,202	67,171,955
1908	60,996,474	3,956,910	1,271,485	63,681,899
1909	67,250,317	4,787,325	1,405,200	70,632,442
1910	72,994,478	4,382,067	1,404,553	75,971,982
1911	78,894,261	3,674,350	1,363,733	81,204,878
1912	78,949,488	3,650,736	1,695,140	80,905,084
1913	83,577,326	4,080,710	1,534,918	86,123,118
1914	80,065,262	4,160,843	1,596,073	82,630,032
1915	70,152,366	3,289,727	1,392,343	72,049,750
1916	77,076,536	3,953,499	1,850,661	79,179,374
1917	93,210,559	3,115,483	2,109,139	94,216,903
1918	60,761,230	1,374,096	....	62,135,326
1919	65,625,796	3,188	....	65,628,984
1920	6,394,897	282,674	....	6,677,571
1921	9,258,500	487	264,000	8,994,987
1922	2,718,534	64	286,000	2,432,598
1923	13,007,432 *	50	303,000	12,704,482
1924	19,813,322 *	53	238,000	19,575,375
1925	35,156,785 *	58	118,000	35,038,843
1926	52,586,654 *	72	294,000	52,292,726
1927	43,876,342 *	70	34,000	43,842,412
1928	40,191,134 *	79	216,000	39,975,213
1929	52,033,170 *	81	383,000	51,650,251
1930	47,402,120 *	43	13,000	47,389,163
1931	34,281,295 *	33	31,000	34,250,328
1932	24,476,341 *	39	14,000	24,462,380

* Includes 25% of alcohol production estimated as used by bootleggers.

TABLE 37.—PRODUCTIONS AND REMOVALS OF WINE, AMOUNTS ON HAND IN BONDED WINERIES, AND REVENUE FROM TAXES ON WINES, FISCAL YEARS 1912 TO 1932, INCLUSIVE

[Statement in wine gallons]

Fiscal year	Total production	Removed, tax paid	Removed as distilling material	Removed as vinegar	Losses	Wine on hand June 30	Revenues from taxes on wine
1912	51,577,000.00						
1913	49,759,000.00						
1914	45,915,000.00						
1915	28,075,000.00						
1916	43,362,000.00						
1917	39,885,508.00						
1918	51,029,821.97					47,159,384.00	$9,124,368.56
1919	55,756,171.00					17,521,147.00	10,521,609.14
1920	20,082,458.49					17,677,370.49	4,017,596.82
1921	20,532,343.19	6,353,731.84	3,642,570.98		933,681.73	27,604,898.76	2,001,779.87
1922	6,357,456.97	3,014,364.88	2,870,268.80	34,475.50	963,463.00	27,069,539.90	1,306,249.72
1923	14,706,495.07	3,697,985.50	3,521,002.85	36,351.50	1,023,618.91	33,383,400.86	1,531,991.38
1924	9,056,170.46	4,194,030.65	4,809,269.60	82,343.00	1,316,774.77	31,905,896.10	1,454,062.88
1925	3,638,446.17	4,817,228.22	2,984,698.20	111,653.00	1,657,053.83	26,290,417.55	1,595,488.63
1926	5,841,095.63	4,973,197.98	2,849,410.34	82,902.00	1,230,416.56	23,393,964.34	1,679,434.38
1927	4,406,564.16	2,223,384.52	1,412,574.03	28,648.50	1,114,283.09	23,283,890.62	795,602.83
1928	4,922,617.03	2,382,644.07	2,326,139.50	55,986.00	1,073,662.28	22,498,714.51	893,408.41
1929	11,381,590.43	3,004,200.65	6,932,139.67	382,889.00	1,286,275.98	23,138,754.99	292,549.93
1930	3,154,866.47	2,512,333.06	1,801,141.30	69,758.50	955,826.47	21,168,643.86	239,383.68
1931	6,658,854.00	2,186,306.50	4,522,794.27	294,587.50	1,146,693.52	20,094,522.06	228,495.06
1932	5,210,453.71	1,917,165.64	3,478,831.50	175,229.80	1,485,359.65	18,659,481.71	186,563.29

TABLE 38.—WINE EXPORTS AS SHOWN BY THE REPORTS OF THE BUREAU OF FOREIGN AND DOMESTIC COMMERCE, FISCAL YEARS 1901 TO 1925, INCLUSIVE

Year	Gallons	Year	Gallons	Year	Gallons	Year	Gallons
1901	1,147,561	1908	457,495	1914	941,357	1920	4,573,587
1902	962,756	1909	427,408	1915	819,310	1921	26,000
1903	693,846	1910	519,234	1916	1,133,274	1922	12,000
1904	914,841	1911	1,394,994	1917	2,245,013	1923	47,000
1905	856,786	1912	957,120	1918	2,765,344	1924	13,000
1906	806,314	1913	1,075,151	1919	4,926,425	1925	14,000
1907	573,359						

TABLE 39.—WINE IMPORTS AS SHOWN BY THE REPORTS OF THE BUREAU OF FOREIGN
AND DOMESTIC COMMERCE, FISCAL YEARS 1901 TO 1931, INCLUSIVE

Year	Still wines	Champagnes and other sparkling wines	Total
	Gallons	Gallons	Gallons
1901	3,907,346	933,234	
1902	4,493,480	995,768	
1903	5,075,818	1,223,832	
1904	5,421,150	1,008,735	
1905	5,440,238	1,115,433	
1906	6,122,563	1,246,182	
1907	7,124,272	1,258,209	
1908	7,329,066	1,100,007	
1909	7,699,639	1,309,884	
1910	9,567,390	1,173,009	
1911	6,602,350	665,485	
1912	5,595,802	843,402	
1913	6,461,523	842,484	
1914	7,405,289	810,006	
1915	5,740,868	343,890	
1916	5,094,113	618,630	
1917	4,770,606	587,142	
1918	3,604,335	372,690	
1919	251,865	27,822	
1920	723,723	96,861	
1921	1,446,809	122,942	1,569,751
1922	645,987	32,652	678,639
1923	161,510	13,959	175,469
1924	90,721	2,209	92,930
1925	79,713	1,926	81,639
1926	63,033	13,946	76,979
1927	33,337	3,545	36,882
1928	33,497	1,911	35,408
1929	34,211	1,298	35,509
1930	28,196	1,238	29,434
1931	26,306	1,343	27,649

TABLE 40.—APPARENT CONSUMPTION OF WINE IN THE UNITED STATES, FISCAL YEARS
1912 TO 1932, INCLUSIVE

Year	Gallons	Year	Gallons
1912	57,059,000	1923	14,812,000
1913	55,988,000	1924	9,112,000
1914	53,189,000	1925	3,701,000
1915	33,341,000	1926	5,890,000
1916	47,942,000	1927	4,434,000
1917	42,999,000	1928	4,974,000
1918	52,242,000	1929	11,416,000
1919	51,110,000	1930	3,200,000
1920	16,329,000	1931	6,680,000
1921	21,989,000	1932	5,243,000
1922	6,538,000		

TABLE 41.—INTERNATIONAL TRADE IN WINES—1930

(In thousands of gallons)

Country	Production	Imports	Exports
France.	1,109,933	352,781	28,914
Italy.	551,737	494	27,263
Spain.	481,585	37	92,200
Algeria.	359,314	838	297,135
Roumania.	221,542	10	34
Argentina.	165,000*	1,512	144
Portugal.	155,669	17	21,603
Hungary.	95,626	23	8,265
Germany.	66,905	21,934	1,394
Bulgaria.	62,417	†	2
Austria.	31,768	9,850	17
Australia.	18,150*	58	470
Switzerland.	16,909	30,902	69

* Estimated.
† Less than 1,000 gallons.
Source: International Yearbook of Agricultural Statistics.

SELECTED BIBLIOGRAPHIES *

YEASTS AND FERMENTATION

AHRENS. Das Gährungs-Problem. In Sammlung Chemischer & Technischer Vortraege. Vol. II. Stuttgart, 1902.

ALLEN, PAUL W. Industrial Fermentation. New York, 1926.

BITTING, K. G. Yeasts and Their Properties. (Purdue University Monograph Series, No. 5.)

BUCHNER, E. H., and M. HAHN. Die Zymase Gährung. München, 1903.

EFFRONT. Biochemical Catalysts. New York, 1917.

GREEN-WINDISCH. Die Enzyme. Berlin, 1901.

GUILLIERMOND, A. The Yeasts. New York, 1920.

HANSEN, E. CHR. Practical Studies in Fermentation. London, 1896.

HARDEN, ARTHUR. Alcoholic Fermentation. London, 1923.

HENRICI, A. T. Molds, Yeasts, and Actinomycetes. New York, 1930.

JÖRGENSEN. Micro-Organisms of Fermentation. London, 1900.

KLOECHER. Fermentation Organisms. London, New York, 1903.

LAFAR. Technical Mycology. London, 1910.

MAERCKER. Handbuch der Spiritusfabrikation. Berlin, 1908.

MATTHEWS, CHAS. G. Manual of Alcoholic Fermentation. London, 1901.

OPPENHEIMER. Dis Fermente. Leipzig, 1913-29.

RIDEAL, SAMUEL. The Carbohydrates and Alcohol. London, 1920.

* The student will find directions to further bibliographies on all of the topics included here, except whiskey, in: West and Berolzheimer. Bibliography of Bibliographies on Chemistry and Chemical Technology. Washington, 1925, 1929, 1932.

DISTILLATION. PRACTICAL AND THEORETICAL

ELLIOT, C. Distillation in Practice. London, 1925.
ELLIOT, C. Distillation Principles. London, 1925.
HAUSBRAND, E. Principles and Practice of Industrial Distillation. Trans. from the 4th German ed. by E. H. Tripp. London, 1925.
ROBINSON, C. S. The Elements of Fractional Distillation. New York, 1930.
YOUNG, SYDNEY. Distillation Principles and Processes. London, 1922.

ALCOHOL

FARMER, R. C. Industrial and Power Alcohol. London, 1921.
FOTH, GEORGE. Handbuch der Spiritus Fabrikation. Berlin, 1929.
FRITSCH, J., and VASSEUX, A. Traite Théoretique et Practique de la fabrication de L'alcool et de Produits Accessoires. Paris, 1927.
MCINTOSH, JOHN G. Industrial Alcohol. London, 1922.
SIMMONDS, CHAS. Alcohol: Its Production, Properties, Chemistry and Industrial Applications. London, 1919.
DE VOL, EVERETT T. A Farmers' Practical Treatise on Fermentation, Distillation and General Manufacture of Alcohol from Farm Products with Subsequent Denaturing. Omaha, 1921.
WRIGHT, FREDERICK B. Alcohol from Farm Products. New York, 1933.

DISTILLED LIQUORS

DE BREVANS, J. La Fabrication des Liqueurs. Paris, 1897.
FOTH, GEORGE. Handbuch der Spiritus Fabrikation. Berlin, 1929.
KULLMANN, OTTO. Die Spirituosen Industrie. Leipzig, 1912.
MACDONALD, AENEAS. Whiskey. Porpoise Press, Edinburgh, 1930.
PIAZ, A. DAL. Die Cognac und Wein Spirit Fabrikation so wie die Trester und Hefebranntweinbrennerei. Wien, 1891.
ROCQUES, X. Les Eaux-de-vie et Liqueurs. Paris, 1898.

ROGERS, ALLEN. Industrial Chemistry. New York, 1926.
SCHEDEL, C. F. B. Der Distillateur. Leipzig, 1921.

WINES

ALWOOD, WILLIAM B. Experiments in cider making applicable to farm conditions and notes on the use of pure yeast in wine making. Bull. 129, U. S. Bur. of Chem., 1909.

BABO, A. VON, and MACH. Handbuch des Weinbaues and der Kellerwirtschaft, 4. Auflage. Berlin, 1910.

BARTH, MAX. Kellerbehandlung der Traubenweine, 2. Auflage. Stuttgart, 1903.

BENEVEGNIN, LUCIEN, and others. Manuel de Vinification. Payot et cie., Lausanne, 1930.

BIOLETTI, FREDERICK T. Defecation of Must for White Wine. California Agricultural Experiment Station, Circular 22. Berkeley, 1906.

BIOLETTI, FREDERICK T. Manufacture of Dry Wines in Hot Countries. California Agricultural Experiment Station, Bulletin 167. 1905.

BIOLETTI, F. T., and CRUESS, W. V. The Practical Application of Improved Methods of Fermentation in California Wines during 1913 and 1914. California Agricultural Experiment Station, Circular 140. 1915.

BIOLETTI, F. T. The principles of wine making. California Agricultural Experiment Station, Bulletin 213. 1911.

BIOLETTI, F. T. The Best Wine Grapes for California. California Agricultural Experiment Station, Bulletin 193. 1907.

BIOLETTI, FREDERICK T. Winery Directions. California Agricultural Experiment Station, Circ. 119. 1914.

BIOLETTI, FREDERICK T. The practical application of improved methods of fermentation in California wineries during 1913 and 1914. California Agricultural Experiment Station, Circ. 140. 1915.

BOULLANGER, EUGENE. Distillerie Agricole et Industrielle Eaux-de-vie de Fruits. Paris, 1924.

DE BREVANS, J. La Fabrication des Liqueurs. Paris, 1897.

COSTE-FLORET, P. Procédés Modernes de Vinification.
1. Vin Rouges. Paris, 1899.
2. Vin Blanc. Paris, 1903.
3. Les Residus de la Vendage. Paris, 1901.
CUNIASSE, L. Memorial du Distillateur Liquoriste. Paris, 1925.
DUBOR, GEORGES DE. Viticulture et Vinification Moderne. Paris, 1894.
EMERSON, E. R. The Story of the Vine. New York and London, 1902.
FRITSCH, J. Nouveau Traité de la Fabrication des Liqueurs. Paris, 1926.
GRAZZI-SONCINI, G. Wine. Classification, Wine Tasting, Qualities and Defects. Translated by F. T. Bioletti. Sacramento, 1892.
GUNTHER. Die Gesetzgebung des Auslandes über den Verkehr mit Wein. Berlin, 1910.
HAYNE, ARTHUR PERONNEAU. Bull. 117, Agricultural Experiment Station. University of California, 1897.
HUSMANN, GEORGE, and others. American Grape Growing and Wine Making. New York, 1919.
KULISCH, P. Sachgemasse Weinverbesserung. Berlin, 1903.
LARRONDE, EUGENE. Vins et Boissons. Paris, 1894.
MEISSNER, RICHARD. Untersuchung der Weinplize. Stuttgart, 1901.
MENOTTI DAL PIAZ, A. Handbuch des Praktischen. Weinbaues, 1908.
NESSLER-WINDISCH. Die Bereitung, Pflege, und Untersuchung des Weines, 8. Auflage. Stuttgart, 1907.
PIQUE, RÉNÉ. Vinification et Alcoolisation des Fruits Tropicaus et Produits Coloniaux. Paris, 1908.
RHEINBERG, H. Herstellung von Schaumwein unter Obst Schaumwein. Leipzig, 1913.
ROGERS, ALLEN. Industrial Chemistry. New York, 1926.
SCHMITTHENNER, F. Weinbau und Weinbereitung. Leipzig, 1911.
SEMICHON, L. Traité des Maladies des Vins. Paris, 1905.

352   SELECTED BIBLIOGRAPHIES

Sheen, James R.   Wines and other Fermented Liquors.   London, 1864.

Simon, Andre L.   Wines and the Wine Trade.   London, 1921.

Simon, Andre L.   Bibliotheca Vinaria; a bibliography of books and pamphlets dealing with viticulture, wine making, distillation, management, sale, taxation, use and abuse of wines and spirits.   G. Richards, Ltd., London, 1913.

Thorpe, Edward.   Dictionary of Applied Chemistry, Vols. 1 to 7.   1921–1927.

Thudichum, J. L. W.   A Treatise on Wines.   London, 1896.

University of California College of Agriculture.   Report of the Viticultural Work during 1887 to 1895.   Sacramento, 1896.

Venta, J.   Les Levures dans la Vinification.   Paris, 1911.

Ventre, Jules.   Traité de Vinification Pratique et Rationnelle.   Montpettier, 1930–31.

Visetelly, Henry.   A History of Champagne with Notes on other Sparkling Wines of France.   New York and London, 1882.

Wiley, Harvey W.   American Wines at the Paris Exhibition in 1900: their composition and character.   Also a monograph on the manufacture of wines in California by Henry Lachmann.   U. S. Bur. of Chemistry.   Bull. 72.   1903.

Windisch, K.   Die chemischen Vorgange beim Werden des Weines.   Stuttgart, 1906.

Wortmann, Julius.   Die wissenschaftlichen Grundlagen der Weinbereitung und Kellerwirtschaft.   Berlin, 1905.

ANALYSIS

Cuniasse, L.   Mémorial du Distallateur Liquorist.   Paris, 1925.

Griffin, John J.   Chemical Testing of Wines and Spirits.   London, 1872.

Leach, A. E.   Food Inspection and Analysis.   New York, 1920.

Prescott, Albert B.   Chemical Examination of Alcoholic Liquors.   New York, 1880.

Rocques, X.   Encyc. Scient. des aide mém.   Sect. de L'ingénieur.   Paris.   Page 198.

ROSENHEIM, OTTO, and SCHIDROWITZ, PHILLIP. On some Analyses of Modern "Dry" Champagne. In the Analyst 25, 6–9, 1900.

Royal Commission on Whiskey and Other Potable Spirits. Report and Appendix. 1908–9.

TOLMAN, L. M., and TRESCOT, T. C. A Study of the Methods for the Determination of Esters, Aldehydes and Furfural in Whiskey. Jour. Am. Chem. Soc. 28, 1619–29 (1906).

WILEY, H. W. Beverages and their Adulteration. Philadelphia, 1919.

STATISTICS

Bureau of Industrial Alcohol. Statistics Concerning Intoxicating Liquors. U. S. Treas. Dept. Washington, 1932.

International Yearbook of Agricultural Statistics. Rome, 1930.

KOLLER, ARNOLD. La Production et la Consommation des Boissons Alcooliques dans les Different Pays. Bur. Internat. Contre L'Alcoolisme. Lausanne, 1925.

Rep. of the Bur. of Foreign and Domestic Commerce. U. S. Dept. Comm. Washington, 1901–date.

INDEX

CPSIA information can be obtained at www.ICGtesting.com
Printed in the USA
BVOW04*0019120314

347352BV00001B/1/P